Introduction to THz Wave Photonics

X.-C. Zhang · Jingzhou Xu

Introduction to THz Wave Photonics

Springer

Xi-Cheng Zhang
Rensselaer polytechnic Institute
School of Science
Center for Teraheriz Research
110 8th Street
Troy NY 12180
USA
zhangxc@rpi.edu

Jingzhou Xu
IMRA America, Inc.
1044 Woodridge Avenue
Ann Arbor MI 48105
USA
jxu@imra.com

ISBN 978-1-4419-0977-0 e-ISBN 978-1-4419-0978-7
DOI 10.1007/978-1-4419-0978-7
Springer New York Dordrecht Heidelberg London

Library of Congress Control Number: 2009939154

Printed on acid-free paper

Springer is part of Springer Science+Business Media (www.springer.com)

Preface

Terahertz (THz) radiation, which is electromagnetic radiation in a frequency interval from 0.3 to 10 THz (1 mm–30 μm wavelength), is the next frontier in science and technology. This band occupies a large portion of the electromagnetic spectrum between the infrared and microwave bands. Basic research, new initiatives, and developments in advanced sensing and imaging technology with regard to the THz band remain unexplored compared to the relatively well-developed science and technology in the microwave and optical frequencies.

Historically, THz technologies were used mainly within the astronomy community for studying the background of cosmic far-infrared radiation, and by the laser-fusion community for the diagnostics of plasmas. Since the first demonstration of THz wave time-domain spectroscopy in the late 1980s, there has been a series of significant advances (particularly in recent years) as more intense THz sources and higher sensitivity detectors provide new opportunities for understanding the basic science in the THz frequency range.

As developments move forward, THz science will not only have an impact on material characterization and identification but also have potential applications in the fields of communications, imaging, medical diagnosis, health monitoring, environmental control, and chemical and biological sensing, as well as security and quality-control applications. Twenty-first century research in the THz band is one of the most promising areas of study for transformational advances in imaging and other interdisciplinary fields. This book intends to explain the fundamental principles of broadband THz wave generation and detection, highlighting recent advances of novel optical materials that have impacted THz wave sensing and imaging applications. Emphasis is placed on the use of nonlinear optical materials with femtosecond laser pulses.

There are many new technologies which have been developed in this field since the first pulsed THz time-domain spectroscopic system was invented more than 20 years ago. Benefiting from those novel technologies, researchers have gradually revealed the unique properties and promising applications of THz waves. In this book, we systematically introduce the various sciences, technologies, and applications of THz waves, with a particular focus on pulsed THz technologies.

This text begins by reviewing the history of THz research, then summarizes state-of-the-art THz wave generation and detection methods, discusses THz wave interaction with matter, and finally introduces unique applications developed for this most-elusive electromagnetic wave. Although this book places a heavy emphasis on pulsed THz technology, it will also provide a brief introduction to continuous-wave THz technologies.

The intention of this book is to provide an overview of THz photonics rather than focusing on specific technologies as other books have done in the past. It introduces the fundamentals of THz photonics and also includes the latest technologies currently in use and those that are on the near-term horizon.

We hope to help readers, especially graduate students and researchers in both academic and industry fields, who are interested in the fundamentals of THz technologies, by providing a guide to the learning of the principles which govern them. We also hope that readers who are new to THz photonics will find this book to be a valuable reference tool.

All data, figures, tables, equations, and papers cited in this book are presented solely to illustrate our discussion. There are many outstanding achievements in THz research which are not mentioned in detail or discussed at length in this book. This is due to the constraints of space and is by no means an indication that the authors have ignored these works or consider them less important.

For readers seeking further information which might not be covered exclusively in this book, please see the section labeled "Additional References" preceding the index. The list of books have been edited or written by experts in the field, and we recommend them highly.

The authors hope to convey the sense of enjoyment found in THz study, research, and development, as well as the awe at what has already been learned about THz waves, and just as importantly, what is yet to be discovered.

We thank many colleagues and friends for their guidance, support, and friendship. Special thanks to Dr. David Auston, Dr. Chi Lee, Dr. Dan Grischkowsky, Dr. Ci-Ling Pan, Dr. Kiyomi Sakai, Dr. Hiromasa Ito, Dr. Dan Mittleman, Dr. Peter Siegel, Dr. Alexander Shkurinov, Dr. Masayoshi Tonouchi, Dr. Qing Hu, Dr. Keith Nelson, Dr. Charles Schmuttenmaer, Dr. Peter Jepsen, Dr. Paul Planken and Dr. Kodo Kawase.

We would like to thank our colleagues from the Center for Terahertz Research at Rensselaer for their support and collaboration. Special thanks to Dr. Michael Shur, Dr. Gwo-Ching Wang, Dr. Roland Kersting, Dr. Ingrid Wilke, and Dr. Masashi Yamaguchi.

We would like to thank those former students who have graduated from our THz group at Rensselaer: Ms. Yahong Jin, Ph.D. 1994; Mr. Xinfa Ma, Ph.D. 1995; Mr. Greg Wagoner, Ph.D. 1997; Mr. Qi Wu, Ph.D. 1997; Ms. Jennifer Riordan, Ph.D. 1998; Mr. Ming Li, Ph.D. 2000; Mr. Pengyu Han, Ph.D. 2000; Mr. F.G. Sun, Ph.D. 2001; Mr. Qin Chen, Ph.D. 2001; Mr. Kwang-Su Lee, Ph.D. 2000; Mr. Shaohong Wang, Ph.D. 2003; Ms. Fatemah M. Al-Douseri, Ph.D. 2005; Mr. Kai Liu, Ph.D. 2005; Mr. Haibo Liu, Ph.D. 2006; Mr. Tao Yuan, Ph.D. 2006; Ms. Hua Zhong, Ph.D. 2006, Mr. Xu Xie, Ph.D. 2007, Mr. Brian Schulkin, Ph.D. 2008; Mr. Nick Karpowicz, Ph.D. 2009, and Mr. Jian Chen, Ph.D. 2009. Many selected materials used in this book come from their research.

We thank our senior researchers (postdoctoral associates, research associates, research assistant professors, and visiting scholars): Dr. Zhiping Jiang, Dr. Gyu Cheon Cho, Dr. Masahiko Tani, Dr. Bradley Ferguson, Dr. Samuel Mickan, Dr. Xuguang Huang, Dr. Yuchuang Chen, Dr. Jiayin Qin, Dr. Abdellah Menikh, Dr. Zhengguo Lu, Dr. Yunqing Chen, Dr. Albert Redo, Dr. Jianming Dai, Dr. Jin Wook Choi, Dr. Abul Azad, Dr. Ying Hu, Dr. Wei Liang, Dr. Yuguang Zhao, Dr. Cunlin Zhang, Dr. Guozhong Zhao, Dr. Yan Zhang, Dr. Hongwei Zhao, Dr. Hongkyu Park, and Dr. Etienne Gagnon. They are the key players in our group and have dedicated their time and effort to advancing the THz research.

Many people have helped to complete this book. Particular thanks to Mr. Christopher Thorne, Mr. Benjamin Clough, Ms. I-Chen Ho, Ms. Xiaofei Lu, Mr. Xiaoyu Guo, Ms. Gurpreet Kaur, Ms. Yuting Chen, Mr. Jingle Liu, Mr. David Brigada and Dr. Yimin Sun. Without their tireless help and support, this book is not possible.

Troy, NY X.-C. Zhang
Ann Arbor, MI Jingzhou Xu
July 2009

Contents

Chapter 1
Terahertz Radiation

Introduction to THz Radiation

Various frequencies are spaced along the frequently used electromagnetic spectrum, including microwaves, infrared radiations, visible lights, and X-rays. Between the microwave and infrared frequencies lies terahertz (THz) radiation (Fig1.1). In the electromagnetic spectrum, radiation at 1 THz has a period of 1 ps, a wavelength of 300 μm, a wave number of 33 cm^{-1}, a photon energy of 4.1 meV, and an equivalent temperature of 47.6 K. In the same way that visible light can create a photograph, radio waves can transmit sound, and X-rays can see shapes within the human body, terahertz waves (also called as T-rays) can create pictures and transmit information.

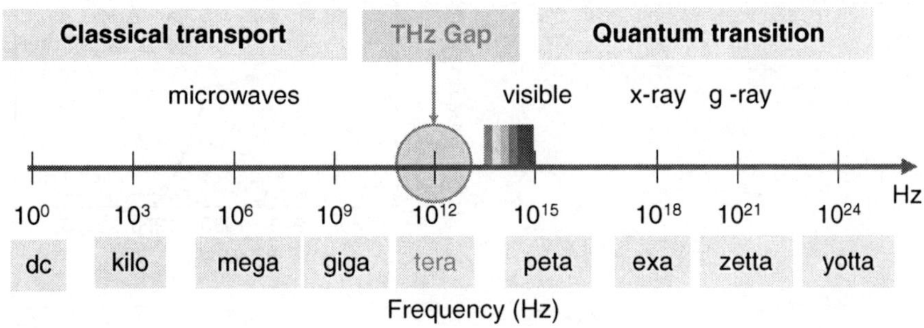

Fig. 1.1 The electromagnetic spectrum. The development of efficient emitters and detectors within each of the spectral regimes has resulted in the birth of numerous industries. The search for potential applications using THz radiation is steadily intensifying as materials research provides improved sources and detectors

Plenty of THz radiation sources surround us, from cosmic background radiation to blackbody radiation from room temperature objects (Fig.1.2). Most of those THz sources are incoherent and can hardly be utilized. Until recently, however, the very large THz portion of the spectrum has not been particularly useful because there were neither suitable emitters to send out controlled THz signals nor efficient sensors to collect them and record information. As a result, the THz portion of

X.-C. Zhang, J. Xu, *Introduction to THz Wave Photonics*,
DOI 10.1007/978-1-4419-0978-7_1,

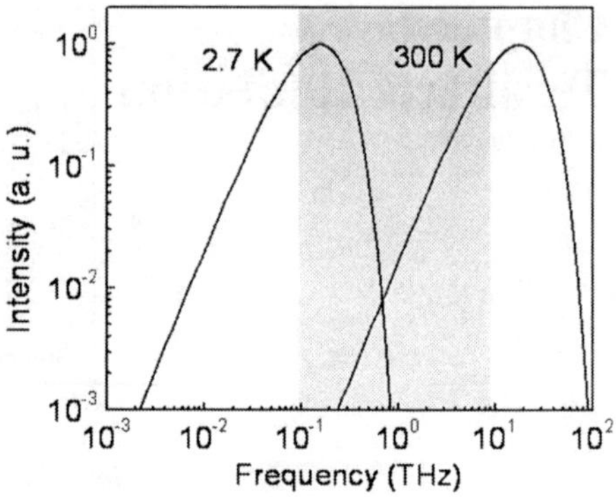

Fig. 1.2 Cosmic background radiation and blackbody radiation from room temperature object. *Shadow area* indicates THz band

the electromagnetic spectrum was called the THz gap. Tremendous effort has been made in order to fill in the THz gap. Figure 1.3 shows the road map of development of THz sources using various technologies. Recent developments in time-domain THz spectroscopy and related THz technologies now lead us to view the world in a new way. As a result of developing research, THz radiation now has widespread potential applications in medicine, microelectronics, agriculture, forensic science, and many other fields.

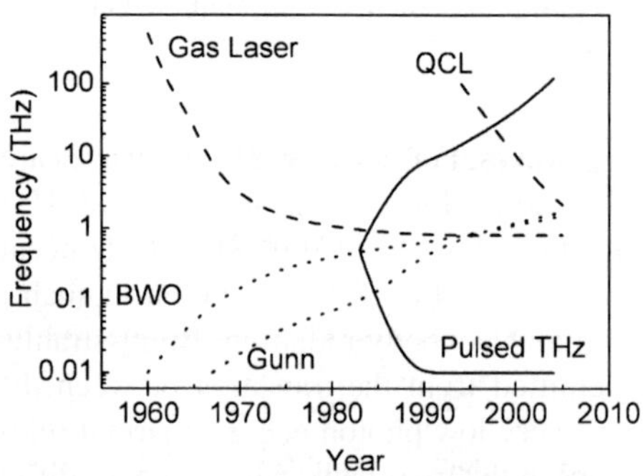

Fig. 1.3 Road map of coherent THz source development

Being electromagnetic waves, the properties and behavior of THz waves are governed by the Maxwell equations, just like the other waves. However, due to their specific location on the electromagnetic spectrum, THz waves are much more difficult to handle than waves adjacent to them. Historically, research on electromagnetic waves was divided into two different fields: optics and electromagnetics. Figure 1.4 compares mechanisms in the generation of electromagnetic waves and optical waves. The principles, methods, and tools used to deal with optical and electrical waves are very different, as presented in Table 1.1. THz waves, however, do not fit simply into either category, since approximations used for optical waves or electrical waves may not be still true in the THz waves' regime. New principles, new technologies, and new tools must be developed in order to understand and utilize

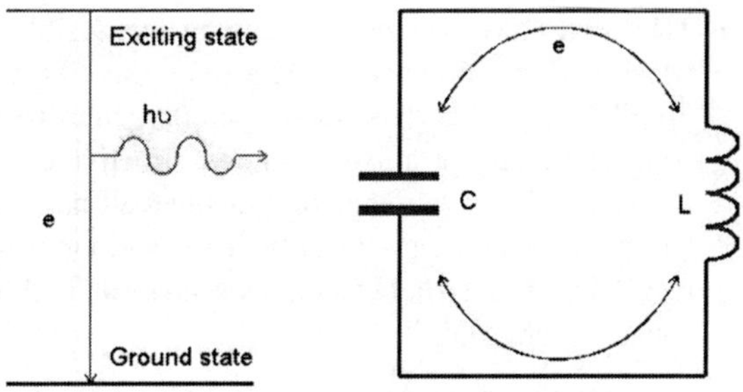

$$\omega = \frac{\Delta E}{\hbar} \qquad \omega = \frac{1}{\sqrt{LC}}$$

Fig. 1.4 Electron transition (**a**), classical movement (**b**), and electromagnetic wave emitted thereby

Table 1.1 Comparing of optics and electromagnetics

	Electromagnetics	Optics
Name of waves	Electromagnetic waves	Optical waves
Principle	Maxwell equations	Schrodinger equations
Emission	Classical movement	Quantum transition
Measurement	Electric field	Intensity
Tools	Circuit, antenna, waveguide	Lens, mirror, fiber
Approximation	Uniform field	Uniform medium

THz waves. For instance, electric waves are usually generated from the drift motion of carriers. However, most traditional RF sources cannot generate high-frequency radiation above several hundred GHz because the motion of the carrier cannot follow the speed of THz oscillation in such a device. Except for a few mechanisms such as blackbody radiation, Bremsstrahlung, or synchrotron radiation, most light is emitted from the transition between different energy states. Since THz waves have very low photon energy, thermal relaxation can easily obscure the distinction between the two states, whose energy difference is equal to a THz photon. Another example where optical and electromagnetic techniques may be incompatible is in waveguides; metallic waveguides used to carry microwaves generally result in high loss for THz waves, while dielectric waveguides and optical fibers suffer from high loss and dispersion.

Despite these difficulties, the unique nature of THz waves has stimulated researchers to develop this frequency band for various applications:

- THz waves have low photon energies and thus cannot lead to photoionization in biological tissues as can X-rays. As a result, THz waves are considered safe for both the samples and the operator. Due to extreme water absorption, THz waves cannot penetrate into the human body like microwaves can. Therefore, even if THz waves do cause any harm, it is limited to skin level.

- THz waves have longer wavelengths than visible and IR waves. This means THz waves are less affected by Mie scattering. THz waves are transparent to most dry dielectric materials, such as cloth, paper, wood, and plastic. THz waves are considered very promising in nondestructive evaluation applications. Figure 1.5 shows a THz wave image of a racquetball bat in a black plastic bag. The wavelength of the THz waves is sufficiently short to provide submillimeter level spatial resolution. If near-field techniques are used, higher spatial resolution down to nm could be achieved.

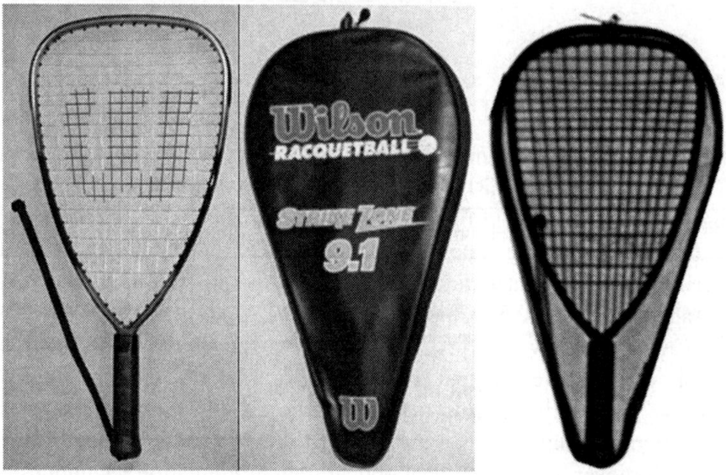

Fig. 1.5 Photo of racquetball bat (**a**), the bat in a plastic bag (**b**), and THz wave (0.6 THz) image of the bat in a plastic bag (**c**)

- At THz frequencies, many molecules exhibit strong absorption and dispersion due to dipole-allowed rotational and vibrational transitions. These transitions are specific to the molecule and therefore enable spectroscopic fingerprinting in the THz range. Combined with imaging technology, inspection using THz waves provides both profile and composition information of the target. Figure 1.6 gives a comparison between the electromagnetic wavelength and the size of common objects, as well as photon energy and chemical bonds.
- Coherent THz signals can be detected in the time domain by mapping the transient electric field in amplitude and phase. This gives access to absorption and dispersion spectroscopy.

Coherent THz time-domain spectroscopy with an ultrawide bandwidth provides a new method to characterize the electronic, vibronic, and compositional properties of solids, liquids, and gases as well as flames and flows. In theory, as mentioned above, many biological and chemical compounds have distinct signature responses to THz waves due to their unique molecular vibrations and rotational energy levels, implying that their chemical compositions might be examined using a THz beam. Such a capability could be applied to the diagnosis of a disease, detection of pollutants,

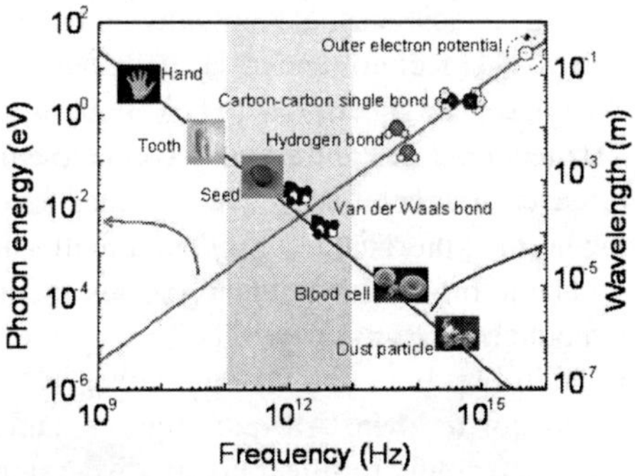

Fig. 1.6 Wavelength and photon energy of electromagnetic wave with general comparison to size of common object and chemical bond energy. *Shadow area* indicates THz band

sensing of biological and chemical agents, and quality control of food products. It is also quite possible that plastic explosives could be distinguished from suitcases, clothing, common household materials, and equipment based on molecular structure. Detecting the binding state of genetic materials (DNA and RNA) by directly using THz waves, without requiring markers, allows for a label-free method for genetic analysis for future bio-chip technologies.

THz wave imaging would produce images with *component contrast* enabling an analysis of the water content and composition of tissues in biological samples. Such a capability presents tremendous potential to identify early changes in composition and function as a precursor to specific medical investigations and treatment. Moreover, in conventional optical trans-illumination techniques that use near-infrared pulses, large amounts of scattering can spatially smear out the objects to be imaged. THz wave imaging techniques, due to their longer wavelengths, can provide significantly enhanced contrast as a result of low scattering.

It has been several decades since electromagnetic radiation in the THz band was first scientifically studied. In the beginning, such radiation was called far-infrared waves or millimeter/submillimeter waves. It was through the development of ultrafast optoelectronics, which led to the successful generation and detection of THz pulses in the middle 1980s, that interest in this particular electromagnetic wave widely bloomed. Early researchers were focused on developing technologies in generating, detecting, and manipulating THz waves. This was followed by utilizing THz waves in fundamental research. Lately, interest has become more focused on the real-world application of THz waves.

In the decades since the first pulsed THz system was invented, a wealth of new technologies have been developed in this field. Benefiting from those novel technologies, researchers, little by little, have revealed unique properties and promising applications of THz waves. Several books, which deeply discuss frontier THz research in various topics, have been published. These books are very helpful for researchers who are interested in related topics. However, they are less favorable for new comers to THz research, such as graduate students and researchers from outside the THz community who are interested in the fundamentals of THz technologies or in viewing an overall picture of this field. In this book, we systematically introduce the various sciences, technologies, and applications of THz waves, with emphasis on pulsed THz technologies at a level appropriate to graduate students and researchers. We hope that our effort will help readers to catch hold of the principles that govern THz technologies and provide them a panoramic view of THz research. In addition, sciences and technologies related to THz waves, with electromagnetic waves being the latest explored, are introduced in this book.

The aim of this book is to introduce fundamental knowledge of THz wave photonic technologies rather than to review or summarize the history of THz research. In addition, this book will not cover THz electronics, which is truly deserving of a book of its own. All data, figures, tables, and equations cited in this book are presented solely to illustrate our discussion. There are numerous key works in THz research which are not mentioned in this book. This does not mean we ignored those works or consider them as less important. We have simply limited acknowledgment to those sources from which data are specifically cited within the text.

THz Research – History and Present

Pulsed THz systems refer commonly to those which generate, propagate, and detect transient electromagnetic pulses. The pulse information is taken in time domain, and it may finally be transformed into frequency domain by Fourier transform. In order to have the electromagnetic energy power spectrum in the THz range, the duration of the electromagnetic oscillation must be within the picoseconds range. Early work began in late 1970s and early 1980s with the study of the response of photoconductors (PC) to laser pulses [1–4]. In the late 1980s, the use of PC antennas generated the first prototype of a pulsed THz wave emitter and detector [5, 6], and such devices remain widely used for emission and detection of THz radiation. Figure1.7 shows the schematic concept of PC antennas [22]. Various mechanisms have been exploited to generate THz radiation, including photo-carrier acceleration in PC antennas, second-order nonlinear effects in electro-optic (EO) crystals (optical rectification), plasma oscillations, and electronic nonlinear transmission lines. Photoconduction and optical rectification are the two most common approaches for generating broadband, pulsed THz beams. Actual conversion efficiencies for these techniques are very low, and average THz powers tend to be in the nW to μW range while the average power of the femtosecond optical source is in the region of tens of mW to tens of W.

Optical rectification is based on the inverse process of the EO effect [7]. This technique was first demonstrated for generating far-infrared radiation using $LiNbO_3$ and ZnTe [8] and considerable research has been employed toward optimizing THz generation through investigating the EO properties of different materials including traditional semiconductors such as GaAs, organic crystals such as the ionic salt 4-dimethylamino-*N*-methyl-4-stilbazolium-tosylate (DAST), and many others [9–14]. Generally, materials with a large second-order nonlinear coefficient and low absorption for both optical and THz waves are good candidates for a THz source

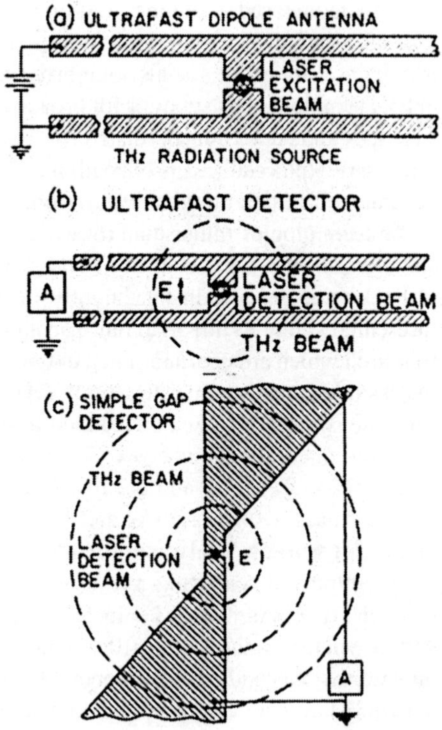

Fig. 1.7 Dan Grischkowsky antenna being used as THz wave emitter (**a**) and detector (**b**) and (**c**). (**a**) Ultrafast dipole antenna, (**b**) ultrafast detector, and (**c**) a simple gap detector (courtesy of Dr. Grischkowsky)

[15]. Currently, ZnTe is the material of choice due to its high nonlinear coefficient, high laser damage threshold, and best phase-match condition with a Ti:sapphire laser [16]. Compared to traditional solid-state lasers, fiber lasers provide more compact, higher efficiency, and higher power. The most popular wavelengths for fiber lasers are around 1.04 μm for an Ytterbium-based system or around 1.56 μm for an Erbium-based system. New materials need to be used in order to optimize the phase-match condition for those wavelengths. GaP was used to develop a high-power THz source pumped by an Yb-doped fiber laser [17]. GaAs is believed to be the most favorable EO crystal for the Er-doped fiber laser [18]. Tilting the wave front of the pumping beam [19] or using a periodical polarized material [20] can increase the phase-matching length in some crystals, which have high nonlinear coefficients but used to suffer from poor phase matching. Intense THz waves can be generated from those phase-match-improved emitters.

Photoconductive antennas are biased THz wave emitters [21, 22] and are capable of relatively large average THz powers in excess of 40 μW [23] and bandwidths as high as 10 THz [24]. The most widely used materials for PC antennas are semi-insulating (SI) GaAs [23, 25] and low-temperature-grown (LTG) GaAs [26, 27]. Low bandgap semiconductors such as InGaAs are used to replace GaAs when a longer wavelength laser source such as a fiber laser is used due to its lower band gap

[28, 29]. Compared with other traditional THz wave emitters such as the ZnTe crystal, GaAs-based PC antennas do not have phonon absorption around 5 THz, which is quite advantageous for gap-free broadband spectroscopy up to 7 THz. However, optical rectification is superior for broadband THz radiation for the bandwidth of the THz spectra is determined only by the laser pulse duration, and not by the response time of the material [30, 31]. With an ultrashort pulse laser, the bandwidth of the generated THz radiation could be as high as 100 THz [32].

Another type of unbiased THz wave emitter is based on the THz surface emission by the ultrafast transport of charged carriers. THz transients arise either due to the charge separation by the built-in electrical field at the semiconductor surface [33] or due to the difference in the electron and hole diffusion coefficients (photo-Dember effect) [34]. The study of the THz radiation mechanism from bulk materials induced a new category of THz science: THz emission spectroscopy [35]. The analysis of THz pulses irradiated from bulk semiconductors can help to understand the materials themselves. For efficient THz radiation, it is desirable to have rapid photocurrent rise and decay times. Thus semiconductors with small effective electron masses such as InAs and InP are attractive, as well as GaAs. The maximum drift velocity is also an important material parameter; it is generally limited by the intraband scattering rate or by intervalley scattering in direct semiconductors such as GaAs [36, 37]. Since the radiating energy mainly comes from stored surface energy in the form of the static bias field, the THz radiation energy scales up with the bias and optical fluency [38, 39]. The dielectric breakdown field of the material is another important parameter as this determines the maximum bias that may be applied [40]. Other materials such as semiconductor quantum wells [41–43], super-lattices [44, 45], superconductors [46–48], dye molecules [49], and even magneto-molecules [50, 51] have also been studied for THz radiation. The acceleration of electron bunches from free-electron lasers is a brand new source of pulsed THz wave [52–54]. This method provides a powerful pulsed source which offers the feasibility of nonlinear THz research.

Recently, it has been demonstrated that THz radiation can be generated using air [55]. Air does not show a second-order nonlinear coefficient, but it has a third-order nonlinear coefficient instead. By generating plasma with a pulsed laser, it is possible to mix a fundamental wave and its second harmonic through the third-order nonlinearity leading to efficient THz generation. This process opens up the possibility of using THz techniques for remote sensing by propagating a laser beam far from the source, generating THz locally, and transforming the reflected THz radiation into an optical beam again and returning it to the point of origin.

The traditional detection techniques in pulsed THz technology are EO sampling and the use of PC antennas. PC antennas, which are widely used for pulsed THz generation, can also be used to detect THz pulses. Rather than applying a bias voltage to the electrodes of the antenna, a current amplifier and ammeter are used to measure the transient current generated by an optical pulse and biased by the instantaneous THz field. The biased current is proportional to the applied THz field [5, 56–58]. It is possible to use a PC antenna for broadband THz wave detection by properly selecting the substrate materials. Ultrahigh bandwidth detection has been demonstrated using photoconductive antenna detectors made of LTG-GaAs with detectable frequencies in excess of 60 THz [58]. The apertures of the PC antennas are usually in the micron range, and the optical alignment is more difficult compared to free-space EO sampling.

Free-space EO sampling utilizes second-order nonlinear crystals to detect the polarization variation of the probe pulse interfering with the THz pulse. Fundamentally, the EO effect is a coupling effect between a low-frequency electric field (THz pulse) and a laser beam (optical pulse) in the sensor crystal [59–61]. The frequency response for detection is the same as for generation and also the phase-matching conditions and crystal absorption play an essential role in the detector's performance. Thus, thinner crystals provide broader frequency response but then the interaction distance is shorter, resulting in reduced sensitivity. Among many EO crystals such as GaP [62], GaAs [63], organic crystals [12], and EO polymers [13], ZnTe [64] is currently the best candidate for THz wave detection using laser pulses with wavelengths around 800 nm. The phase-matching condition of the sensor is same as the emitter, thus material and laser wavelength match in the same way. The use of an extremely short laser pulse (< 15 fs) and a thin sensor crystal (<30 μm) allows EO detection of signals into the mid-infrared range. Extremely high detection bandwidths in excess of 100 THz have been demonstrated using thin sensors [65].

Continuous-wave (CW) THz generation can be considered as the extreme case of pulsed emission when the duration of the pulse is infinite. Generally, a CW system can only produce a single frequency at a time, but in some particular cases broadband sources may be used. Regarding the detectors, either broadband or narrowband is available. The frequency range that CW systems offer is less than those operating in pulsed mode. On the other hand, narrowband sources are very important for high-resolution spectroscopy applications and also have broad potential applications in telecommunications. CW sources also offer higher average power than pulsed systems and they could be less affected by water vapor absorption if the frequency is selected at the air windows. CW systems are very interesting for field applications because they are in a better position than pulsed systems to achieve small, reliable, and portable devices. A multitude of techniques are under development, including up-conversion of electronic RF sources, down-conversion of optical sources, lasers, and backward-wave oscillators (BWO).

The most commonly employed technique for generating low-power (from 100 μW to 20 mW) CW THz radiation up to 0.6 THz is through up-conversion of lower frequency microwave oscillators such as voltage-controlled oscillators (VCO) and dielectric-resonator oscillators. The most widely used oscillators are Gunn oscillators and IMPATT diodes. The Gunn effect was discovered while testing the behavior of n-GaAs samples under high electric fields applied within the bulk of the material. These samples showed microwave radiation emission for a certain range of DC bias voltage, which did not depend on the junction properties but on the properties of the semiconductor itself [66]. A typical frequency limit for commercially available devices is about 0.6 THz, but frequencies as high as 2.7 THz have been reported [67]. Important research is oriented to increase the frequency of Gunn and IMPATT diodes by using alternate semiconducting structures and improved fabrication techniques [68].

A BWO is another commonly used lower frequency THz source. The basic structure of a BWO is a magnetically focused electron beam passing through a corrugated structure carrying an RF signal. BWO sources are capable of output powers of tens of mW in the sub-THz range and available devices can operate above 1 THz [69].

THz waves can be emitted from a gas laser, where a CO_2 laser pumps a low pressure gas cavity, which emits radiation at its emission-line frequencies that lie within the THz range [70, 71]. This gas is usually methanol or hydrogen cyanide. Gas lasers can provide a broad range of frequencies and are tunable to distinct lines by varying the composition and pressure of the gas in the laser cavity and the frequency of the CO_2 pump laser.

Free-electron lasers with energy-recovering linear accelerators, by forcing bunches of high-speed electrons in vacuum to go through regions where magnetic field variations are very strong (Fig.1.8), generate extremely high-powered THz radiation [52, 72]. However, they may operate CW or pulsed and provide average brightnesses more than six orders of magnitude higher than typical PC antenna emitters. Free-electron lasers have significant potential in applications where high-power sources are essential or in the investigation of nonlinear THz spectroscopy.

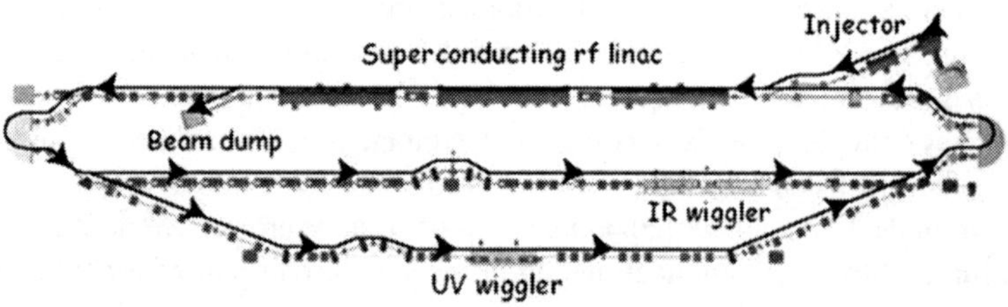

Fig. 1.8 Free-electron laser in Jefferson Lab (courtesy of Dr. Williams, http://www.jlab.org/FEL/felspecs.html)

More recently, long-wavelength quantum-cascade lasers (QCL) have evolved very rapidly and appear to be a promising technique to generate THz. The first QCL was demonstrated in 1994 at Bell Laboratories based on a structure of a series of coupled quantum wells constructed using molecular beam epitaxial (MBE) [73]. Some previous work was done with semiconductor lasers more than 20 years ago using lightly doped p-type germanium with hole population inversion induced by crossed electric and magnetic fields [74]. These lasers are tunable by adjusting the magnetic field or external stress. THz lasing in germanium has also been demonstrated by applying a strong uniaxial stress to the crystal to induce hole population inversion [75]. Recent advances in semiconductor deposition techniques such as MBE allow the construction of multiple quantum well semiconductor structures and THz QCL becomes feasible (Fig.1.9). Light is produced in a QCL by intersubband transitions in the serial of quantum wells. In other words, an upper level acts as the injector, which provides free carriers for the active region, where the transition happens and light is emitted, of the next quantum well and the process is repeated again

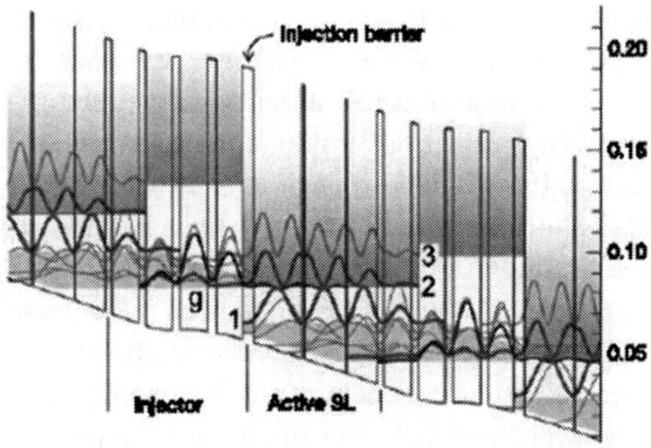

Fig. 1.9 Concept of THz QCL (courtesy of Dr. Tredicucci)

[76]. THz QCLs can be operated in both CW and pulsed modes, and the latter usually works at higher operation temperature, producing a longer lasing wavelength and higher peak power. Recent results reported QCL power above tens of mW [77]. THz QCL is used to work at low operating temperature. As the understanding of charge transport inside these materials improves, the design of QCL also allows higher operating temperature. At this time, QCL can be operated well-above liquid nitrogen temperatures [78] and comes very close to the temperature accessible by thermoelectric cooling (~70 K below room temperature).

Optical techniques have also been pursued for generating narrowband THz radiation. Original efforts began in the 1960s using nonlinear photo-mixing of two laser sources but struggled with low conversion efficiencies [79]. In this technique, two CW lasers with slightly differing centre frequencies are combined in a material exhibiting a high second-order optical nonlinearity such as DAST. Tunable CW THz radiation has been demonstrated by mixing two frequency-offset lasers in LTG GaAs [80] and by mixing two frequency modes from a single multi-mode laser. Further techniques utilize optical parametric generators and oscillators where a Q-switched Nd:YAG laser pump beam generates a second idler beam in a nonlinear crystal and the pump and idler signal beat to emit THz radiation [81–83]. Optical techniques provide broadly tunable THz radiation and are relatively compact due to the availability of solid-state laser sources; output powers in excess of 100 mW (pulsed) have been demonstrated [84]. Optical down-conversion is a rich area for materials research as molecular beam epitaxial and other materials advances allow for engineered materials with improved photo-mixing properties [85].

Far-infrared interferometry techniques using incoherent detectors such as bolometers [86] have also been used to perform detection but lose the coherency that EO sampling and PC antennas provide. Superconductor research has yielded extremely sensitive bolometers based on the change of state of a superconductor such as niobium, and a single-photon detector for THz photons has been demonstrated [87]. This detector offers unparalleled sensitivity using a single-electron transistor consisting of a quantum dot in a high magnetic field. Although detection speeds are currently limited to 1 ms, high-speed designs are proposed, and this has the potential to revolutionize the field of THz detection. The bolometer requires a liquid-helium-cooled environment but, more recently, the use of a Golay cell detector provides a new and convenient approach for interferometry detection because it does not require a low ambient temperature [88]. The responsivity of the Golay is very high but it has a poor time response and there is a maximum input power that usually is in the order of tens of μW. If a higher power is applied, the gas may expand too much and the gold film will be damaged. The beam must be modulated around 10 Hz for detection and to reduce noise effects. Both the bolometer and Golay cell, which detect power rather than the electric field, cannot provide a coherent measurement. At room temperature, semiconductor structures may also be used. The most common is the Schottky diode used as direct detector or with a heterodyne approach. For higher sensitivity, cryogenic cooling is used for heterodyne superconductor detectors. Several superconductor structures can be used and have been for over 20 years. The most widely used is the superconductor–insulator–superconductor tunnel junction mixer [89]. High-temperature superconductors such as YBCO are under investigation for their potentially higher bandwidth operation. Alternative narrowband detectors such as electronic resonant detectors, based on the fundamental frequency of plasma waves in field effect transistors, have been demonstrated up to 0.6 THz [90].

The maximum frequency that commercially available Schottky diodes can achieve is around 0.8 THz using direct detection, but the frequency can go higher if the heterodyne technique is used and then the Schottky diode works as a mixer. A planar Schottky diode mixer has been operated successfully at 2.5 THz for space sensing applications [91]. The requirement to implement a heterodyne approach is that another source with a different frequency is required and the difference between the two frequencies must be within the detection range of the mixer, which can be within the MHz or GHz range.

Time-domain spectroscopy (TDS) is the workhorse of most of the spectroscopic measurements performed in the THz range. It combines the pulse generation by using a femtosecond laser onto a nonlinear crystal or photoconductive antenna with a coherent detection by using an EO crystal or also a PC antenna [92]. Figure 1.10 shows an example of THz wave TDS, the spectrum of water vapor. Although THz power is usually fairly weak in a TDS system, the measurement dynamic range could be as high as 10^6 benefiting from the nature of time gating and coherent detection, thus the background noise is dramatically suppressed. Since the electric field is directly measured, TDS provides both absorptive and dispersive properties of the target [93]. The latter is difficult to directly measure in a traditional optical spectrometer. Another advantage of the THz-TDS is its intrinsic temporal resolution, which allows it to provide ps temporal resolution in a dynamic spectroscopy. The recording speed in TDS is mainly dependent upon time delay scanning in the sampling process. A mirror with circular involute shape can be used in a fast time delay line, where an optical beam is retro-reflected by the involute curve and a linear temporal delay is generated by rotation of the mirror [94, 95]. Driving the mirror using a powerful motor can boost the recording speed of a TDS system up to hundreds of hertz. An alternative to high-speed THz-TDS is spectroscopy with a chirped pulse [96, 97]. This technique allows the full THz waveform to be measured

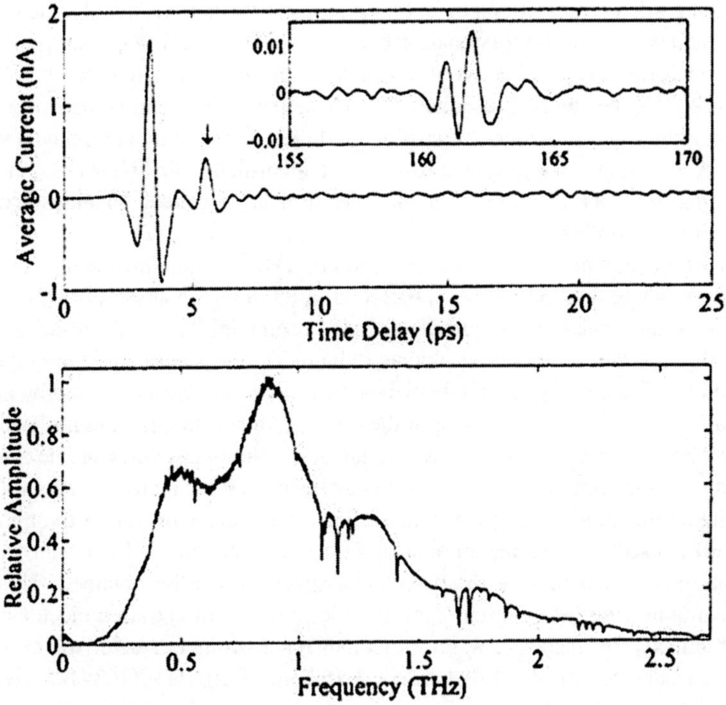

Fig. 1.10 THz waveform after transmission through water vapor (a) and time-domain spectroscopy of water vapor (courtesy of Dr. Grischkowsky)

simultaneously rather than scan the temporal pulse with a stepped motion stage. This technique derives from real-time picosecond optical oscilloscopes [98], and the geometry is similar to the standard THz-TDS.

Explosives and explosive-related compound (ERC) sensing and identification have been a very important topic recently for most security and defense organizations. In particular, a non-contact, fast technique is required in order to be used in real applications. THz-TDS offers several advantages beyond identifying the fingerprints of the explosives. THz waves can, to some extent, penetrate cloth, plastic, paper, and many other non-metallic and dry materials and, thus, provide the possibility to identify the explosives behind optically opaque materials. The identification can be performed either in transmission or in reflection. For a real application, the reflection geometry, especially diffuse reflection, is the most realistic approach. The spectra of ERC samples, such as TNT, RDX, HMX, can be measured with THz-TDS [99–102], and can also be simulated based on density function theory. Petroleum-product analysis has been performed with several spectroscopic techniques such as ultraviolet absorption, infrared absorption, Raman scattering, mass spectroscopy, emission spectroscopy, nuclear magnetic resonance, and chromatography. THz-TDS can also be used in this field to analyze gasoline, diesel, and lubricating oil. The absorption of THz waves in liquid phase is mainly due to vibrational modes that can be torsional or bending. The real and imaginary part of the dielectric permittivity, refractive index, and absorption coefficient can be obtained with THz-TDS because of its coherency. These parameters can be used to identify various grades and providers of gasoline and diesel. Aromatic compounds such as *o*-xylene, *m*-xylene, and *p*-xylene can also be identified with THz-TDS. THz spectroscopy also has applications in the identification and chemical dynamics of organic and biological molecules.

Imaging is another important application of THz waves, which can be classified as passive where no THz source is included in the imaging system or active where the THz source needs to be provided. Passive imaging is based on the detection of the thermal radiation that all objects radiate. The distribution of the radiation is governed by Planck's law and it describes the radiation intensity emitted by a blackbody at a given temperature. Most of the background cosmic radiation is within the THz range, radiated by cool (30 K) stellar dust. The oldest form of THz imaging is passive submillimeter sensing, which has been used for many decades for space imaging applications. In these systems, a heterodyne detector (on Earth or aboard a satellite) is used to sense the amount of THz radiation emitted by distant galaxies. By tuning the frequency of the detector, a spectrum can be obtained. This spectrum contains vital information regarding the presence of certain molecules in that distant galaxy. For instance, water molecules have strong characteristic absorption resonances at 0.557, 0.752, 1.097, 1.113, 1.163, and 1.207 THz [103–105]. By comparing the amplitude of the received THz power at these frequencies relative to the background radiation, astronomers can determine whether water is likely to exist on distant planets. This is a vital tool in the search for extraterrestrial life. Other molecules that can be easily identified using this technique include oxygen, carbon monoxide, and nitrogen [106].

Similarly, passive THz imaging principles have recently been employed in terrestrial applications. This type of imaging system is aided by the fact that a wide variety of common materials, such as plastics, cloth, paper, and cardboard, which are totally opaque in the optical spectrum, have very low absorption coefficients at THz frequencies and thus appear transparent to THz imaging systems. This led groups such as the European Space Agency (ESA) [107] to invest in the development of a passive CCD camera operating at THz frequencies. This project focused on combining micro-machined THz antennas with a silicon photonic band gap back plane to form an imaging array. THz emission by the object, because of its temperature, enables passive imaging but it is also a severe source of noise. For this reason, passive THz imaging methods have had most success in space, where the detector can be mounted on a satellite, away from the strong thermal background that exists on Earth and directed solely at the target of interest.

In active imaging, the target is illuminated with the radiation of interest and then the reflected or transmitted radiation is measured. Active imaging systems can use pulsed or CW illumination. Early THz-imaging systems used gas THz lasers to illuminate the target and thermal detectors [108, 109] or pyroelectric cameras [110] to record the image, but thermal background noise was and still is a common problem. One way to overcome this problem is using a source more powerful than the background. For pulsed systems, the illumination power is compressed into a short pulse width (a few picoseconds) causing a very high-peak illumination power. This high-peak power together with coherent detection techniques allows the use of much lower average power sources while providing the same signal-to-noise ratio (SNR).

Pulsed THz imaging was first demonstrated by Hu and Nuss at Bell Laboratories in 1995 [111] (Fig.1.11), and many applications have been developed including microchips imaging [112], skin burn severity evaluation [113], leaf moisture content [114], tooth cavities [115], and skin cancer [116]. THz imaging has been used for target screening in plastic and low refractive index materials, biomedical imaging, security scanning, and microscopic imaging. Due to its low photon energy (4 meV @ 1 THz), THz wave imaging offers the advantage of being noninvasive and nonionizing, thus representing little harm to biological tissue. In comparison, a typical X-ray photon has an energy in the KeV range and causes ionization and other potentially harmful effects on biological tissue [117, 118]. The wavelength is substantially smaller than that offered by microwave radiation and can provide much higher spatial resolution. When used in a tomography application, THz radiation is also significantly less susceptible to scattering than infrared light, thus allowing improved reconstruction fidelity. THz imaging systems are a recent addition to the wide array of available imaging modalities, but the unique properties of THz radiation allow THz imaging to fill niches that are unreachable using other techniques.

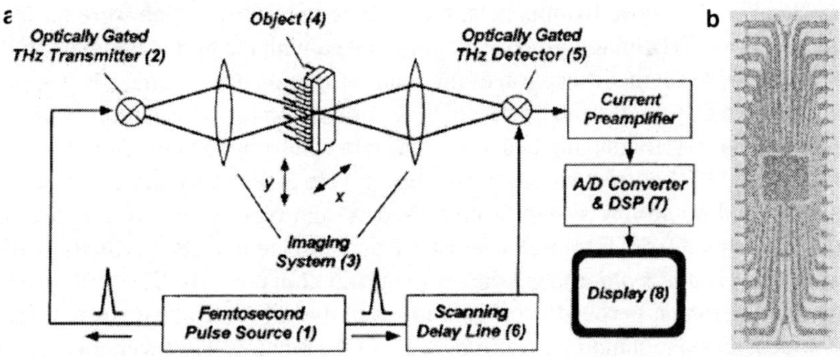

Fig. 1.11 THz wave imaging setup (**a**) and image of an IC chip (**b**) (courtesy of Dr. Nuss)

Classic THz wave imaging works by raster scanning the sample, where the THz beam is focused onto the target. Information on each spot on the target is recorded when scanning the THz beam across the target. Although the raster scanner method secured SNR in the measurement, imaging speed is usually the bottle neck of THz imaging. A telecentric beam scanning technique has been developed, which bursts the raster scanning imaging into multiframes per second [119]. An alternative method to generating real-time THz wave image is to use a so-called 2D imaging method. In this technique, a THz wave is illuminated on the entire target, transmitted or reflected THz waves are imaged by image optics (lenses or mirrors) onto a 2D extended THz wave sensor, which is either a large area EO crystal [120], or a micro-bolometer array [121]. An image of the entire target is recorded spontaneously. Several types of lenses can be designed and fabricated, with the spherical, Fresnel [121–124], and hyperbolic being the most common. The design of a spherical lens is simple, but it results in a thicker lens with prominent aberrations at large apertures. A Fresnel lens is thinner but the focal length depends on the frequency and still shows prominent aberrations at large apertures. A hyperbolic lens corrects for the spherical aberration providing a better resolution, but the thickness is similar to the spherical lens.

Regarding the applications, security inspection was first proposed using far-infrared radiation in the 1970s [125] and it has been extended to THz especially at frequencies below 1 THz. The interest to use THz for security screening is very high because it could perform a similar function to X-ray screening, but with lower cost and health risk. More recently, THz wave imaging has shown up as a very valuable NDE tool and has found some key applications. For instance, a CW compact imaging system has been designed and built to perform scanning of the foam insulation attached to the Space Shuttle fuel tank. The tragedy of the Space Shuttle Columbia, which happened on February 1, 2003 due to the detachment of a foam panel during lift-off, drove NASA to study and analyze possible solutions and technology that could be used to inspect those panels and avoid the detachment. The detachment is mainly due to the presence of defects (voids and delaminations) that may appear during its fabrication. Several technologies and techniques were tested to catch defects in those foam panels, the most promising of which were backscattered X-ray and THz imaging. THz imaging ranked with the best techniques studied [126] and as a result the inspection of the foam insulation has become a key application for THz technology as an NDE tool. The sprayed-on foam insulation (SOFI) is a good subject for THz imaging because it has a low absorption coefficient and index of refraction [127] at frequencies lower than 1 THz. The inspection can be carried out with a pulsed or CW system. The pulsed system requires working at a central frequency around 0.5 THz and the substrate and surface must be perfectly perpendicular to the beam in order to get the pulse information properly. The CW system is more tolerant to non-perpendicular conditions of the substrate and the roughness of the surface but some standing pattern may affect the pictures. However, this standing wave pattern can be reduced resulting in a picture where the defects and the general structure of the foam show up clearly [128].

The use of THz to perform tomography has always been in the mind of researchers. Aside from target reconstruction, it can also provide spectroscopic information of a target, whereas microwave and X-ray imaging modalities produce only density pictures. THz wave tomography of a target can be reconstructed using a Fresnel lens, which is a Fresnel zone plate with phase or amplitude patterns formed by a series of concentric ring structures. The main focal length of this lens is proportional to the frequency of the radiation [129]. Thus an image carried with different frequency responds to the target by locating at a different depth. Although this concept has been exposed using THz wave pulses, it is also applicable to any tunable narrowband imaging beam. Unlike other transmitted THz tomography techniques, which require the rotation of the target [130], this tomography image is achieved without rotating or moving the target.

An alternative method of performing tomography without using a Fresnel lens is THz computed tomography (THz wave CT) which is similar to conventional computed tomography techniques such as X-ray CT. In a THz wave CT measurement, THz wave passes through the target following different paths described by certain horizontal, vertical displacement and angle is recorded and a 3D image of the target is reconstructed using the inverse Radon transformation.

Prospective

Pulsed systems based on optical rectification and photo-conducting antennas have been well developed in recent years and have revealed a large amount of scientific information in the THz region. Figure1.12 summarizes growth of scientific publications in THz research. It has triggered many academic and industrial applications, such as security scanning, defect characterization, label-free gene and protein analysis, bio-chemicals identification, gas and liquid measurement, and dynamics study of semiconductors. In particular, THz-TDS has been very successful in the identification of small molecules such as explosives, medicines, and bio-molecules, because many of the rotational and vibrational peaks of these small molecules are located between 0.3 and 3 THz. However, there are still some issues in these applications which may make further developments difficult:

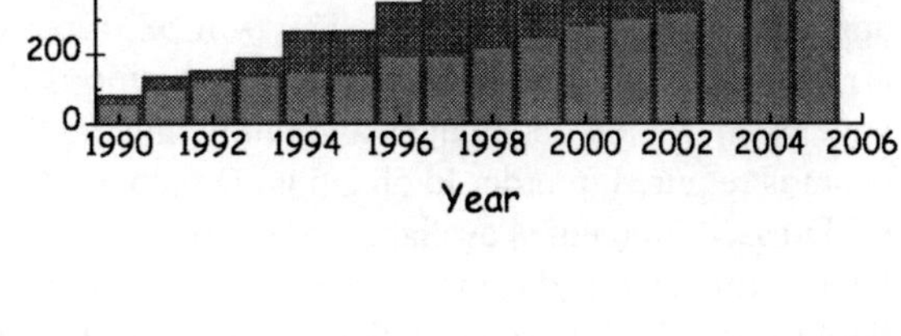

Fig. 1.12 Publications in THz research

- Current pulsed THz wave emitters offer extremely low-conversion efficiency from the input laser power into THz wave power, and a lock-in amplifier has to be used to increase the SNR and time constants around ms are usually set.
- The high water-vapor absorption significantly weakens the THz wave signal during its propagation in air and, therefore, it is a challenge to accomplish remote sensing with THz waves in air over several meters.

- Thick samples, or samples with a high absorption coefficient in the THz range, greatly attenuate the transmitted THz wave and in some cases only the reflected or scattered THz wave signal can be collected and analyzed, although the signal is very weak.

All these issues come down to a common solution of improving the systems with large SNR. A real-world application utilizing a THz-TDS or THz imaging system with a large SNR will provide quick scanning or even snapshot data acquisition, as such a system with a large SNR needs less averaging time for the lock-in amplifier. A powerful source will provide enough THz power to compensate for both water-vapor absorption during propagation and attenuation through the transmitted object, and a very sensitive detector will enable the detection of a severely attenuated or scattered signal from the noise. For some applications, semiconductors such as GaAs and ZnTe are widely used as the THz emitter and detector. However, these materials cannot satisfy some of the applications which have critical high SNR requirements. A better THz emitter and detector would increase the SNR, reduce scanning time significantly, and would make it possible to perform long-distance and scattering spectroscopy or imaging. The new devices would provide better measurement precision, higher sensitivity and larger suitability than that which can be reached currently. With CW applications, water-vapor absorption can be reduced by selecting frequencies within the transmission windows available. Nevertheless, in both cases there are regions of the spectra useless because of its high absorption.

As metals are not transparent to THz, they can be an issue for imaging applications regarding security screening and inspection. THz cannot be used to scan metal packages or regular packages with a high content of metal parts but it can still be used for scanning envelopes and other small mail packages looking for explosives of biological agents, which cannot be done with X-ray. However, it is still an attractive technology because it is safer and more reliable than the X-ray technology widely used today and can be used to screen people looking for weapons and explosives without posing a health risk which is also a key application compared to X-ray technology.

Acquisition speed is a very important technological problem especially for those applications involving imaging. This will be solved enormously when 2D detector arrays become available. Most of the detection is done by a single device, so for raster scan or other techniques that imply moving the sample, the transceiver or the beam is required in order to obtain a 2D picture of the target. Even though the 2D-CCD-based imaging is available for the pulsed system, its use is still limited to the lab environment, and some engineering is required to come up with a compact and field use device. There is a major ongoing effort to create an array of detectors and finally build a THz camera.

Another problem is that the cost in generating THz photons is relatively high compared with its closer technologies. Figure1.13a shows the number of photons generated by each femto-dollar of microwave, infrared, and optical technologies compared with THz. Figure 1.13b compares electrical to photon energy conversion coefficient for waves at different frequencies. It can be seen that THz is the most

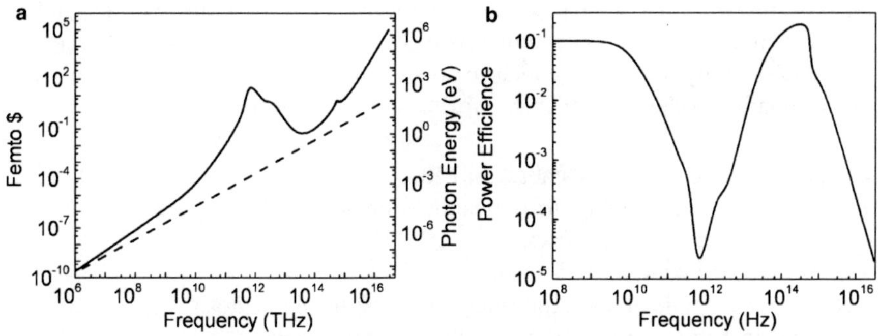

Fig. 1.13 "Photon dollar" (a) and energy conversion coefficient (b) of present coherent light sources in different bands

expensive of all, and in some cases by several orders of magnitude. However, the cost is going down as the research and interest of the industry increases, and it could reach a similar cost as the optical technologies.

Current applications of THz waves are summarized in Fig. 1.14. The future applications of THz technology can be divided into three time periods: short term, medium term, and long term. Short term implies applications that are already using THz technology as a tool and are in the process of optimization and perfection. Medium-term applications are those for which the proof of principle has been demonstrated and that are being developed to move from the laboratory environment to a more realistic environment within 3–5 years. Long-term applications are beyond 5 years and are applications that are envisioned as feasible, but they still show important constraints and issues. Examples of short-term applications are those related to NDE and spectroscopy. Medium-term applications are long-distance sensing and

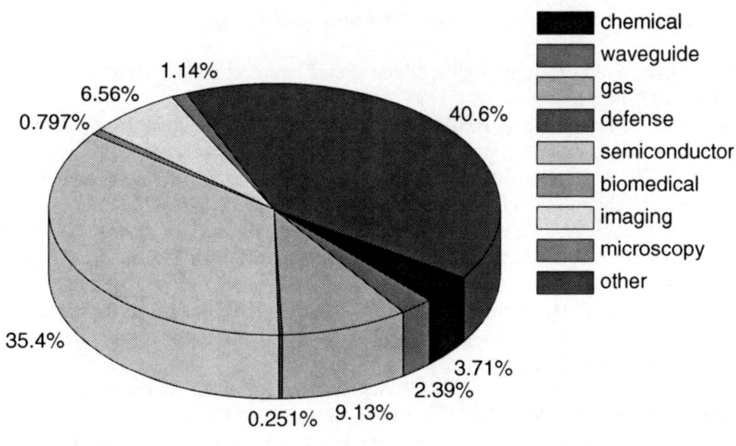

Fig. 1.14 A pie chart of current THz research areas

identification together with imaging. Long-term applications involve the biology and medical fields. The momentum THz technology gained recently will bring it to a level of maturity of its own similar to optics and radio technologies in a few years. Our estimation of the most relevant projects in the future are

- Complete spectroscopic database of ERC.
- THz sensing and imaging instrument development.
- THz camera and spectrometer.
- THz real-time imaging and sensing.
- Biomedical applications: skin cancer and superficial diagnosis.
- Biomedical applications: in vivo analysis and diagnosis.

The main problem facing the application of THz waves within the biomedical field is the high water content of samples. This makes it difficult to use THz for in vivo inspection. However, it may be possible in the future to introduce the THz radiation with some kind of THz fiber and perform imaging or sensing by introducing the probe inside the body.

References

1. C. H. Lee, "Picosecond optoelectronic switching in GaAs," *Appl. Phys. Lett.* **30**, 2, 84–86 (1977).
2. C. H. Lee, and V. K. Mathur, "Picosecond photoconductivity and its applications," *IEEE J. Quantum Electon.* **17**, 10, 2098–2112 (1981).
3. G. A. Mourou, C. Stancampiano, A. Antonetti, and A. Orszag, "Picosecond microwave pulses generated with a subpicosecond laser driven semiconductor switch," *Appl. Phys. Lett.* **39**, 4, 295–365 (1981).
4. J. A. Valdmanis, G. A. Mourou, and C. W. Gabel, "Subpicosecond electrical sampling," *IEEE J. Quantum Electron.* **19**, 4, 664–667 (1983).
5. C. Fattinger, and D. Grischkowsky, "Terahertz beams," *Appl. Phys. Lett.* **54**, 6, 490–494 (1989).
6. C. Fattinger, and D. Grischkowsky, "Point source terahertz optics," *Appl. Phys. Lett.* **53**, 16, 1480–1482 (1988).
7. M. Bass, P. A. Franken, J. F. Ward, and G. Weinreich, "Optical rectification," *Phys. Rev. Lett.* **9**, 11, 446–448 (1962).
8. K. H. Yang, P. L. Richards, and Y. R. Shen, "Generation of far-infrared radiation by picosecond light pulses in $LiNbO_3$," *Appl. Phys. Lett.* **19**, 9, 320–323 (1971).
9. T. Yajima and N. Takeuchi, "Spectral Properties and Tunability of Far-Infrared Differencerequency Radiation Produced by Picosecond Laser Pulses," *Jpn J. Appl. Phys.* 10, 907–915 (1971).
10. X.-C. Zhang, X. F. Ma, Y. Jin, T.-M. Lu, E. P. Boden, P. D. Phelps, K. R. Stewart, and C. P. Yakymyshyn, "Terahertz optical rectification from a nonlinear organic crystal," *Appl. Phys. Lett.* **61**, 26, 3080–3082 (1992).
11. X.-C. Zhang, Y. Jin, K. Yang, and L. J. Schowalter, "Resonant nonlinear susceptibility near the GaAs band gap," *Phys. Rev. Lett.* **69**, 15, 2303–2306 (1992).
12. A. Rice, Y. Jin, X. F. Ma, X.-C. Zhang, D. Bliss, J. Larkin, and M. Alexander, "Terahertz optical rectification from <110> zinc-blende crystals," *Appl. Phys. Lett.* **64**, 11, 1324–1326 (1994).

13. P. Y. Han, M. Tani, F. Pan, and X.-C. Zhang, "Use of the organic crystal DAST for terahertz beam applications," *Opt. Lett.* **25**, 9, 675–677 (2000).
14. A. M. Sinyukov, and L. M. Hayden, "Generation and detection of terahertz radiation with multilayered electro-optic polymer films," *Opt. Lett.* **27**, 1, 55–57 (2002).
15. A. M. Sinyukov, M. R. Leahy, M. Hayden, M. Haller, J. Luo, A.K-Y. Jen, L. R. Dalton, "Resonance enhanced THz generation in electro-optic polymers near the absorption maximum," *Appl. Phys. Lett.* **85**, 24, 5827–5829 (2004).
16. A. Leitenstorfer, S. Hunsche, J. Shah, M. C. Nuss, and W. H. Knox, "Detectors and sources for ultrabroadband electro-optic sampling: Experiment and theory," *Appl. Phys. Lett.* **74**, 11, 1516–1518 (1999).
17. J. P. Hornak, "Imaging technology and systems section", in *The Encyclopedia of Imaging Science and Technology*, vol. 2, pp. 1393–1404, Editor L. Hornak, John Wiley and Sons, New York (2002).
18. G. Chang, C. J. Divin, C.-H. Liu, S. L. Williamson, A. Galvanauskas, T. B. Norris, "Power scalable compact THz system based on ultrafast Yb-doped fiber amplifier," *Opt. Express* **14**, 17, 7909–7913 (2006).
19. M. Nagai, K. Tanaka, H. Ohtake, T. Bessho, T. Sugiura, T. Horosumi, M. Yoshia, "Generation and detection of terahertz radiation by electro-optical process in GaAs using 1.56 μm fiber laser pulses," *Appl. Phys. Lett.* **85**, 18, 3974–3976 (2004).
20. J. Hebling, G. Almasi, I. Z. Kozma, and J. Kuhl, "Velocity matching by pulse front tilting for large area THz-pulse generation," *Opt. Express* **10**, 21, 1161–1166 (2002).
21. G. Imeshev, M. E. Fermann, K. L. Vodopyanov, M. M. Fejer, X. Yu, J. S. Harris, D. Bliss, and C. Lynch, "High-power source of THz radiation based on orientation-patterned GaAs pumped by a fiber laser," *Opt. Express* **14**, 10, 4439–4444 (2006).
22. P. R. Smith, D. H. Auston, and M. C. Nuss, "Subpicosecond photoconducting dipole antennas," *IEEE J. Quantum Electron.* **24**, 2, 255–260 (1988).
23. M. van Exter, C. Fattinger, and D. Grischkowsky, "High-brightness terahertz beams characterized with an ultrafast detector," *Appl. Phys. Lett.* **55**, 4, 337–339 (1989).
24. G. Zhao, R. N. Schouten, N. Van-der-Valk, W. T. Wenckebach, and P. C. M. Planken, "Design and performance of a THz emission and detection setup based on a semi-insulating GaAs emitter," *Rev. Sci. Instrum.* **73**, 4, 1715–1719 (2002).
25. M. Tani, M. Nakajima, S. Kono, and K. Sakai, "Generation of ultra broadband terahertz radiation with a photoconductive emitter," *IEEE/LEOS Annu. Meeting Conf. Proc.* Galsgow, UK, **2**, 532–533 (2002).
26. M. Nakajima, M. Takahashi, and M. Hangyo, "Strong enhancement of THz radiation intensity from semi-insulating GaAs surfaces at high temperatures," *Appl. Phys. Lett.* **81**, 8, 1462–1464 (2002).
27. H. M. Heiliger, M. Vosseburger, H. G. Roskos, H. Kurz, R. Hey, and K. Ploog, "Applications of liftoff low-temperature-grown GaAs on transparent substrates for THz signal generation," *Appl. Phys. Lett.* **69**, 19, 2903–2905 (1996).
28. M. Tani, S. Matsuura, K. Sakai, and S. Nakashima, "Emission characteristics of photoconductive antennas based on low-temperature-grown GaAs and semi-insulating GaAs," *Appl. Opt.* **36**, 30, 7853–7859 (1997).
29. C. Baker, I. S. Gregory, W. R. Tribe, I. V. Bradley, M. J. Evans, M. Withers, P. F. Taday, V. P. Wallace, E. H. Linfield, A. G. Davies, and M. Missous, "Terahertz pulsed imaging with 1.06 μm laser excitation," *Appl. Phys. Lett.* **83**, 20, 4113–4115 (2003).
30. M. Suzuki, and M. Tonouchi, "Fe-implanted InGaAs terahertz emitters for 1.56 μm wavelength excitation," *Appl. Phys. Lett.* **86**, 051104 (2005).
31. P. Y. Han, and X.-C. Zhang, "Coherent, broadband mid-infrared terahertz beam sensors," *Appl. Phys. Lett.* **73**, 21, 3049–3051 (1998).
32. R. Huber, A. Brodshelm, F. Tauser, and A. Leitenstorfer, "Generation and field-resolved detection of femtosecond electromagnetic pulses tunable up to 41 THz," *Appl. Phys. Lett.* **76**, 22, 3191–3193 (2000).

33. K. Reimann, R. P. Smith, A. M. Weiner, T. Elsaesser, and M. Woerner, "Direct field-resolved detection of terahertz transients with amplitudes of megavolts per centimeter," *Opt. Lett.* **28**, 6, 471–473 (2003).
34. X.-C. Zhang, and D. H. Auston, "Optoelectronic measurement of semiconductor surfaces and interfaces with femtoseconds optics," *J. Appl. Phys.* **71**, 1, 326–338 (1992).
35. T. Dekorsy, H. Auer, H. J. Bakker, H. G. Roskos, and H. Kurtz, "THz electromagnetic emission by coherent infrared-active phonons," *Phys. Rev. B* **53**, 7, 4005–4014 (1996).
36. C. A. Schumttenmaer, "Exploring dynamics in the far-infrared with terahertz spectroscopy," *Chem. Rev.* **104**, 4, 1759–1779 (2004).
37. E. Gornik, and R. Kersting, "Coherent THz emission in semiconductors", *Ultrafast Physical Processes in Semiconductors*, vol. 67 of semiconductors and semimetals, chapter 8, Academic Press, New York (2001).
38. A. Leitenstorfer, S. Hunsche, J. Shah, M. C. Nuss, and W. H. Knox, "Femtosecond high-field transport in compound semiconductor," *Phys. Rev. B* **61**, 24, 16642–16652 (1999).
39. J. T. Darrow, X.-C. Zhang, and D. H. Auston, "Power scaling of large-aperture photoconducting antennas," *Appl. Phys. Lett.* **58**, 1, 25–27 (1991).
40. J. T. Darrow, Z.-C. Zhang, D. H. Auston, and J. D. Morse, "Saturation properties of large-aperture photoconducting antennas," *IEEE J. Quantum Electron.* **28**, 6, 1607–1616 (1992).
41. D. You, R. R. Jones, and P. H. Bucksbaum, D. R. Dykaar, "Generation of high-power sub-single-cycle 500-fs electromagnetic pulses," *Opt. Lett.* **18**, 290–292 (1993).
42. P. C. M. Planken, M. C. Nuss, I. Brener, K. W. Goossen, M. S. C. Luo, S. L. Chuang, and L. Pfeiffer "Terahertz emission in single quantum wells after coherent optical excitation of light hole and heavy hole excitons," *Phys. Rev. Lett.* **69**, 26, 3800–3803 (1992).
43. H. G. Roskos, M. C. Nuss, J. Shah, K. Leo, D. A. Miller, A. M. Fox, S. Schmitt-Rink, and K. Kohler, "Coherent Submillimeter-wave emission from charge oscillations in a double-well potential," *Phys. Rev. Lett.* **68**, 14, 2216–2219 (1992).
44. N. Sekine, K. Yamanaka, K. Hirakawa, M. Vosseburger, P. Haring-Bolivar, and H. Kurz, "Observation of terahertz radiation from higher-order two-dimensional Plasmon modes in GaAs/AlGaAs single quantum wells," *Appl. Phys. Lett.* **74**, 7, 1006–1008 (1999).
45. X.-C. Zhang, J. T. Darrow, B. B. Hu, D. H. Auton, M. T. Schmidt, P. Tham, E. S. Yang, "Optically induced electromagnetic radiation from semiconductor surfaces," *Appl. Phys. Lett.* **56**, 22, 2228–2230 (1990).
46. Y. Shimada, K. Hirakawa, and S.-W. Lee, "Time-resolved terahertz emission spectroscopy of wide miniband GaAs/AlGaAs superlattices," *Appl. Phys. Lett.* **81**, 9, 1642–1644 (2002).
47. M. Hangyo, S. Tomozawa, Y. Murakami, M. Tonouchi, M. Tani, Z. Wang, K. Sakai, and S. Nakashima, "Terahertz radiation from superconducting $YBa_2Cu_3O_{7-\delta}$ thin films excited by femtosecond optical pulses," *Appl. Phys. Lett.* **69**, 14, 2122–2124 (1996).
48. C. Jaekel, H. G. Roskos, and H. Kurz, "Emission of picosecond electromagnetic pulses from optically excited superconducting bridges," *Phys. Rev. B* **54**, 10, R6889–R6892 (1996).
49. H. Wald, C. Steigmeier, P. Seidel, S. Nashima, M. Tonouchi, and M. Hangyo, "Comparison of the electromagnetic pulse emission from $YBa_2Cu_3O_{7-\delta}$ and $Y_{0.7}Pr_{0.3}Ba_2Cu_3O_7$ excited by femtosecond laser pulses," *Physica C* **341**, 3, 1899–1900 (2000).
50. M. C. Beard, G. M. Turner, and C. A. Schmuttenmaer, "Measuring intramolecular charge transfer via coherent generation of THz radiation," *J. Phys. Chem. A* **106**, 6, 878–883 (2002).
51. N. Kida, and M. Tomouchi, "Terahertz radiation from magnetoresistive $Pr_{0.3}Ca_{0.3}MnO_3$ thin films," *Appl. Phys. Lett.* **78**, 26, 4115–4117 (2001).
52. N. Kida, and M. Tonouchi, "Reversible and bistable terahertz radiation from magnetoresistive $Pr_{0.3}Ca_{0.3}MnO_3$ thin films," *Appl. Phys. Lett.* **82**, 20, 3412–3414 (2003).
53. G. L. Carr, M. C. Martin, W. R. McKinney, K. Jordan, G. R. Neil, and G. P. Williams, "High-power terahertz radiation from relativistic electrons," *Nature* **420**, 6912, 153–156 (2002).
54. G. L. Carr, M. C. Marting, W. R. McKinney, K. Jordan, G. R. Neil, and G. P. Williams, "Very high power THz radiation sources," *J. Bio. Phys.* **29**, 2-3, 319–325 (2003).

55. I. Wilke, A. M. MacLeod, W. A. Gillespie, G. Berden, G. M. H. Knippels, and A. F. G. van-der-Meer, "Single-shot electron-beam bunch length measurements," *Phys. Rev. Lett.* **88**, 12, 124801 (2002).
56. X. Xie, J. Dai, and X.-C. Zhang, "Coherent control of THz wave generation in ambient air," *Phys. Rev. Lett.* **96**, 7, 075005 (2006).
57. D. H. Auston, and P.R Smith, "Generation and detection of millimeter waves by picosecond photoconductivity," *Appl. Phys. Lett.* **43**, 7, 631–633 (1983).
58. S. Kono, M. Tani, and K. Sakai, "Ultrabroadband photoconductive detection: Comparison with free-space electro-optic sampling," *Appl. Phys. Lett.* **79**, 7, 898–900 (2001).
59. S. Kono, M. Tani, and K. Sakai, "Coherent detection of mid-infrared radiation up to 60 THz with an LT-GaAs photoconductive antenna," *IEE Proc. Optoelectron.* **149**, 3, 105–109 (2002).
60. Q. Wu, and X.-C. Zhang, "Free-space electro-optic sampling of terahertz beams," *Appl. Phys. Lett.* **67**, 24, 3523–3525 (1995).
61. A. Nahata, A. S. Weling, and T. F. Heinz, "A wideband coherent terahertz spectroscopy system using optical rectification and electro-optic sampling," *Appl. Phys. Lett.* **69**, 16, 2321–2323 (1996).
62. P. U. Jepsen, C. Winnewisser, M. Schall, V. Schyja, S. R. Keiding, and H. Helm, "THz electromagnetic emission by coherent infrared-active phonons," *Phys. Rev. E* **53**, 4, 3052–3054 (1996).
63. Q. Wu, and X.-C. Zhang, "7 terahertz broadband GaP electro-optic sensor," *Appl. Phys. Lett.* **70**, 14, 1784–1786 (1997).
64. M. Vossebrger, M. Brucherseifer, G. C. Cho, H. G. Roskos, and H. Kurz, "Propagation effects in electro-optic sampling of terahertz pulses in GaAs," *Appl. Opt.* **37**, 15, 3368–3371 (1998).
65. Q. Wu, and X.-C. Zhang, "Ultrafast electro-optic field sensors," *Appl. Phys. Lett.* **68**, 12, 1604–1606 (1996).
66. A. Brodschelm, F. Tauser, R. Huber, J. Y. Sohn, and A. Leitenstorfer, "Amplitude and phase resolved detection of tunable femtosecond pulses with frequency components beyond 100 THz," *Springer Ser. Chem. Phys.* **66**, 215–217 (2000).
67. R. S. Robert, *An Introduction to Guided Waves and Microwave Circuits*, 1st edn., Prentice-Hall, Englewood Cliffs, NJ, ISBN 0-13-481052-X (1993).
68. F. Maiwald, S. Martin, J. Bruston, A. Maestrini, T. Crawford, and P. H. Siegel, "2.7 THz waveguide tripler using monolithic membrane diodes," *IEEE MTT-S Int. Microwave Symp. Digest* **3**, 1637–1640 (2001).
69. V. Ryzhii, I. Khmyrova, and M. Shur, "Terahertz photomixing in quantum well structures using resonant excitation of plasma oscillations," *J. Appl. Phys.* **91**, 4, 1875–1881 (2002).
70. A. Dobroiu, M. Yamashita, Y. N. Ohshima, Y. Morita, C. Otani, and K. Kawase, "Terahertz imaging system based on a backward-wave oscillator," *Appl. Opt.* **43**, 30, 5637–5646 (2004).
71. E. Mueller, "Submillimeter wave lasers", *Wiley Encyclopedia of Electrical and Electronics Engineering*, vol. 20, pp. 597–615, Editor J. G. Webster, John Wiley & Sons Inc, New York (1999).
72. R. Densing, A. Erstling, M. Gogolewski, H. P. Gemund, G. Lundershausen, and A. Gatesman, "Effective far-infrared laser operation with mesh couplers," *Infrared Phys.* **33**, 3, 219–226 (1992).
73. G. P. Williams, "Far-IR/THz radiation from the Jefferson Laboratory, energy recovered linac, free electron laser," *Rev. Sci. Instrum.* **73**, 3, 1461–1463 (2002).
74. J. Faist, F. Capasso, D. L. Sivco, C. Sirtori, A. L. Hutchinson, and A. Y. Cho, "Quantum cascade laser," *Science* **264**, 5158, 553–556 (1994).
75. S. Komiyama, "Far-infrared emission from population-inverted hot-carrier system in p-Ge," *Phys. Rev. Lett.* **48**, 4, 271–274 (1982).

76. Y. P.Gousev, I. V. Altukhov, K. A. Korolev, V. P.Sinis, M. S. Kagan, E. E. Haller, M. A. Odnoblydov, I. N. Yassievich, and K. A. Chao, "Widely tunable continuous wave THz laser," *Appl. Phys. Lett.* **75**, 6, 757–759 (1999).
77. R. Kohler, A. Tredicucci, F. Beltram, H. E. Beere, E. H. Linfield, A. G. Davies, D. A. Ritchie, R. C. Iotti, and F. Rossi, "Terahertz semiconductor-heterostructure laser," *Nature* **417**, 6885, 156–159 (2002).
78. L. Mahler, A. Tredicucci, R. Kohler, F. Beltram, H. E. Beere, E. H. Linfield, and D. A. Ritchie, "High-performance operation of single-mode terahertz quantum cascade lasers with metallic gratings," *Appl. Phys. Lett.* **87**, 181101 (2005).
79. A. Wade, G. Fedorov, D. Smirnov, S. Kumar, B. S. Williams, Q. Hu, and J. L. Reno, "Magnetic-field-assisted terahertz quantum cascade laser operating up to 225 K," *Nat. Photonics* **3**, 41–45 (2008).
80. F. Zernike Jr. and P. R. Berman, "Generation of Far Infrared as a Difference Frequency," *Phys. Rev. Lett.* 15, 999–1001(1965).
81. J. R. Morris, and Y. R. Shen, "Theory of far-infrared generation by optical mixing," *Phys. Rev. A* **15**, 3, 1143–1156 (1977).
82. E. R. Brown, K. A. McIntosh, K. B. Nichols, and C. L. Dennis, "Photomixing up to 3.8 THz in low-temperature-grown GaAs," *Appl. Phys. Lett.* **66**, 3, 285–287 (1995).
83. K. Kawase, M. Sato, T. Taniuchi, and H. Ito, "Coherent tunable THz wave generation from $LiNbO_3$ with monolithic grating coupler," *Appl. Phys. Lett.* **68**, 18, 2483–2485 (1996).
84. J. Shikata, K. Kawase, M. Sato, T. Taniuchi, and H. Ito, "Enhancement of THz-wave output from $LiNbO_3$ optical parametric oscillators by cryogenic cooling," *Opt. Lett.* **24**, 4, 202–204 (1999).
85. K. Imai, and K. Kawase, "A frequency-agile terahertz-wave parametric oscillator," *Opt. Express* **8**, 13, 699–704 (2001).
86. K. Kawase, J. Shikata, K. Imai, and H. Ito, "Transform-limited, narrow-linewidth, terahertz-wave parametric generator," *Appl. Phys. Lett.* **78**, 19, 2819–2821 (2001).
87. C. Kadow, A. W. Jackson, A. C. Gossard, S. Matsuura, and G. A. Blake, "Self-assembled ErAs islands in GaAs for optical-heterodyne THz generation," *Appl. Phys. Lett.* **76**, 24, 3510–3512 (2000).
88. E. L. Dereniak, and G. D. Boreman, *Infrared Detectors and Systems*, John Wiley & Sons Inc, New York, ISBN-0471122092, (1996).
89. S. Komiyama, O. Astaflev, V. Antonov, T. Kutsuwa, and H. Hirai, "A single-photon detector in the far-infrared range," *Nature* **403**, 405–407 (2000).
90. M. R. Stone, M. Naftaly, R. E. Miles, and J. R. Fletcher, D. P. Steenson, "Electrical and radiation characteristics of semi-large photoconductive terahertz emitters," *IEEE Trans. Microwave Theory Tech.* **52**, 10, 2420–2429 (2004).
91. G. J. Dolan, T. G. Phillips, and D. P. Woody, "Low noise 115 GHz mixing in superconductor oxide barrier tunnel junctions," *Appl. Phys. Lett.* **34**, 5, 347–349 (1979).
92. W. Knap, Y. Deng, S. Rumyantsev, J. Q. Lu, M. S. Shur, C. A. Saylor, and L. C. Brunel, "Resonant detection of subterahertz radiation by plasma waves in the submicron field effect transistor," *Appl. Phys. Lett.* **80**, 18, 3433–3435 (2002).
93. M. C. Gaidis, H. M. Pickett, C. D. Smith, S. C. Martin, R.P. Smith, P.H. Siegel, "A 2.5 THz receiver front end for spaceborne applications," *IEEE Trans. Microwave Theory Tech.* **48**, 4, 733–739 (2000).
94. D. Grischkowsky, S. R. Keiding, M. P. van-Exter, and C. Fattinger, "Far-infrared time-domain spectroscopy with terahertz beams of dielectrics and semiconductors," *J. Opt. Society Am. B* **7**, 10, 2006–2015 (1990).
95. D. Grischkowsky, "An ultrafast optoelectronic THz beam system: Applications to time-domain spectroscopy," *Opt. Photonics News* **3**, 22 (1992).
96. J. Xu, and X.-C. Zhang, "Circular involute stage," *Opt. Lett.* **29**, 17 2082–2084 (2004).
97. J. Xu, Z. Lu, and X.-C. Zhang, "Compact involute optical delay line," *Electron. Lett.* **40**, 19, 1218–1219 (2004).

98. Z. Jiang, and X.-C. Zhang, "Electro-optic measurement of THz field pulses with a chirped optical beam," *Appl. Phys. Lett.* **72**, 16, 1945–1947 (1998).
99. Z. Jiang, and X.-C. Zhang, "Single-shot spatiotemporal terahertz field imaging," *Opt. Lett.* **23**, 14, 1114–1116 (1998).
100. A. Galvanauskas, J. A. Tellefsen, A. Krotkus, M. Oberg, and B. Borberg, "Real-time picosecond electro-optic oscilloscope technique using a tunable semiconductor laser," *Appl. Phys. Lett.* **60**, 2, 145–147 (1992).
101. M. C. Kemp, P. F. Taday, B. E. Cole, J. A. Cluff, A. J. Fitzgerald, and W. R. Tribe, "Security applications of terahertz technology," *Proc. SPIE* **5070**, 44–52 (2003).
102. Y. Chen, H. Liu, Y. Deng, D. Veksler, M. Shur, X.-C. Zhang, D. Schauki, M. J. Fitch, and R. Osiander, "Spectroscopic characterization of explosives in the far-infrared region", *Proc. SPIE* **5411**, 1–8 (2004).
103. K. Yamamoto, M. Yamaguchi, F. Miyamaru, M. Tani, M. Hangyo, T. Ikeda, A. Matsushita, K. Koide, M. Tatsuno, and Y. Minami, "Noninvasive inspection of C-4 explosive in mails by terahertz time-domain spectroscopy," *Jpn. J. Appl. Phys.* **43**, 3B, L414–417 (2004).
104. Y. C. Shen, T. Lo, P. F. Taday, B. E. Cole, W. R. Tribe, and M. C. Kemp, "Detection and identification of explosives using terahertz pulsed spectroscopic imaging," *Appl. Phys. Lett.* **86**, 241116 (2005).
105. H. M. Pickett, R. L. Poynter, and E. A. Cohen, *Submillimeter, Millimeter and Microwave Spectral Line Catalog* (2003). Available: http://spec.jpl.nasa.gov
106. H. M. Pickett, R. L. Poynter, E. A. Cohen, M. L. Delitsky, J. C. Pearson, and H. S. P. Muller, "Submillimeter, millimeter and microwave spectral line catalog," *J. Quant. Spectrosc. Rad. Transfer* **60**, 5, 883–890 (1998).
107. R. L. Poynter, and H. M. Pickett, "Submillimeter, millimeter and microwave spectral line catalog," *Appl. Opt.* **24**, 14, 2235–2240 (1985).
108. P. H. Siegel, "Terahertz technology," *IEEE Trans. Microwave Theory Tech.* **50**, 3, 910–928 (2002).
109. C. M. Mann, P. de Maagt, G. McBride, F. van-de-Water, D. Castiglione, A. McCalden, L. Deias, J. O'Neill, A. Laisne, J. T. Vallinas, I. Ederra, and D. Haskett, "Micro-fabrication of 3D terahertz circuitry," *IEEE MTT-S Int. Microwave Symp. Digest* **2**, 739 (2003).
110. N. I. Malykh, A. G. Nagornyi, and E. S. Yampolskii, "Submillimeter-wave imagery," *Instrum. Exp. Tech.* **18**, 1, 182–184 (1975).
111. T. S. Hartwick, D. T. Hodges, D. H. Barker, and F. B. Foote, "Far-infrared imagery," *Appl. Opt.* **15**, 8, 1919–1922 (1976).
112. A. A. Lash, and D. N. Yundev, "Visualization of submillimeter laser radiation with a pyroelectric television camera," *Instrum. Exp. Tech.* **27**, 3, 736–739 (1984).
113. B. B. Hu, and M. C. Nuss, "Imaging with terahertz waves," *Opt. Lett.* **20**, 16, 1716–1718 (1995).
114. D. M. Mittleman, R. H. Jacobson, and M. C. Nuss, "T-ray imaging," *IEEE J. Select. Topics Quant. Electro.* **2**, 3, 679–692 (1996).
115. D. M. Mittleman, M. Gupta, R. Neelamani, R. G. Baraniuk, J. V. Rudd, and M. Koch, "Recent advances in terahertz imaging," *Appl. Phys. B* **68**, 6, 1085–1094 (1999).
116. S. Hadjiloucas, L. S. Haratzas, and J. W. Bowen, "Measurement of leaf-water content using terahertz radiation," *IEEE Trans. Microwave Theory Tech.* **47**, 2, 142–149 (1999).
117. M. Knott, "See-through teeth," *New Scientist* **162**, 2192, 22 (1999).
118. R. M. Woodward, B. Cole, V. P. Wallace, D. D. Arnone, R. Pye, E. H. Linfield, M. Pepper, and A. G. Davies, "Terahertz pulse imaging of in-vitro basal cell carcinoma samples," *Proc. Conf. Lasers Electro-Opt.*, Baltimore, pp. 329–330 (2001).
119. W. S. Smye, J. M. Chamberlain, A. J. Fitzgerald, and E. Berry, "The interaction between terahertz radiation and biological tissue," *Phys. Med. Biol.* **46**, 9, R101–R112 (2001).
120. G. C. Walker, E. Berry, N. N. Zinov'ev, A. J. Fitzgerald, R. E. Miles, J. Chamberlain, and M. A. Smith, "Terahertz imaging and international safety guidelines," *Proc. SPIE Medical Imaging* **4682**, 683–690 (2002).

121. J. Xu, and G. Cho, "A real-time terahertz wave imager," CLEO, CThN2, San Jose, California (2008).
122. Q. Wu, T. D. Hewitt, and X.-C. Zhang, "Two dimensional electro-optic imaging of terahertz beams," *Appl. Phys. Lett.* **69**, 8, 1026–1028 (1996).
123. A. W. M. Lee, and Q. Hu, "Real-time, continuous-wave terahertz imaging by use of a microbolometer focal plane array," *Opt. Lett.* **30**, 19, 2563–2565 (2005).
124. V. Minin, and O. V. Minin, "System of microwave radio-vision of three-dimensional objects in real time," *Proc. SPIE-Int. Soc. Opt. Eng.* **4129**, 616 (2000).
125. E. D. Walsby, S. Wang, B. Ferguson, J. Xu, T. Yuan, R. Blaikie, S. M. Durbin, D. R. S. Cumming, and X.-C. Zhang, "Investigation of a THz Fresnel lenses", *Chemical Physics, Ultrafast Phenomena XIII*, p. 292, Editors R. D. Miller, M. M. Murnane, N. F. Scherer, and A. M. Weiner, Springer, New York (2002).
126. S. H. Wang, and X.-C. Zhang, "Terahertz tomographic imaging with a Fresnel lens," *Optics & Photonics News Special December Issue Optics in 2002*, 59 (2002).
127. T. S. Hartwick, "Far-infrared imaging techniques for law enforcement applications," *Opt. Sec. Law Enforcement (SPIE)* **108**, 139 (1977).
128. NASA's Implementation Plans for Space Shuttle Return to Flight and Beyond, vol. 1, rev. 2, April 26 (2004).
129. J. Xu, H. Zhong, T. Yuan, X. Xu, X.-C. Zhang, R. Reightler, and E. Madras, "T-rays identify defects in insulating materials," CLEO, CMB2, San Jose, California (2004).
130. N. Karpowicz, H. Zhong, C. Zhang, K. Lin, J. S. Hwang, J. Xu, and X.-C. Zhang, "Compact continuous-wave sub-Terahertz system for inspection applications," *Appl. Phys. Lett.* **86**, 054105 (2005).
131. J. Jahns, and S. J. Walker, "Two-dimensional array of diffractive micro-lenses fabricated by thin film deposition," *Appl. Opt.* **29**, 7, 931–936 (1990).
132. S. H. Wang, and X.-C. Zhang, "Tomographic imaging with a terahertz binary lens," *Appl. Phys. Lett.* **82**, 12, 1821–1823 (2003).

Chapter 2
Generation and Detection of THz Waves

Before discussing the nature of THz waves and their applications, it is suitable to introduce how THz waves are generated and detected. As mentioned in Chapter 1, this book will focus on pulsed THz technologies. A typical pulsed THz wave generation and detection system is a pump and probe setup as presented in Fig. 2.1. The most common way that pulsed systems work is by splitting a beam from a femtosecond (fs) laser into two beams: the probe and the pump beams. The pump beam is used to generate the THz pulse, while the probe beam is used to sample and obtain the pulse profile. Detecting of THz field is performed by modulating the probe pulse with the THz field or by accelerating free carriers induced by the probe pulse with the THz field. A mechanical delay line is used to change the time delay between THz pulse and the probe pulse. The THz waveform can be obtained by scanning this time delay. To increase the sensitivity, the pump beam is modulated by an optical chopper, and the THz-induced modulation on the probe beam is extracted by a lock-in amplifier. This pulse information acquired in the time domain is transformed to the frequency domain with a Fourier transform from which spectral information can be obtained.

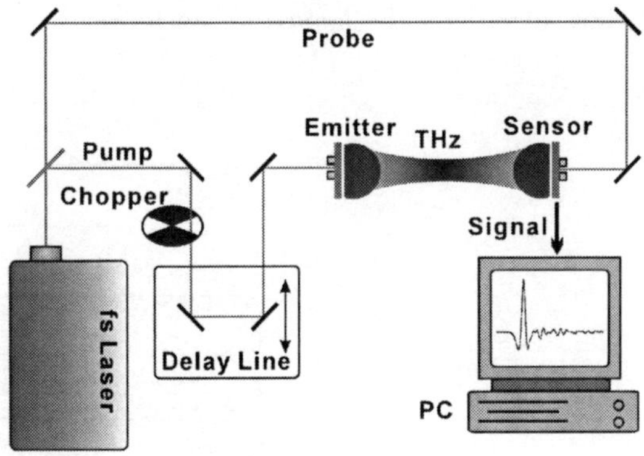

Fig. 2.1 Pulsed THz wave generation and detection setup

X.-C. Zhang, J. Xu, *Introduction to THz Wave Photonics*,
DOI 10.1007/978-1-4419-0978-7_2, © Springer Science+Business Media, LLC 2010

Photoconductive Antenna

The photoconductive (PC) antenna is one of the most frequently used components for THz generation as well as detection. It generates and detects THz pulses by transient photocarriers induced with ultrafast laser pulses. Figure 2.2 gives a sketch of a PC antenna and concept of how to use a PC antenna to generate THz pulses. A PC antenna consists of two metal electrodes that are coated on a semi-insulation semiconductor substrate with a gap between these two electrodes. To generate THz pulses, voltage is applied across the electrodes. Since the substrate is semi-insulating, electric energy is stored in the gap area. Ultrafast laser pulses act like transient switches to open this reservoir of electric energy and release it in the form of THz pulses. Polarization of the THz wave radiated is parallel to the biased field, which is perpendicular to the strip lines in Fig. 2.2. Polarization of THz wave can be altered by switching polarization of bias. By doing so, the laser pulses must have enough photon energy in order to generate photoinduced free carriers in the substrate. Generally, the photon energy of the excitation optical pulse should be higher than band gap of the substrate. Sometimes, multiphoton absorption could be used, and an excitation laser with lower photon energy can also generate free carriers. Free carriers are driven by bias field across the gap and produce photocurrent. Since electrons usually have much higher mobility than holes, the contribution of holes can be ignored in most cases. The current density is described as

$$J(t) = N(t)e\mu E_b, \tag{1}$$

where N is density of photocarriers, e denotes the elementary charge, μ is the mobility of electron, and E_b is the bias electric field. The photocarrier density N is a function of time, whose format is determined by the laser pulse shape and the carrier lifetime. Since the photocurrent varies in time, it generates electromagnetic pulse, whose electric field is approximately

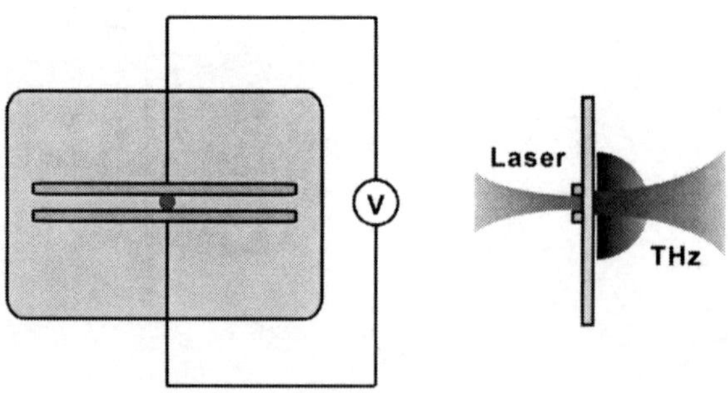

Fig. 2.2 PC antenna and THz emission from PC antenna

$$E_{\mathrm{THz}} = \frac{1}{4\pi\varepsilon_0}\frac{A}{c^2 z}\frac{\partial J(t)}{\partial t} = \frac{Ae}{4\pi\varepsilon_0 c^2 z}\frac{\partial N(t)}{\partial t}\mu E_b, \tag{2}$$

where A is the area in the gap illuminated by the laser light, ε_0 is the vacuum permittivity, c is the speed in vacuum, and z is the distance between the field point and the THz source. To derive Equation (2), the field point is assumed located at normal to the PC antenna and the distance between the field point and the source is much larger than the dimension of the PC antenna.

The energy of the THz pulse comes from the electric energy stored across the gap rather than the optical pulse energy. In principle, the pulse energy of THz radiation is not limited by the pulse energy of excitation laser and optical to THz quantum conversion efficiency greater than one is possible. However, the pulse energy of THz radiation has a very immediate relationship with the excitation laser pulse energy. In fact the excitation acts as a trigger to release the stored energy into THz radiation. The more photocarriers being generated, the more stored energy is converted into THz radiation. Under weak excitation condition, pulse energy of the THz wave is proportional to pulse energy of the excitation laser. In reality, linear relationships between the biased field and THz field, as well as between the excitation pulse energy and THz field, is only true under weak excitation and low bias field. When the substrate of the PC antenna is excited, it is no longer a semi-insulating material, but rather a conductive medium. As a result, the induced field screens the biased field, and the photo current is modified from Equation (1) to [1]

$$J(t) = \frac{\sigma(t)E_b}{\dfrac{\sigma(t)\eta_0}{1+n}+1}, \tag{3}$$

where σ is the conductivity of the substrate, η_0 denotes the impedance of air, which is $\eta_0 = 377\Omega$, and n is the refractive index of the substrate. The substrate conductivity σ is induced by the excitation laser, and can be considered as$\sigma \propto I_0$, where I_0 is the laser intensity. Combining Equation (2) with (3), the THz field is:

$$E_{\mathrm{THz}} \propto \frac{d\sigma(t)}{dt}\frac{1}{\left[1+\dfrac{\sigma(t)\eta_0}{1+n}\right]^2} \propto \frac{I_O}{(1+kI_O)^2}. \tag{4}$$

Here $k = \kappa(t)\eta_0/(1+n)$, where $\kappa(t)$ denotes the ratio between $\sigma(t)$ and I_0. Equation (4) clearly shows that when the excitation laser is strong enough, the impedance of the substrate becomes comparable to the air, and the THz field becomes saturating to the excitation laser power.

Increasing the biased field also has a limitation, since a high electric field may cause dielectric breakdown in the substrate. Breakdown of a PC antenna in THz wave generation can be grouped into two categories: field induced breakdown and thermal induced breakdown. Field induced breakdown happens when the biased field is higher than the breakdown field of the semiconductor material, i.e. 4×10^5 V/cm for GaAs. It usually happens instantly with field-induced avalanche. Thermal induced breakdown is caused by heating of the substrate by photocurrent flow in the substrate (as well as photon absorption of the excitation laser). Heating of the substrate reduces its resistance, which leads to even higher current flow. The thermal induced breakdown is a slow process, and usually takes seconds to even minutes. Most breakdowns of PC antennas in THz wave generation are thermal induced except for those having a very small gap or high bias. The thermal induced breakdown of PC antennas sets a tradeoff between the excitation laser intensity and the maximum biased voltage. Proper treatment or coating on the substrate surface could increase the breakdown field and thus enhance THz wave generation from the PC antenna.

Using a PC antenna as a THz detector is quite similar to using it as a THz emitter. The only major difference is that as a detector, its two electrodes are connected to a current sensor rather than a power supply. In the THz generation and detection system presented in Fig. 2.1, by controlling the time delay between the THz pulse and the optical probe pulse, the electric field across the stripline of the PC antenna at any given point in time can be sampled by the optical probe pulse which serves to generate transient photocarriers in the substrate at that specific time. Since the THz pulses and laser pulses remain for a certain time delay, the photoinduced carriers see a steady electric field, and are driven by this field to form current between the two electrodes. The THz field induced current is

$$\overline{J} = \overline{N} e \mu E(\tau). \tag{5}$$

Here $\overline{N}$ denotes the average electron density, and τ is the temporal delay between probe pulse and the THz pulse. By scanning the temporal delay, the THz pulse waveform as a function of τ is recorded.

Figure 2.3 shows a typical waveform of THz pulse. The period of the THz oscillation is about 1 picosecond (ps), and typical THz pulses are a sub cycle to a few

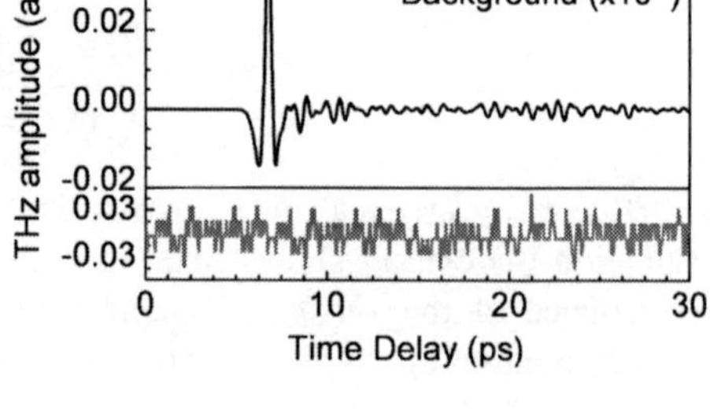

Fig. 2.3 Temporal waveform of a THz pulse

cycles of oscillation. As shown in Equation (5), the detection of THz pulses directly records its field rather than intensity. The measurement records not only amplitude, but also phase information of the THz pulse, while the latter is rather difficult to be directly measured in optics.

Performance of a PC antenna depends mainly on the following factors: the substrate material, geometry of the active area, geometry of antenna, and the excitation laser pulse. Materials with a short carrier lifetime, such as LT-GaAs or doped silicon, are usually selected as the substrate in order to increase the response speed of the PC antenna. The response speed is essential to generate and detect THz pulses containing high frequency components. Higher carrier mobility is also a desired character since it results in the high efficiency of THz wave generation. High dark resistance is required for the substrate in order to have sufficient breakdown voltage across the antenna. The active area of the PC antenna, which is the gap area with laser excitation, is another crucial aspect of a PC antenna, since that is the active area to generate and detect THz waves. Careful design of the field distribution in the gap can increase the breakdown field and allow the PC antenna to generate more intense THz pulses [2]. Distribution of excitation light within the gap is also important. It is arguable that concentrating the excitation beam close to the anode, where the electric potential has the highest slope, leads to higher THz wave generation [3]. It is also arguable that concentrating the excitation beam on the high field area may also result in lower breakdown voltage. PC antennas with smaller gaps are more sensitive, especially when low excitation power is used. However, larger gaps allow higher excitation power and bias voltage to be applied on the PC antenna, thus generating higher power THz waves. Larger active area can also help to lower the screening effect. Shape of the antenna is crucial to optimize coupling of THz waves between the device and free space. In terms of frequency response, various antennas are divided into two groups: resonant and nonresonant antennas. The former has a resonant frequency, which emits THz waves around a certain central-frequency. The dipole-antenna is the most widely used resonant antenna, which emits THz wave with a central wavelength of $\lambda_n = 2L/m$. Here λ_n is wavelength in the substrate, and the response wavelength in the free space is $\lambda = \lambda_n \times n$, where n is refractive index of the substrate. L denotes the width of the antenna from one side of the anode to the other side of the cathode. m can be any positive integer. A nonresonant antenna has a variable width, and leads to a broader frequency response range. Geometries of a nonresonant antenna include bowtie, spiral, and logarithmic periodic antennas. Optics, such as hyperhemispherical silicon lens, can also be used to enhance the coupling coefficient.

PC antennas are not only used to generate and detect THz pulses. The similar device can also be used to generate and detect CW THz waves. In such as system, two CW laser beams with different frequencies illuminate at the same spot on the PC antenna. Beating between these two laser beams results in an oscillation of the laser intensity at the overlapped spot. This further induced the oscillation of photocurrent, which emits electromagnetic wave. The PC antenna acts as a frequency mixer, and narrow band CW THz wave is generated when the beating frequency lies in the THz band.

Built-In Field in Semiconductor

The surface states in some semiconductors, such as GaAs, can be used for THz generation. The Fermi level of a surface state may be different than in the bulk material. This difference of Fermi levels induces bending of energy bands just beneath the surface. The built-in surface field is formed by the band bending area. Figure 2.4 presents the band bending and surface field in an n-type GaAs wafer. The Fermi level of n-type GaAs is close to the conduction band, thus higher than the Fermi level of the surface state, which is closer to center of the band gap. The surface field drives free electrons drifting toward the inside of the bulk material. As a result, the free electron density in this layer with surface field is much lower than the bulk material, and it is so called the depletion layer.

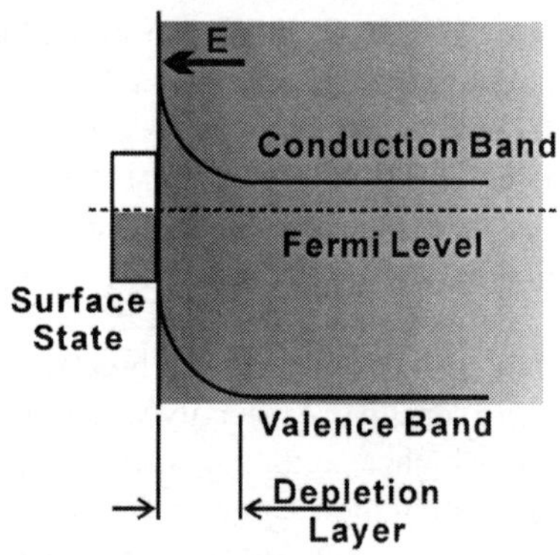

Fig. 2.4 Schematic of band bending and surface field of a n-type GaAs wafer

Without excitation, a balance is reached between the drift and the diffusion of free carriers in semiconductor, including the depletion layer. Therefore, net charge movement is observed in a macro scale. When a laser pulse is absorbed in the ablation layer, the photo induced electron hole pairs will be accelerated by the existing electric field just like what happens in a PC antenna. In n-type GaAs, for instance, electrons are driven toward the inside of the wafer, while holes are driven in the opposite direction. Dipole oscillations occur until a new balance is reached. THz radiation can be estimated using dipole radiation

$$E_{\text{THz}} \propto \frac{\partial N(t)}{\partial t} e\mu E_S \sin\theta, \quad (6)$$

where E_s is the built-in surface field, and θ is the angle between the radiation direction and the normal of the dipole oscillation, which is perpendicular to surface of the semiconductor. Since the surface field in a p-type GaAs wafer points to the opposite direction than a n-type wafer, THz pulses generated from a p-type GaAs wafer have a reversed polarity compared with those generated from a n-type wafer. Equation (6) shows that the radiation has a nonuniform angular distribution with the maximum radiation angle parallel to surface of the semiconductor wafer. This angular distribution is not favorable for THz wave generation, especially because semiconductors mostly have a large refractive index. For example, the refractive index of GaAs is 3.6 for THz waves, resulting in large portions of generated THz radiation that cannot be coupled into the free space, without additional coupler, such as index matching lens or prism, due to total internal reflection.

If the semiconductor wafer is illuminated with a laser beam and the illumination area is much larger than THz wavelength, interference guides THz waves toward the same direction where the optical beam propagates as well as reflects off the semiconductor surface. According to Equation (6), THz radiation is more efficient when optical beam incident angle is larger. Figure 2.5 shows the amplitude of THz radiation as a function of the optical beam incident angle. The maximum radiation is obtained when the incident angle is close to the Brewster angle. A larger angle does not result in higher generation efficiency, which is caused by the Fresnel loss. Although a large incidence angle improves the coupling coefficient of THz radiation from the semiconductor surface, since semiconductors usually have high refractive index, the real angle with respect to normal of the dipole oscillation is very small. For instance, when GaAs is used, the maximum θ is only 16° even at grazing incident, which gives less than 30% coupling efficiency. When a magnetic field is applied parallel to the semiconductor surface, movement of carriers is bent by the magnetic field and coupling coefficient is dramatically increased.

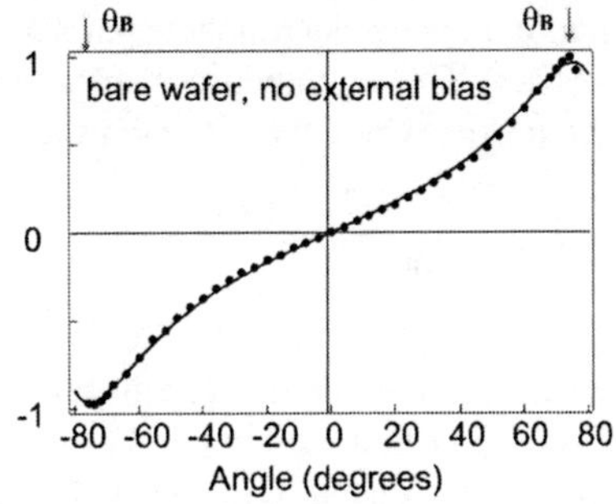

Fig. 2.5 Amplitude of THz radiation generated from semiconductor surface field as a function of excitation beam incident angle. Excitation beam is p polarized

According to Equation (6), the THz field is linearly proportional to the built-in field. Increasing the built-in field will lead to stronger THz wave generation. To increase the surface field, one can either increase the Fermi level difference between the surface state and bulk state, or decrease the thickness of depletion layer. This can be accomplished by growing a very thin layer of low-temperature-grown GaAs (LT-GaAs) film on n-type GaAs wafer. Since LT-GaAs film has a very high density of defect state, the Fermi level of the bulk GaAs is pinned to the defect state, which is located close to center of the band gap. If LT-GaAs is doped with boron, the boron ions lower the Fermi level in defect state and thus increase the surface field. The change of Fermi level in defect state is [4]

$$E_F - E_d = kT \ln \left\{ \frac{\exp(w/kT) - \exp(fw/kT)}{g[\exp(fw/kT) - 1]} \right\}, \quad (7)$$

where E_F and E_d denotes the Fermi level and bottom of the defect state, respectively, k is the Boltzmann constant, w is the width of the defect band, f indicates the ionization possibility of deep donors, and g is a degeneracy factor. Doping boron in LT-GaAs can increase the difference of surface level from 0.65 to 0.75 eV and enhance THz wave radiation.

Built-in fields not only exist on the surface of semiconductor wafers, but also at boundaries between different layers of materials, especially if there are strong built-in fields existing in junctions. By investigating THz wave radiation from built-in fields in a semiconductor material, one can study the properties of the material.

Photo-Dember Effect

Even if there is no built-in field or only a very weak built-in field exists, exciting semiconductor materials with ultrafast laser pulses may also generate THz pulses through the photo-Dember effect. Figure 2.6 shows concept of photo-Dember effect. When a laser pulse with photon energy higher than the band gap illuminates a semiconductor wafer, free electron hole pairs are generated due to absorption of laser photons. The laser beam is strongly absorbed by semiconductor material, so that photo induced electron hole pairs have very inhomogeneous distribution close to the surface. The asymmetric distribution causes electrons and holes to diffuse toward the inside, where the diffusion speed is

$$\frac{\partial N}{\partial t} = D \frac{\partial^2 N}{\partial z^2}, \quad (8)$$

where z is the coordinate unit toward the inside of the semiconductor, D is the diffusion constant, which can be obtained from Einstein relationship as $D = k_B T \mu$. Since electrons have higher mobility than holes, they are able to diffuse faster. The different diffusion speed between electrons and holes leads to a charge separation

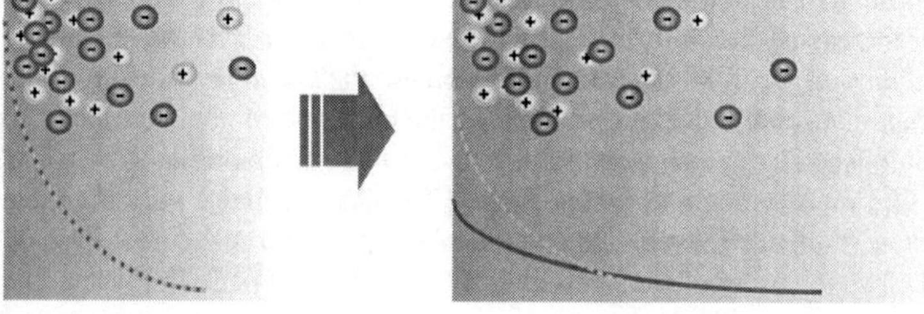

Fig. 2.6 Photo-Dember effect on InAs surface

in the semiconductor and generates a transient photo-Dember field. Emission of the transient photo-Dember field generates THz radiation.

Due to its high electron mobility, InAs is considered a promising THz emitter among other narrow band-gap semiconductors. According to the discussion in the previous section, THz wave generated from n- and p-type GaAs wafers should have opposite polarity. This prediction has been verified in experiment; however, when InAs is used to generate THz pulses, both n- and p-type InAs generate THz pulses with the same polarity. The polarity for both materials is the same as what is generated from an n-type GaAs wafer. This phenomenon is difficult to explain by surface-field-induced THz wave generation, while it agrees with the prediction given by the photo-Dember effect. In fact, both surface field acceleration and the photo-Dember effect exist when ultrafast laser pulses excite the semiconductor surface. Which mechanism dominates THz wave generation depends on properties of the excitation laser pulse as well as the semiconductor material. Table 2.1 compares properties of InAs and GaAs. The following factors make THz wave generation from GaAs and InAs due to different mechanisms. First, GaAs has a larger band gap than InAs, so that it has higher surface field. A higher surface field is favorable for generating THz waves through surface-field-induced photocurrent. Second, InAs has stronger absorption of the excitation light than GaAs; as a result it has a larger slope in photo carrier distribution. And finally, since InAs has a lower band gap, the free electrons have residue energy after being excited, which leads to a higher carrier temperature. The latter two factors give InAs stronger photo-Dember effect properties than GaAs. Similar to surface-field-induced THz wave generation, photocarrier oscillations in the photo-Dember process are also perpendicular to the semiconductor surface and thus have small external coupling efficiency.

Table 2.1 Comparing properties of GaAs and InAs

	Band gap (eV)	Electron mobility ($cm^2V^{-1}s^{-1}$)	Absorption depth (nm)	Residue energy (eV)
GaAs	1.43	8,500	1,000	0.05
InAs	0.35	40,000	150	0.5

When the dynamics of optical excitation and doping is taken into account, Equation (8) can be further modified to [5]

$$\frac{\partial N_i(z,t)}{\partial t} = G(z,t) + \frac{\partial}{\partial z}\left(D_i(z,t)\frac{\partial N_i(z,t)}{\partial z}\right) \pm \frac{\partial}{\partial z}[\mu_i(z,t)E(z,t)N_i(z,t)], \tag{9}$$

where N_i is carrier density, with $i{=}e,h$. The electric field E can be derived from Maxwell equation as

$$\frac{\partial E(z,t)}{\partial z} = \frac{e}{\varepsilon_s\varepsilon_0}[N_h(z,t) - N_e(z,t)]. \tag{9a}$$

The first item on the right side of the Equation (9) gives gain of photocarrier density due to laser excitation and it can be described as

$$G(Z,t) = I(t)\alpha(1 - R)e^{-\alpha z}, \tag{9b}$$

where α is the absorption coefficient, and R is the reflectance of the laser beam on InAs surface. The second term indicates the diffusion as presented in Equation (8), and the third item is the drift caused by the electric field, whose sign is determined by the sign of carrier selected. Equation (9) describes drift and diffusion of free carriers in semiconductor, which include photocarriers and doping-induced free carriers. Solving Equation (9) results in THz pulse generation from the photo-Dember effect. Figure 2.7 shows calculation and experimental results for the amplitude of generated THz pulses as a function of doping density in n- and p-type InAs wafers [6]. Doping-induced free carriers screen THz generation from the InAs material, and as a result, higher doping density leads to lower THz power. Since electrons have higher mobility than holes, n-type InAs gives a higher screening effect than p-type InAs with same doping level, and hence generate weaker THz radiation.

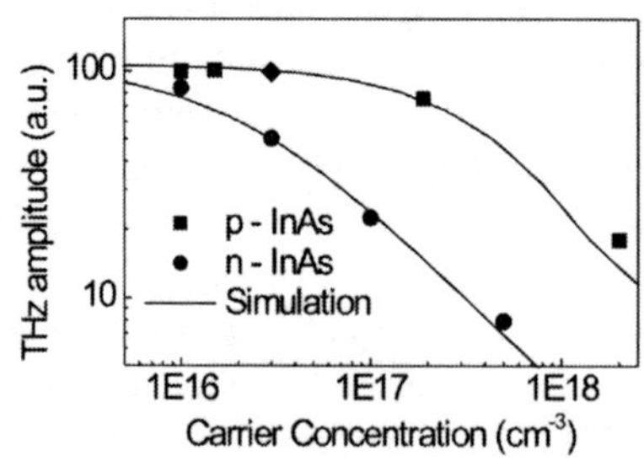

Fig. 2.7 Amplitude of THz field generated via photo-Dember effect as a function of doping. *Solid curves* are calculated using Equation (9). *Dots* are experimental results, where *circles* represent n-type (100) crystal, squares represent p-type (100) crystal, and diamonds represent p-type (111) crystal

Optical Rectification

THz wave generation from the above mechanisms utilize real photoinduced carriers. Here we discuss several THz wave generation methods using nonlinear optical process, which can be considered as virtual carriers induced by optical excitation. Although many different physical principles, such as surge current, Bloch oscillation, and coherent phonon and plasma oscillation, can be used to generate freely propagating electromagnetic waves in the THz region, THz generation by optical rectification has the unique advantage of extremely broad spectral bandwidth.

Optical rectification is a second-order nonlinear optical effect. It is basically a difference-frequency generation with the frequency difference close to zero. Typically, femtosecond laser pulses are used to generate THz from EO crystals via optical rectification. Because a femtosecond pulse contains many frequency components, any two frequency components contribute to the difference-frequency generation, and the overall result is the weighted sum of all the contributions. One femtosecond laser pulse is enough to stimulate optical rectification radiation, which makes the experiment very simple.

Mathematically, the polarization P can be expanded into a power series of the electric field E

$$\begin{aligned}\mathbf{P}(\mathbf{r},t) = \chi^{(1)}(\mathbf{r},t)\,\mathbf{E}(\mathbf{r},t) + \chi^{(2)}(\mathbf{r},t):\mathbf{E}(\mathbf{r},t)\,\mathbf{E}(\mathbf{r},t)\\ +\chi^{(3)}(\mathbf{r},t):\mathbf{E}(\mathbf{r},t)\,\mathbf{E}(\mathbf{r},t)\,\mathbf{E}(\mathbf{r},t) + \ldots\end{aligned}, \tag{10}$$

where $\chi^{(n)}(\mathbf{r},t)$ is the nth-order nonlinear susceptibility tensor. Optical rectification comes from the second term of Equation (10). If the incident light is plane wave, then E can be expressed as

$$\mathbf{E}(t) = \int_0^{+\infty} E(\omega)\exp(-i\omega t)\,\mathrm{d}\omega + \text{c.c.} \tag{11}$$

By substituting Equation (10) into Equation (11) the polarization for optical rectification is given by

$$\begin{aligned}\mathbf{P}_{\mathrm{OR}}^{(2)}(t) &= 2\chi^{(2)}:\int_0^{\infty}\int_0^{\infty}\mathbf{E}(\omega_1)\,\mathbf{E}^*(\omega_2)\exp[-i(\omega_1-\omega_2)\,t]\,\mathrm{d}\omega_1\mathrm{d}\omega_2\\ &= 2\chi^{(2)}:\int_0^{\infty}\int_0^{\infty}\mathbf{E}(\omega+\Omega)\,\mathbf{E}^*(\omega)\exp[-i\Omega t]\,\mathrm{d}\Omega\,\mathrm{d}\omega,\end{aligned} \tag{12}$$

where Ω is the frequency difference of two optical frequency components. In the far field, the radiated electric field $E_r(t)$ is proportional to the second derivative of $\mathbf{P}_{\mathrm{OR}}^{(2)}(t)$ with respect to time t,

$$\mathbf{E}_r(t) \propto \frac{\partial^2}{\partial^2 t}\mathbf{P}_{\mathrm{OR}}^{(2)}(t)\,. \tag{13}$$

The susceptibility tensor $\chi^{(2)}$ depends on the crystal structure. Given a crystal structure and incident light, Equation (13) can be used to calculate the far-field waveform of the radiation. Many factors, such as materials, crystal orientation, thickness, absorption and dispersion, diffraction, phase matching, and saturation, affect the radiation efficiency, waveform, and frequency distribution.

Phase matching is the most important factor for a nonlinear process such as THz generation from optical rectification. Phase matching requires conservation of energy and momentum in the nonlinear process, which is described by:

$$\begin{cases}\omega_{O1}-\omega_{O2}=\Omega_{\mathrm{THz}}\\ k_{O1}-k_{O2}=k_{\mathrm{THz}}\end{cases}, \tag{14}$$

where ω_{01}, ω_{02} and k_{01}, k_{02} are frequencies and wave vectors of optical waves involved in THz wave generation, respectively. Ω_{THz} and k_{THz} are the frequency and wavelength of the generated THz wave. Only when phase matching is satisfied, all three waves participating in the optical rectification process can keep in phase and lead to maximum energy conversion coefficient along the light propagation. Phase mismatch leads to a phase walk off while propagating. The coherence length is defined by the interaction length when the phase change reaches π

$$\delta k L_C = \pi, \tag{15}$$

where $\delta k = k_{01} - k_{02} - k_{\mathrm{THz}}$. To generate THz efficiently from a bulk crystal, the thickness of crystal cannot be selected longer than the coherent length in order to avoid conversion cancellation due to phase mismatch. Since THz frequency is much lower than optical frequency, Equations (14) can be simplified through dividing the first equation by the second one

$$\frac{\partial \omega_O}{\partial k_O} = \frac{\Omega_{\mathrm{THz}}}{k_{\mathrm{THz}}}. \tag{16}$$

According to electromagnetic principle, Equation (16) gives

$$v_{G,O} = v_{\mathrm{Ph,Thz}}. \tag{17}$$

This means phase matching is satisfied in THz wave generation when the group velocity of the optical beam equals phase velocity of the THz beam. Now we can understand the phase matching condition in a more straightforward way. Since the optical pulse has a much higher frequency than the THz pulse, the THz pulse only sees the profile of the optical pulse rather than the oscillations. To have maximum energy conversion, the optical pulse should have a constant temporal delay according to the THz pulse along the entire interaction length. Equation (17) gives the collinear phase matching condition, where phase matching occurs when the excitation and THz beams collinearly propagate through the nonlinear crystal. This type of phase matching not only gives a high generation coefficient due to long interaction length, but also generates THz radiation with very good beam quality.

According to the excitation wavelength and properties of the nonlinear material, optical rectification can happen under three different conditions. If the photoenergy of the excitation beam is higher than the band gap of the nonlinear material, the optical beam will be absorbed within a fairly short distance. In this case, usually phase matching is not very important, since the interaction range is much shorter than the coherence length. Even with a short absorption distance, there is typically strong THz generation due to a resonating enhanced nonlinear process. When the photon energy is less than the band gap of the nonlinear material, the excitation laser is able to propagate through the nonlinear crystal for a long distance. THz generation is also different with different phase matching conditions. If phase matching is satisfied in optical rectification, the generated THz field will continually increase along the entire depth of the nonlinear crystal. Therefore, strong THz wave generation is expected. If phase matching is not satisfied, generated THz waves will be canceled after each coherence length and THz radiation efficiency will be low.

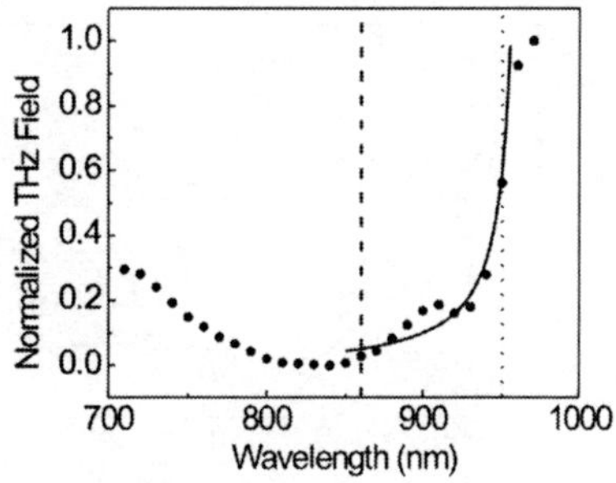

Fig. 2.8 THz generation from a (110) orientation CdTe crystal via optical rectification. *Dots* are experimental results; the *solid curve* is calculated according to the phase matching condition. The *dashed line* indicates crystal bandwidth. The *dotted line* indicates where the coherence length equals the crystal thickness

Only THz waves generated within a very short range close to both surfaces of the nonlinear crystal may not have been fully canceled due to velocity mismatch between the optical and THz pulse. In this case, two THz pulses with reverse polarity may be observed. Figure 2.8 shows the amplitude of THz pulses generated from a CdTe crystal via optical rectification [7]. Due to changes in the excitation beam wavelength, THz waves are generated through all three conditions.

When selecting nonlinear crystals for THz wave generation, three major factors need to be considered: nonlinearity of the material, absorption of both optical and THz waves in the material, and the coherence length of the optical rectification process. After considering these factors, ZnTe crystal was found to be most favorable to generate THz waves excited by fs laser pulses with central wavelength around 800 nm. To obtain maximum conversion efficiency from pump light to THz radiation, it is important to select proper crystal cutting and orientation. ZnTe is a zincblende crystal, and the only nonzero tensor elements are $\chi_{14} = \chi_{25} = \chi_{36}$. Solving Equation (12) gives the generated THz fields from different cuts of ZnTe crystals. When only normal incidence is considered, the THz field generated [8] from a (100) oriented crystal is

$$|E_{\mathrm{THz}}| = 0. \tag{18a}$$

From a (110) orientated crystal it is

$$|E_{\mathrm{THz}}| \propto d_{14}E^2[\sin^2\theta(1 + 3\cos^2\theta)]^{1/2},$$

$$\phi = \arctan(2\cot\theta) \tag{18b}$$

and from the (111) orientation crystal it is

$$|E_{\mathrm{THz}}| \propto d_{14}E^2,$$

$$\phi = -2\theta. \tag{18c}$$

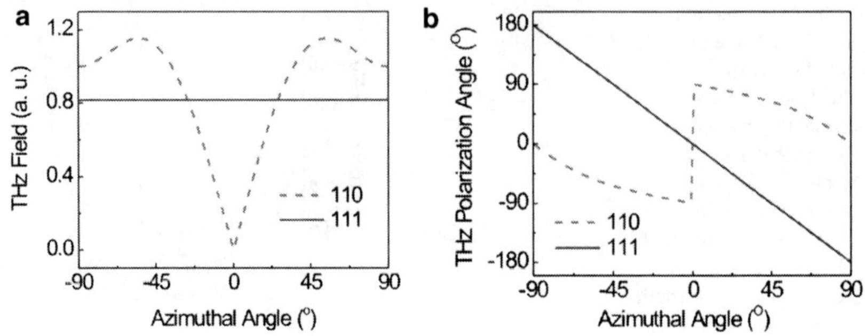

Fig. 2.9 THz wave generation via optical rectification. (**a**) THz wave amplitude and (**b**) THz wave polarization

Here θ denotes the angle between laser polarization and the reference axis in the crystal, and ϕ is the angle between THz wave polarization and the reference axis. In (110) crystals, the reference axis is defined as [001], and is [-1-12] in a (111) crystal. Figure 2.9a shows the calculated THz amplitude as a function of excitation optical beam polarization azimuthal angle, and 2.9b gives the polarization of THz wave.

Electro-optical Sampling

Electro-optical (EO) sampling can be considered as a reciprocal process of the optical rectification. In EO sampling, the THz field is measured by modulating a probe laser beam inside an EO crystal, where it changes the polarization ellipsoid of the refractive index of the EO crystal. The linearly polarized probe beam co-propagates inside the crystal with the THz beam, and its phase is modulated by the refractive index change induced by the electric field of the THz pulse. The existance of the THz field changes the birefringence of the EO crystal, i.e. causing the refractive index difference for polarizations along different axes of the crystal. The electric field induced birefringence changes polarization of the probe beam. This polarization change is converted to intensity change by an analyzer, for example a Wollaston prism. Usually a pair of balanced photodiodes is used to suppress the common laser noise while the signal is doubled.

For a zincblade crystal, such as ZnTe, when an electric field is applied, its ellipsoid of the refractive index is

$$\frac{x^2+y^2+z^2}{n_0^2}+2\gamma_{41}E_xyz+2\gamma_{41}E_yzx+2\gamma_{41}E_zxy=1, \tag{19}$$

where n_0 is refractive index of the crystal without electric field, x, y, z are coordinate units of the ellipsoid, and E_x, E_y, E_z are applied electric field along corresponding axes, respectively. γ_{41} is EO coefficient of the crystal. A phase delay Γ can be calculated according to change of the refractive index

$$\Gamma = \frac{2\pi d}{\lambda}\Delta n, \tag{20}$$

where d is thickness of the EO crystal and Δn is difference between long and short axes of the ellipsoid. Here the process is assumed to be phase matched. If only normal incidence is considered. The phase delay in (100), (110), and (111) orientation ZnTe crystal is

$$\Gamma = 0\,(\text{for}\,(100)\,\text{crystal})] \tag{21a}$$

$$\Gamma = \frac{\pi d n_0^3 \gamma_{41} E}{\lambda}\sqrt{1 + 3\sin^2\phi}\,(\text{for (110) crystal})], \tag{21b}$$

$$\Gamma = \frac{\pi d n_0^3 \gamma_{41} E}{\lambda}\sqrt{\frac{8}{3}}\,(\text{for (111) crystal})]. \tag{21c}$$

In linear EO processes, the field induced phase delay is proportional to the applied electric field; thus, the EO coefficient of certain EO crystals can be represented by half wave field E_π of the crystal, which is defined as the minimum electric field which gives π phase delay in certain EO crystal with unit thickness. As a result, the maximum phase delay a THz field may generate in the EO crystal is

$$\Gamma = d\pi\frac{E_{\mathrm{THz}}}{E_\pi}. \tag{22}$$

Table 2.2 summaries properties of five mostly used zincblade EO crystals [9].

Table 2.2 Properties of 5 EO crystals with Zincblade structure

	ZnTe	GaAs	InP	GaP	ZnS
E_π ($l = 1$ mm) (kV/cm)	89.0	161	153	252	388
Field sensitivity (mV/cm$\sqrt{\mathrm{Hz}}$)	3.20	5.80	5.51	9.07	12.2
NEP (10^{-16}W/Hz)	0.27	0.89	0.80	2.2	5.2
V_{TO} (THz)	5.3	8.0	10.4	11	10.8
$\sqrt{\varepsilon}$	3.18	3.63	3.54	3.34	2.88
N	2.85	3.63	3.54	3.18	2.32
Phase-matching wavelength (nm)	822	1,405	1,230	1,030	470
Frequency of TO photon (THz)	5.3	7.6	10.0	10.8	9.8

There are two most commonly used methods to measure phase delay of the probe laser beam. They are cross and balanced measurement. The latter one gives higher signal and directly measures the field of the THz field, while the former method is simpler in experimental setup. Figure 2.10 shows the concept of balanced measurement. A linearly polarized probe beam is modified to elliptical polarization through the EO process. A quarter-waveplate is used to bias the polarization of the probe beam, which can be put either in front of or after the EO crystal. An analyzer is used to split the biased probe beam into s and p polarization components. A pair

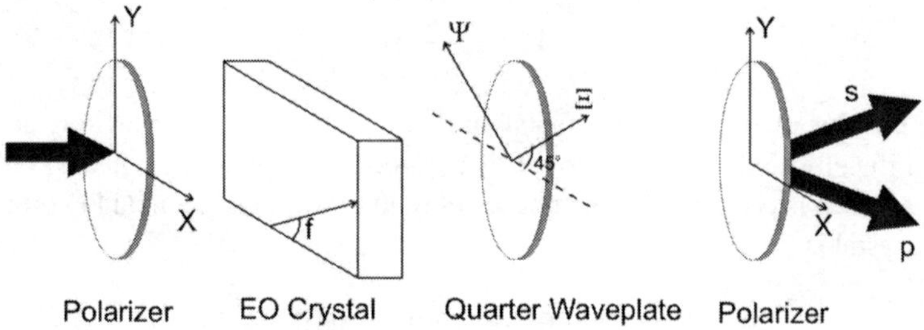

Fig. 2.10 Balanced detection method

of balanced photo detectors is used to measure difference in s and p polarization components. When no THz field is applied, s and p polarization components will have the same intensity after the analyzer; therefore, the balanced detector gives no signal. The presence of a THz electric field changes polarization of the probe beam, generating a measurable signal in the balance detector. If the input beam is x polarized, then the output light can obtained by the following expression

$$\begin{pmatrix} E_x \\ E_y \end{pmatrix} = \begin{pmatrix} \cos\frac{\pi}{4} & -\sin\frac{\pi}{4} \\ \sin\frac{\pi}{4} & \cos\frac{\pi}{4} \end{pmatrix} \begin{pmatrix} \exp(i\Gamma) & 0 \\ 0 & 1 \end{pmatrix} \begin{pmatrix} \cos\frac{\pi}{4} & \sin\frac{\pi}{4} \\ -\sin\frac{\pi}{4} & \cos\frac{\pi}{4} \end{pmatrix} \begin{pmatrix} E_0 \\ 0 \end{pmatrix}. \quad (23)$$

The signal of the balanced detector is

$$S = I_0 \sin 2\phi \sin\Gamma \approx I_0\Gamma \sin 2\varphi, \quad (24)$$

where I_0 is intensity of the probe laser and ϕ is the angle between probe beam polarization and the long axis of ellipsoid induced by THz field. Equation (24) indicates that, in a balanced measurement, the signal is linearly proportional to electric field of the THz radiation. Figure 2.11 shows phase delay as a function of azimuthal angle in (110) and (111) orientated ZnTe crystal.

A cross measurement is similar to balanced measurement except there is no quarter-wave-plate. The analyzer is set cross-polarized to the polarizer located

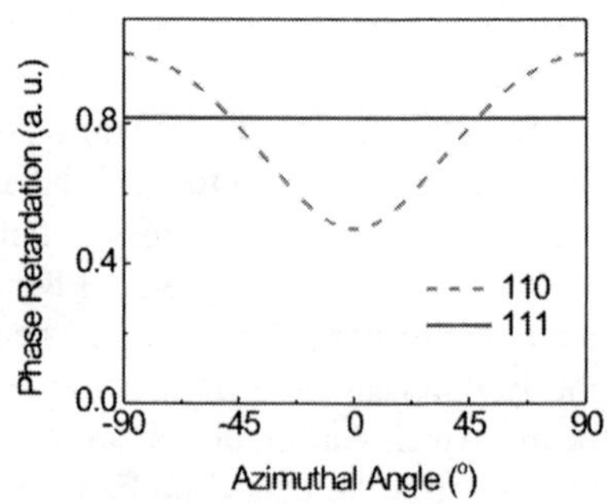

Fig. 2.11 Phase delay in EO sampling with balanced detection

before the EO crystal. Probe beam leaking through the analyzer is detected using a single optical detector. Without THz field, in principle, probe beam does not leak through the analyzer and the recorded signal is 0. If a THz field is applied to modify the polarization of the probe beam, the *leaked* signal will become larger. In an ideal case, the measured signal is

$$S = I_0 \left(\sin 2\phi \sin \frac{\Gamma}{2} \right)^2 \approx \frac{1}{4} I_0 \Gamma^2 \sin^2 2\phi. \tag{25}$$

Equation (25) shows that the measured signal is proportional to the intensity of THz radiation. In reality, however, the EO crystal is not perfectly homogenous. Residual stress in the crystal induces an ellipsoid of the refractive index in the EO crystal, which also generates a phase delay Γ_0. With existence of Γ_0, Equation (25) becomes

$$S = \frac{1}{4} I_0 (\Gamma + \Gamma_0)^2 \sin^2 2\varphi. \tag{26a}$$

If $\Gamma_0 >> \Gamma$, then Equation (26a) is

$$\Delta S = \frac{1}{2} I_0 \Gamma \Gamma_0 \sin^2 2\varphi. \tag{26b}$$

where ΔS represents the signal difference with and without the presence of a THz field. The recorded signal is then proportional to the electric field of THz wave rather than its intensity.

It is worth noticing that the above discussion on EO sampling is based on the steady electric field assumption. For a transient electric field such as a THz pulse, phase matching should be considered. Being the reciprocal process of optical rectification, EO sampling shares the same phase-matching condition.

Broad Band Generation and Detection

Two main factors limit the bandwidth of THz radiation in optical rectification and EO sampling: the pulse duration of the excitation laser pulse and phase matching conditions. Another factor that may not be as important as the previous two, but also limits the useful spectrum is LO phonon absorption; it burns holes in the broad spectrum. Roughly speaking, a laser pulse can generate a THz pulse with bandwidth twice that of the laser pulse bandwidth. Therefore, shorter laser pulses are expected to extend the bandwidth of the THz radiation. With the development of the ultrafast lasers, laser pulse durations of less than a fs already exist; the bandwidth is well above 100 THz. The limiting factor continues to be proper phase matching. Because the frequency extent of the THz pulses is so broad, it is practically impossible to select an EO material that fulfills phase matching requirements for all frequency components. It turns out that one effective way is to decrease the thickness of the material. Since phase matching in optical rectification and EO sampling can be considered as an optical pulse maintaining the same temporal delay as the THz oscillation as discussed before, phase mismatch can be considered as the temporal delay change between optical and THz pulses as they propagate through the EO crystal. A temporal displacement between the optical pulse and THz pulse can be defined as [10]

$$\tau(\Omega) = \frac{n_g(\lambda_0) - n(\Omega)}{c} d. \tag{27}$$

Here λ_0 is the central wavelength, and n_g is its group index of optical beam; n(Ω) is the phase index of THz radiation. In EO sampling, for example, the temporal displacement indicates that the probe optical pulse does not stay at the same THz field along the entire EO sampling process. The modulation of probe pulse by THz field is an average effect within the temporal displacement

$$\begin{aligned} G(\Omega) &= \frac{t(\Omega)}{\tau(\Omega)} \int_0^{\tau} e^{i2\pi\Omega t} dt \\ &= t(\Omega) \frac{e^{i2\pi\Omega\tau(\Omega)} - 1}{i2\pi\Omega\tau(\Omega)}. \end{aligned} \tag{28}$$

Here $2\pi\Omega\tau(\Omega)$ is the phase mismatching factor in the EO process. Smaller value if the phase mismatching factor gives better phase match. The phase mismatch factor is linearly proportional to the crystal thickness *d*. In order to have broad bandwidth, it is essential to decrease the thickness *d*. The phase mismatch factor is also dependent on the difference between $n_g(\lambda_0)$ and $n(\Omega)$. The group index of the laser pulse is fixed when excitation laser is selected, while it is the phase index of THz wave that determines which frequency of the THz wave has high efficiency in the EO process. The phase index of a THz wave as a function of frequency can be described using optical phonons of the EO crystal [11]

$$n(\Omega) = \sqrt{\left[1 + \frac{(\hbar\Omega_{LO})^2 - (\hbar\Omega_{TO})^2}{(\hbar\Omega_{TO})^2 - (\hbar\Omega)^2 - i\hbar\gamma\Omega}\right] \times \varepsilon_\infty}, \tag{29}$$

where $\hbar$ is the reduced Planck's constant, Ω_{TO} and Ω_{LO} are transverse and longitudinal optical phonon frequency respectively, γ indicates delay of oscillation, and ε_∞ is high frequency permittivity of the EO crystal. Table 2.3 compares properties of two commonly used EO crystals, ZnTe and GaP.

The efficiency of EO sampling is affected not only by the phase match between optical and THz pulses but also by the strength of the EO coefficient, which, unfortunately, is also a function of THz frequency,

Table 2.3 Properties of ZnTe crystal and GaP crystal

	$\hbar\Omega_{TO}$ (cm^{-1})	$\hbar\Omega_{LO}$ (cm^{-1})	γ (cm^{-1})	ε_∞	C	N_g @ 835 nm
ZnTe	177	206	3.01	6.7	–0.07	3.224
GaP	367.3	403.0	4.3	9.075	–0.47	3.556

$$r_{41}(\Omega) = r_e \times \left[1 + C(1 - \frac{(\hbar\Omega)^2 - i\hbar\Omega\gamma}{(\hbar\Omega_{TO})^2})^{-1} \right], \tag{30}$$

where r_e is a constant that is independent of frequency and indicates the pure EO coefficient of the crystal in mid-long IR range and C is the Faust-Henry factor of the EO crystal, which gives the ratio between ion effect and electron effect in the DC EO effect. The actual EO effect is calculated by combining Equation (28) with (30). Figure 2.12 shows the frequency response of ZnTe crystal with different thickness when exited with 800 nm laser.

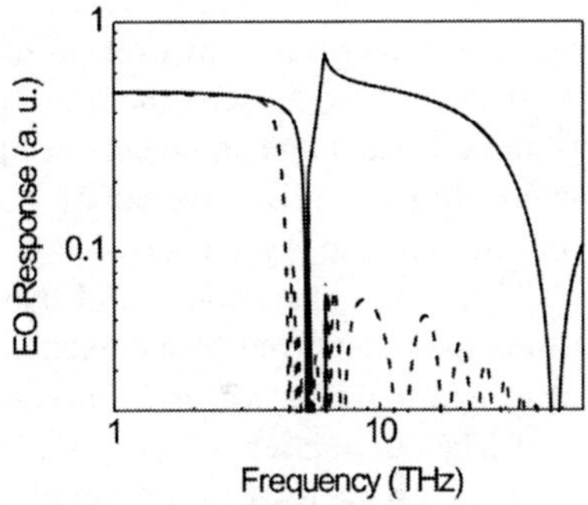

Fig. 2.12 The frequency response of a ZnTe crystal in EO sampling. Thickness of the ZnTe crystal is 10 μm (solid) and 100 μm (dash)

Tilted Pulse Front

THz wave generation from a ZnTe crystal has fairly good collinear phase matching for all polarizations. However, there are still other crystals such as GaSe and $LiNbO_3$ which have high nonlinear coefficients, but do not automatically have collinear phase matching in optical rectification. Different methods need to be used in order to generate high efficiency THz waves from such nonlinear crystals. Some crystals, i.e. GaSe, have high birefringence. By selecting the polarization of the optical and THz beams along different directions of the ellipsoid of the refractive index, collinear phase matching can be satisfied. There are still other nonlinear crystals, such as $LiNbO_3$, which do not have sufficient birefringence to match the index of the optical wave with the low frequency components of the THz wave. Phase matching can also be satisfied by selectively setting the optical and THz beams to propagate in

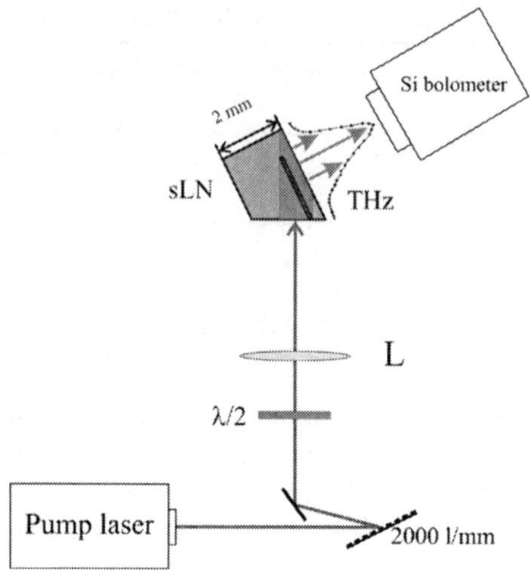

Fig. 2.13 Setup to generate THz waves from a $LiNbO_3$ crystal using a tilted-pulse-front excitation beam (Courtesy of Dr. Nelson)

different directions. However, in this configuration, the THz wave and NIR beam "walk away" from each other in a short distance. The energy conversion coefficient is limited. The large angle between the THz and optical beams also brings difficulty in coupling the THz wave out of the nonlinear crystal, having to avoid total internal reflection on the crystal surface.

THz wave generation from $LiNbO_3$ can satisfy the phase matching condition by tilting the pulse front of the excitation beam. Figure 2.13 shows a schematic setup of generating THz waves from $LiNbO_3$ by optical pulses with a tilted pulse front [12]. The pump laser is incident on a grating, which is used to tilt the intensity front of the pump pulses. Control the tilting angle γ, so that the group velocity of optical beam has the relationship with THz phase velocity,

$$v_{G,O} \cos \gamma = v_{\mathrm{Ph,THz}}. \tag{31}$$

Tilting the pulse front causes phase matching along the THz wave propagation direction. Shown in Fig. 2.13 although the propagation direction of THz wave and optical wave are different, the tilted intensity front of the pump pulses travels collinearly with THz wave with same speed. As a result, both phase and velocity matching are satisfied.

$LiNbO_3$ is an attractive nonlinear crystal that has been widely used due to its high nonlinearity. An additional advantage of $LiNbO_3$ as a THz generator is that it has a higher band gap than other commonly used EO materials. This makes $LiNbO_3$ less susceptible to multiphoton absorption of the excitation laser pulses. Multiphoton absorption not only gives a higher possibility for optical damage to the crystal, but also generates free carriers in the nonlinear material that screen the THz generation. This is considered the major cause of saturation in THz generation through optical rectification when high excitation laser power is used. The higher band gap allows

$LiNbO_3$ crystals to hold higher laser power, and thus delivers stronger THz radiation. A drawback of $LiNbO_3$ crystal is that it has a higher absorption coefficient. This limits the thickness of the $LiNbO_3$ crystal that can be used.

Quasi-Phase-Match

Bulk nonlinear materials always have limited coherence length. Momentum conservation can be improved by introducing an additional wave vector coming from a periodic nonlinear medium with a selected period [13]. This method is called quasi-phase-matching (QPM). The period of the nonlinear medium can be set equal to one coherence length. The polarity of the nonlinear medium reverses alternately from domain to domain, which causes the sign of the phase to flip after one coherence length. Figure 2.14 compares nonlinear processes in bulk material as well as in quasi-phase-matched material. In bulk material, energy flow from pump frequency to signal frequency breathes within a period of each coherence length due to the alternating sign of the phase in each period. The periodic poling of the nonlinear crystal in the quasi-phase-match medium gives a π phase shift after each coherence length. This results in a continually accumulated energy flow from the pump frequency toward the signal frequency.

To make a quasi-phase-matched medium for THz generation, one can stack multiple layers of EO material with alternating pole and thickness of each layer equal to the coherence length of optical rectification. The most popular QPM materials

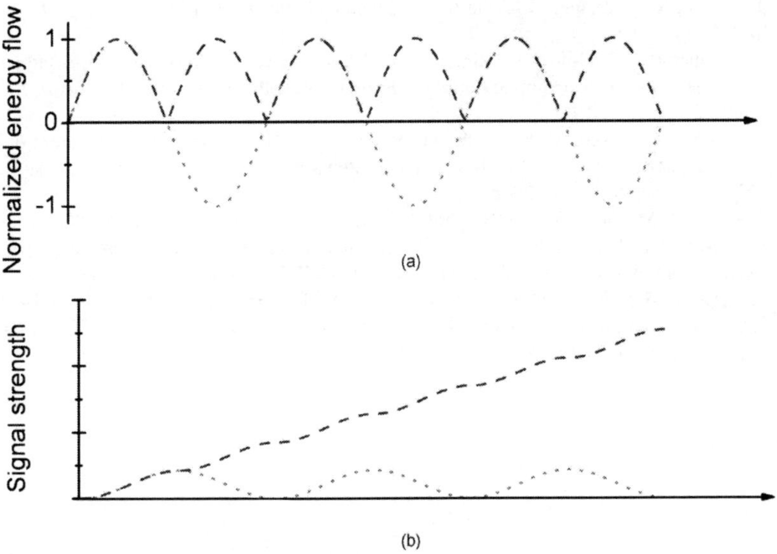

Fig. 2.14 Nonlinear process in bulk material (*dotted*) and QPM material (*dashed*) with increasing of interaction length. (**a**) Energy flow and (**b**) signal strength

include periodically poled lithium niobate and periodically poled gallium arsenide [14]. It is worthwhile to notice that, QPM structures increases the phase matching length in the EO crystal, while it does not correct the temporal delay induced by the velocity mismatch between the optical excitation beam and THz beam. Therefore, using QPM may generate a long THz pulse with many cycles of oscillations, even if a narrow optical pulse is used.

References

1. J. T. Darrow, X.-C. Zhang, D. H. Auston, and J. D. Morse, "Saturation properties of large-aperture photoconducting antennas," *IEEE J. Quantum Electron.* **28**, 1607 (1992).
2. D. S. Kim, and D. S. Citrin, "Coulomb and radiation screening in photoconductive terahertz sources," *Appl. Phys. Lett.* **88**, 161117–161119 (2006).
3. S. E. Ralph, and D. Grischkowsky, "Trap-enhanced electric field in semi-insulators: the role of electrical and optical carrier injection," *Appl. Phys. Lett.* **59**, 1972 (1991).
4. Y. H. Chen, Z. Yang, Z. G. Wang, and R. G. Li, "Temperature dependence of the Fermi level in low-temperature-grown GaAs," *Appl. Phys. Lett.* **72**, 1866 (1998).
5. T. Dekorsy, T. Pfeifer, W. Kutt, and H. Kurz, "Subpicosecond carrier transport in GaAs surface-space-charge fields," *Phys. Rev. B* **47**, 3842 (1993).
6. K. Liu, J. Xu, T. Yuan and X. "C. Zhang, "Terhertz radiation from InAs induced by carrier diffusion and drift," *Phys. Rev. B.* **73**, 1607 (1992).
7. X. Xie, J. Xu and X. "C. "Terahertz wave generation and detecion from a CdTe crystal charactersised by different excitation wavelengths" *Opt. Lett.*. **31**, 978 (2006).
8. Q. Chen, M. Tani, Z. Jiang, and X.-C. Zhang, "Electro-optic transceivers for terahertz-wave applications," *J. Opt. Soc. Am. B* **18**, 823 (2001).
9. Q. Wu, and X.-C. Zhang, "Design and characterization of traveling-wave electrooptic terahertz sensors," *IEEE J. Select. Top. Quantum. Electron.* **2**, 693 (1996).
10. Q. Wu and X.-C. Zhang, "7 terahertz broadband GaP electro-optic sensor," *Appl. Phys. Lett.* **70**, 1784 (1997).
11. A. Leitenstorfer, S. Hunsche, J. Shah, M. C. Nuss, and W. H. Knox "Detectors and sources for ultrabroadband electro-optic sampling: Experiment and theory," *Appl. Phys. Lett.* **74**, 1516 (1999).
12. J. Hebling, K.-L. Yeh, M. C. Hoffmann, B. Bartal, and K. A. Nelson, "Generation of high-power terahertz pulses by tilted-pulse-front excitation and their application possibilities," *J. Opt. Soc. Am. B* **25**, B6 (2008).
13. Y.-S. Lee, T. Meade, V. Perlin, H. Winful, T. B. Norris, and A. Galvanauskas, "Generation of narrow-band terahertz radiation via optical rectification of femtosecond pulses in periodically poled lithium niobate," *Appl. Phys. Lett.* **76**, 2505–2507 (2000).
14. G. Imeshev, M. E. Fermann, K. L. Vodopyanov, M. M. Fejer, X. Yu, J. S. Harris, D. Bliss, and C. Lynch, "High-power source of THz radiation based on orientation-patterned GaAs pumped by a fiber laser," *Opt. Express* **14**, 4439 (2006).

Chapter 3
THz Spectroscopy and Imaging

Once appropriate THz wave sources and detectors become available, researchers will be interested in developing application technologies, in order to utilize the THz waves. By far, most of the applications using THz wave can be sorted into the following two categories, namely, THz wave spectroscopy and THz wave imaging. When a pulsed THz wave is used, the mainly employed spectroscopic technique is THz wave time-domain spectroscopy. In this chapter we briefly introduce the concept of THz wave time-domain spectroscopy and THz wave imaging. Then we discuss a few spectroscopic and imaging techniques.

THz Time-Domain Spectroscopy

In a pulsed THz system, the waveform of THz pulse *E(t)* is detected by temporally sampling the THz pulse using probe laser pulses. As discussed previously, an ultrafast laser pulse is split into pump and probe beams, where the former is used to generate THz pulses and the latter is used to detect those pulses. Coming from the same source, the pump and probe pulses have a defined temporal relationship. Propagating along an optical delay line, the probe pulse samples the THz pulse and records its electric field as a function of delay time. The Fourier transform of the temporal waveform *E(t)* gives spectral distribution of the THz pulse in the frequency domain

$$\tilde{E}(\omega) \equiv A(\omega)e^{-i\phi(\omega)} = \int dtE(t)e^{-i\omega t}. \qquad (1)$$

The THz field in the frequency domain is in principle a complex value, which consists of amplitude and phase. A THz pulse usually contains only a few cycles of oscillation; therefore, its spectral bandwidth can be more than one octave. Different THz sources emit THz radiation with different bandwidths varying from 0.1 THz all the way up to and beyond 100 THz. Recording the waveform of a THz pulse after its interaction with a target, then extracting the spectrum of the THz pulse through Equation (1), one can obtain the THz response of the target within the

X.-C. Zhang, J. Xu, *Introduction to THz Wave Photonics*,
DOI 10.1007/978-1-4419-0978-7_3,

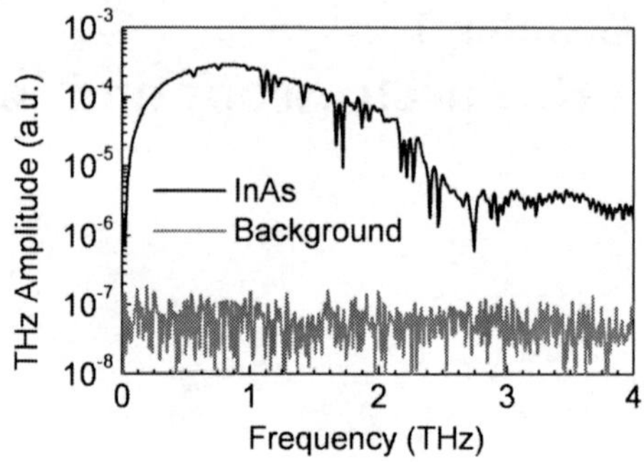

Fig. 3.1 Spectrum of THz pulses generated from an InAs emitter. Background noise is shown as a comparison

THz-pulse-covered bandwidth. Since the above spectroscopic measurement is carried out recording the THz waveform in the time-domain, this technique is called time-domain spectroscopy (also named TDS) [1]. Figure 3.1 shows the spectrum of a THz pulse compared to the background noise.

The spectral resolution of THz-TDS, $\delta\omega$, is determined by the temporal scanning range T. The frequency range of the spectrometer is limited by the response of the THz source and detector, while mathematically the spectrum is significant within a bandwidth $\Delta\Omega$, which is related to the temporal sampling interval δt. The bandwidth and spectral resolution of THz-TDS are given by

$$\begin{aligned} \delta\omega &= \frac{2\pi}{T}, \\ \Delta\Omega &= \frac{2\pi}{\delta t}. \end{aligned} \tag{2}$$

When a fast Fourier transform is used, the spectrum is symmetric about $\Omega = 0$. Therefore, the frequency range spans from $-1/(2\Omega)$ to $1/(2\Omega)$. In order to have a smooth calculated THz spectrum, 0 padding technique may be applied. The 0 padding method adds several 0 values on one side or both sides of the THz waveform. Padding with 0's mathematically increases the temporal scanning range T, and thus gives more intense data points in the THz spectrum. However, padding with 0's does not provide any additional information and does not improve the spectral resolution.

To measure the spectral response of a target, one should first record the THz waveform of a reference sample, i.e. free space. The THz waveform of the reference sample is called the reference waveform. The THz waveform transmitted through the target is called the signal waveform. Fourier transform of the reference and signal waveforms gives the reference spectrum, $A_R(\omega)e^{-i\phi_R(\omega)}$ and the signal spectrum, $A_S(\omega)e^{-i\phi_S(\omega)}$, respectively. The spectral properties of the target can be extracted by comparing the signal spectrum with the reference spectrum

$$\alpha = \frac{1}{d}\ln\frac{A_R}{A_S}$$
$$n = 1 + \frac{[\phi_S(\omega) - \phi_R(\omega)]c}{d\omega}, \tag{3}$$

where α is absorption coefficient, n is refractive index, d is thickness of the target and c is the speed of light in vacuum. In a real measurement, one can use two samples with the same composition but different thickness as the sample and the reference. In this case, the Fresnel loss at the surfaces of the samples will be canceled. Equation (3) gives the result in transmission spectroscopy. The complex refractive index of samples can also be measured by other types of spectroscopies, such as reflection spectroscopy or diffuse scattering spectroscopy. The exact mathematic expression may differ from Equation (3). THz-TDS measures electric field of the THz pulse, which has amplitude and phase information; therefore, it solves both the absorptive and refractive properties of the target. THz-TDS directly measures the complex refractive index of the target, then obtains its complex permittivity without using the Kramers–Kronig (K–K) relationship.

THz-TDS has certain advantages, compared to other spectroscopies. THz-TDS provides coherent spectroscopic detection in a wide range at THz, which is difficult to access using other methods. The THz pulse has ps pulse duration, thus it has intrinsic high temporal resolution. It is thus very suitable for measuring dynamic spectroscopy. THz-TDS utilizes coherent detection methods, which can be used to measure coherent processes of carriers. Additionally, THz-TDS uses time-gating in sampling the THz pulses. This method dramatically suppresses background noise. As a result, THz-TDS usually has a very high signal-to-noise ratio. It is especially useful to measure spectroscopy with high background radiation which is comparable or even stronger than the signal [1].

THz-TDS also presents challenges. THz-TDS is usually slow, due to the temporal sampling of the THz pulses. Techniques have been developed in order to improve the speed of THz-TDS measurements. Today, THz-TDS can perform a single measurement in less than one second with a fairly high SNR. THz-TDS also suffers from poor spectral resolution due to the limited temporal scanning range T in real measurements. In principle, one can scan a THz pulse as long as it is needed. However, a longer scan not only takes more data acquisition time, but also reduces dynamic range of the spectrometer. The relationship between dynamic range and scanning length will be discussed in detail in the next section. The limited spectral resolution is not a problem when measuring the spectrum of a target in a condensed state, whose spectral features often have a bandwidth of a few tenths of THz. The spectral resolution could be insufficient when gases are measured. For such a measurement, a cw-THz source with narrow line width is preferred.

Both THz-TDS and FTIR are spectroscopic techniques in the mid- and far-infrared. THz-TDS uses THz pulses, which are generated using a fs laser, as the light source, while FTIR typically uses a thermal light source. Both use broadband sources, and first record signal based on time delay and present it in the time-domain, and then convert the signal into the frequency domain using Fourier transform. However, there are clear differences between these two spectroscopy techniques.

A standard THz-TDS system covers the frequency range 0.1–3 THz. It can also reach beyond 10 THz if a shorter laser pulse is used, while a 100 THz bandwidth can also be achieved if an even shorter laser pulse is used with a thinner EO crystal as emitter and sensor. However, the measurement of the dynamic range is often limited for a THz-TDS system with a very broad bandwidth. The light source used in an FTIR system can cover a broader band than common THz sources. The band limitation of an FTIR spectrometer depends on its detector, which usually has a lower response for low frequency radiations, and the optics in the FTIR spectrometer have also limited the usable spectral range. Based on current technologies, FTIR spectroscopy usually gives better results with frequencies in excess of 10 THz, while THz-TDS is preferred when the relevant frequencies are below 3 THz. These two techniques give a comparable performance between 3 and 10 THz.

The light source for THz-TDS is a THz pulse; therefore, it is very suitable for time-resolvable spectroscopy, with a temporal resolution in ps scale. The FTIR spectrometer can also measure time-resolvable spectroscopy; however its temporal resolution can hardly reach beyond the ns scale, which is limited by the speed of the flash light source or optical modulator. THz-TDS directly measures the electric field of the THz pulses; therefore, it directly measures the absorption and refraction index of the sample. FTIR measures intensity of the light, and can only obtain amplitude information. The Kramers-Kronig transform is required in order to obtain the refractive index of the sample. Table 3.1 compares THz-TDS and FTIR spectroscopy.

Table 3.1 Comparing of THz-TDS and FTIR

	THz-TDS	FTIR
Bandwidth	0.1–100 THz	Full spectrum
Advanced range	0.1–10 THz	>10 THz
Measurable	Electric field	Intensity
Temporal resolution	ps	ns
Coherent	Yes	No

Dynamic Range of THz-TDS

The measurement of dynamic range $D(\omega)$ is essential in spectroscopic measurements. If the THz source in a spectrometer gives an electric field $E(\omega)$ and the noise equivalent field of detector is $N(\omega)$, the measurement dynamic range of this spectrometer is $D(\omega) = E(\omega)/N(\omega)$. When this spectrometer is used to measure the spectrum of a target with a thickness of d, the reduction of the THz wave caused by the target can be divided into two categories, according to whether or not the loss is associated with the spectral features of the target. To simplify the discussion, we assume that the loss associated with the spectral feature of interest of the target is frequency independent. The spectral-feature-associated loss can be described using the absorption coefficient of the target, $\alpha\ (\omega)$. Since the THz TDS usually directly measures electric field rather than power of the THz beam, here we define a(w) as the absorption coefficient of electric field. To further simplify the discussion, we assume the target has only one absorption line, and this absorption line is a delta function located at $\omega = \omega_0$. The detected THz field of the spectrometer is:

$$\begin{cases} E_D(\omega) = E(\omega)L & \omega \neq \omega_0 \\ E_D(\omega) = E(\omega)e^{-\alpha d}L & \omega = \omega_0 \end{cases}, \tag{4}$$

where L denotes the frequency-independent loss. In order to identify the absorption feature, modulation of the THz field caused by this absorption line must be higher than the noise-equivalent field of the system.

$$E(\omega_0)L\left[1 - e^{-\alpha d}\right] > N(\omega_0)]. \tag{5}$$

When $\alpha d << 1$, Equation (5) yields

$$D(\omega_0) \equiv \frac{E(\omega_0)}{N(\omega_0)} > \frac{1}{\alpha d}\frac{1}{L}. \tag{6}$$

Equation (6) shows the importance of the measurement dynamic range. Only with sufficient dynamic range is the THz wave imager able to identify spectral features of the target.

Another critical parameter of THz-TDS is its signal-to-noise-ratio (SNR), which is defined as the maximum amplitude of THz waveform over noise in the detection system. As we discussed several times already, by using the time-gating technique, THz-TDS usually has very high signal-to-noise-ratio (SNR). However, the SNR in time-domain measurements is not always equal to the actual dynamic range of the THz-TDS system. It is important to understand the relationship between SNR in the time-domain measurement and the dynamic range of the spectrometer.

There are two primary noise sources in the pulsed THz system; those that originate in the probe laser beam N_b and those that originate in the THz pulses N_{THz}. N_{THz} is proportional to THz field $E(t)$ and can be written as:

$$N_{\mathrm{THz}}(t) = R(t)E(t), \tag{7}$$

where $R(t)$ is a unitless factor giving the relationship between the THz wave related noise and the THz field. Both N_b and $R(t)$ can be considered as random functions of time. Standard deviation of N_b and $R(t)$ denoted as respectively, σ_B and σ_R, define the noise level in the THz time-domain measurement. To distinguish the background noise and the noise carried by THz wave, we call the ratio between the THz amplitude A and background noise standard deviation σ_B the time-domain measurement dynamic range (D), and we call $1/\sigma_R$ the measurement SNR. It is worth to notice that the definition of SNR differs from what was defined at the beginning of the previous paragraph. In a common pulsed-THz system, the noise carried by THz wave usually dominates the background noise. Therefore, both definitions give very similar results.

If the temporal resolution in THz-TDS is δt, and the entire scanning range is T, then the spectral range and resolution can be calculated based on Equation (2). In most cases, both σ_B and σ_R are not dependent on the THz spectrum. Those noises, as functions of frequency in the extracted spectrum, are solely due to the measurement. Most likely, they are a combination of white noise plus 1/f noise. Since the detailed distribution of noise does not affect the basis of the discussion, we consider that both of these noise sources are frequency independent. The THz source has a pulse width of T' and a bandwidth of Ω'. In a TDS measurement, one always has $T \geq T'$ and $\Omega \geq \Omega'$, and consequently the noise level in the THz spectrum is

$$\begin{aligned} \rho_{\mathrm{THz}} &= \sqrt{\frac{\delta t}{2\pi}}\sigma_R A\kappa, \\ \rho_B &= \sqrt{\frac{\delta t T}{2\pi}}\sigma_B. \end{aligned} \tag{8}$$

Here $\kappa = A^{-1}\sqrt{\int |E(t)|^2\, dt}$ is the root mean square of the THz field normalized by its amplitude. Equation (8) indicates that, in the frequency domain, the THz wave-carried noise is not related to the temporal scanning range T, while the background noise is proportional to the square root of T.

The measured dynamic range of the THz spectrometer $D(\omega)$, is written as:

$$D(\omega) = \frac{k(\omega)}{\sqrt{\frac{\delta t}{2\pi}\left(\frac{\kappa}{S}\right)^2 + \frac{T\delta t}{2\pi}\frac{1}{D^2}}}, \tag{9}$$

where $k(\omega) = E(\omega)/A$ is the normalized spectrum of the THz pulse. Table 3.2 gives the expression of $k(\omega)$ and κ^2 which can be used to present several typical THz waveforms. Figure 3.2 shows the measured dynamic range of the spectrometer

Table 3.2 $k(\omega)$ and κ^2 for typical THz waveforms

Emitter	Description	Waveform	$k(\omega)$	κ^2
PC antenna	Mono-polar	$2A\frac{t}{\tau^2}e^{-t^2/\tau^2}$	$\frac{\omega\tau}{\sqrt{2}}e^{-\omega^2\tau^2/4}$	$\frac{\sqrt{2\pi}}{2\tau}$
Surface field	Bi-polar	$\frac{2A}{\tau^2}e^{-t^2/\tau^2} - 4A\frac{t^2}{\tau^4}e^{-t^2/\tau^2}$	$\frac{\omega^2\tau}{\sqrt{2}}e^{-\omega^2\tau^2/4}$	$\frac{3\sqrt{2\pi}}{2\tau^3}$
Optical rectification	Damped oscillation	$A\sin(\omega_0 t)e^{-at}$ $t > 0$; 0 $t < 0$	$\sqrt{\frac{\omega_0^2}{[a^2+(\omega_0-\omega)^2][a^2+(\omega_0+\omega)^2}}$	$\frac{\omega_0^2}{4a(a^2+\omega_0^2)}$

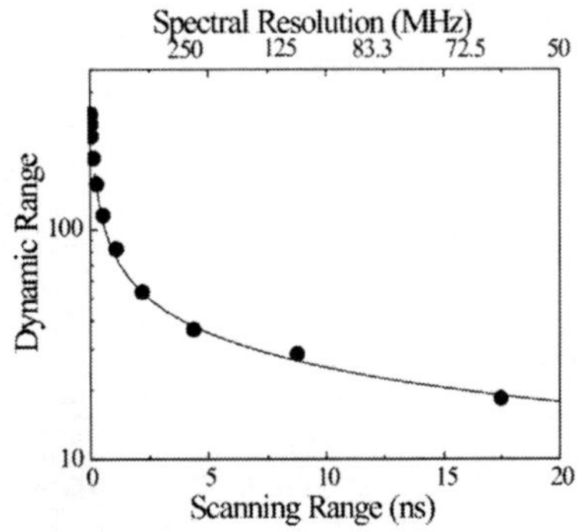

Fig. 3.2 Dynamic range of a THz spectrometer as a function of its spectral resolution

as a function of spectral resolution. Clearly, a higher spectral resolution leads to a lower dynamic range. Equation (9) also indicates that the measured dynamic range is related to the sampling step width in the time-domain. Equation (2) shows that the sampling step width δt decides the total frequency range in the spectrum. When $1/\delta t$ is larger than the bandwidth of the THz pulses, a narrower sampling step width will not lead to a more useful broad bandwidth; however, it does provide higher dynamic range in the frequency domain.

Equation (6) suggests that to identify a sample through its spectral features, the THz wave spectrometer has to give sufficient dynamic range in the frequency domain. Combining Equation (6) with Equation (9) leads us to conclude that the possible spectral resolution, which a THz-TDS system may provide, is limited by the dynamic range of the time-domain measurement,

$$\delta \nu \geq \frac{1}{2\Omega}\left(\frac{1}{\alpha dk(\omega)De^{-\beta d}}\right)^2. \tag{10}$$

The relationship between the measured dynamic range and possible spectral resolution can be easily understood. The THz pulse energy is distributed across its entire bandwidth. A higher spectral resolution means that one needs to detect energy contained in a narrower bandwidth, thus containing less energy. As a result, the dynamic range of a spectrometer is inversely proportional to its spectral resolution. Since it is the electric field, rather than intensity, that is measured in THz-TDS, the possible spectral resolution is thus inversely proportional to the square of the dynamic range in the time-domain measurement. A finer sampling step-width acts to average the measurement, thus it enhances the measured dynamic range in the frequency domain.

Detection of a THz Waveform Using a Single Laser Pulse

Sampling a THz waveform usually requires a series of THz pulses and probe pulses with various temporal delays. However, if the THz field is strong enough, one can catch the entire THz waveform using a single laser pulse. Various methods could

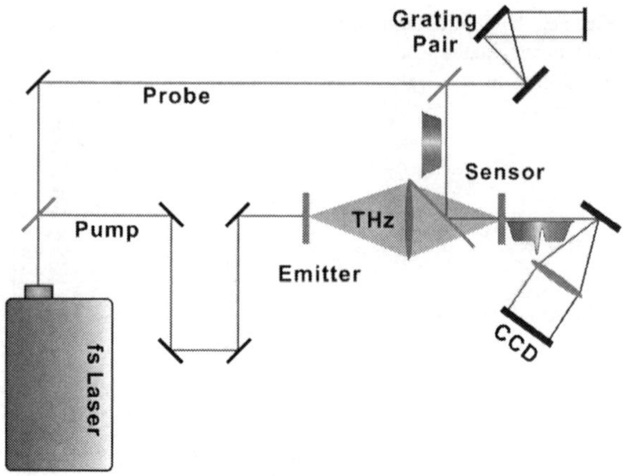

Fig. 3.3 Optical setup of a chirped-pulse detection system

be applied to detect the THz waveform using a single laser pulse, which include evaluating the THz waveform by the spatial distribution of the probe pulse or by the spectral distribution of the probe pulse. The single laser pulse detection technique provides a significant reduction in the acquisition time and greatly extends the applicability of THz systems in situations where the sample is dynamic or moving.

Figure 3.3 exhibits the concept of measuring the THz waveform using a chirped probe pulse. The setup is similar to a standard THz-TDS measurement, where an ultrafast laser beam is split into a pump and probe beam, while the pump beam is used to generate THz pulses and the probe beam is used to detect the THz pulses. However, there is no delay scanning instrument in the chirped pulse detection system. Instead, the optical probe pulse is frequency-chirped and time-stretched with a grating pair from sub-picoseconds to a few tens of picoseconds. The negative chirp of the grating makes the blue component lead the red component. The output from the grating is a pulse with a longer pulse duration and a wavelength that varies linearly with time. The chirped probe pulse is modulated by the THz pulse when they are mixed in the EO crystal. Conceptually, the chirped probe pulse can be seen as a succession of short pulses each with a different wavelength. Each of these wavelength components encodes a different portion of the THz pulse. A spectrometer spatially separates the different wavelength components and thus reveals the temporal THz pulse. The spatial signal output from the spectrometer is measured using a CCD. Figure 3.4 gives the CCD recorded probe pulses with and without the THz pulse. The difference between those two signals gives the waveform of the THz pulse. For maximum image acquisition speed, the THz pulse and probe pulse may be expanded in the vertical dimension using cylindrical lenses. The CCD is then able to capture both the THz temporal waveforms and several hundred vertical pixels simultaneously and only a single translation stage is required for spectroscopic image acquisition.

It is assumed that the probe pulse originally has a Gaussian distribution in both the time and frequency domains, and its central frequency is ω_0. If only

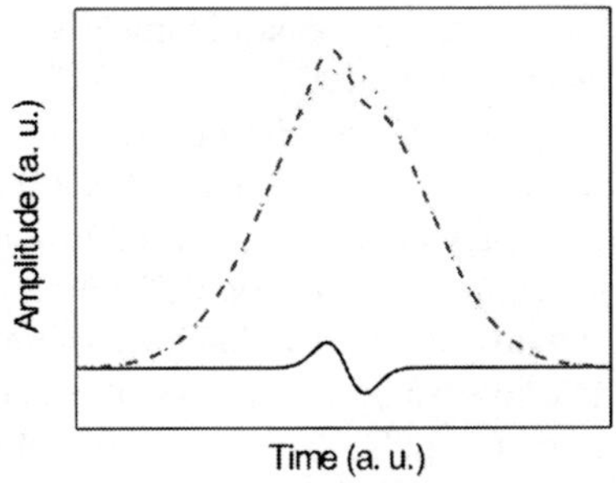

Fig. 3.4 THz waveform extracted using a chirped-pulse measurement

second-order dispersion is taken into account, while ignoring third and higher order dispersion of the grating pair, the electric fields of the probe pulse before and after chirping are:

$$\begin{aligned} E_0(t) &= \exp\left(-\frac{t^2}{T_0^2} - i\omega_0 t\right), \\ E_C(t) &= \exp\left(-\frac{t^2}{T_C^2} - i\alpha t^2 - i\omega_0 t\right), \end{aligned} \tag{11}$$

where T_0 and T_C are the pulse widths of the probe pulse before and after chirping, respectively, and α is the so-called chirping rate of the pulse. Once the chirped pulse is modulated by THz pulses, its electric field is

$$E_m(t) = E_C(t)[1 + kE_{\mathrm{THz}}(t-\tau)], \tag{12}$$

k is a modulation factor, which indicates electric field of the probe pulse being affected by the THz field. τ denotes the relative temporal delay between the THz pulse and the probe pulse. In most cases kE_{THz} is much smaller than 1. The modulation of the electric field in the different frequency components of the probe beam is

$$\begin{aligned} N(\omega) &\equiv \frac{I(\omega)|_{\mathrm{THzOn}} - I(\omega)|_{\mathrm{THzOff}}}{I(\omega)|_{\mathrm{THzOff}}} \\ &= \frac{\int g(\omega-\varpi)2kE(t_\varpi-\tau)\exp(-2t_\varpi^2/T_C^2)d\varpi}{\int g(\omega-\varpi)\exp(-2t_\varpi^2/T_C^2)d\varpi} \\ &\propto 2kE_{\mathrm{THz}}(t_\omega-\tau). \end{aligned} \tag{13}$$

Here $I(\omega)$ is the intensity of the probe pulse in the frequency domain, which can be defined as intensity with or without THz field as presented in the foot notes. $g(\omega-\varpi)$ is the spectral function of the spectrometer. t_ω is defined as $t_\omega \equiv (\omega_0 - \omega)/2\alpha$. Equation (13) indicates that the modulation of the probe pulse intensity in the frequency domain is linearly proportional to the electric field of THz pulse in the time domain.

According to Heisenberg's uncertainty principle, higher temporal resolution requires more bandwidth. When the probe pulse is chirped and only a part of the frequency component is used to sample a temporal window in the THz waveform, the temporal resolution will not be as high as the original pulse. If the original pulse is transform-limited, its time-bandwidth production should be a constant $T_0\ \Omega_0 = k$, whose value depends on the actual pulse shape, i.e. 0.44 for a Gaussian pulse. Once the pulse is chirped, its pulse width becomes T_C while keeping the same bandwidth. The best temporal resolution can be obtained if all frequency components used in the THz waveform measurement remain transform-limited, which is the component with bandwidth $\delta\Omega$ providing a temporal resolution of $\delta T = k/\delta\Omega$. In this case, the temporal resolution δT is:

$$\begin{aligned}\delta T &= \frac{T_C}{\Omega/\delta\Omega} \\ &= \sqrt{T_0 T_C}.\end{aligned} \tag{14}$$

Equation (14) gives the finest temporal resolution possible for the chirped pulse measurement method. The temporal resolution is linearly proportional to the square root of the pulse width for both the original pulse and chirped pulse. A larger chirped pulse width results in lower temporal resolution.

THz Differential Spectroscopy

Lock-in technology is usually used in THz-TDS measurement in order to suppress the background noise. To use lock-in technology, the detected signal is modulated at a certain frequency. The modulation frequency is used as the reference frequency of the lock-in amplifier. Only a certain component within the input, which has the same frequency and certain phase shift with the reference is amplified and recorded. All other components are blocked from recording. Thus tremendous noise is screened due to frequency and phase selection. Using the lock-in technique may suppress noise level down to 10^{-6} of the original noise level. Employing lock-in technology into the THz-TDS measurement by modulating the THz source, one can efficiently reduce the background noise generated in the detection system, such as that which is associated with the probe laser beam. However, such a technique cannot reduce noise associated with THz waves since noise is also modulated at the same frequency. When the target is a thin film or tracing material, it only gives weak modulation to the THz field. This weak modulation may be buried under the noise associated with the THz field. In this condition, one can use the differential spectroscopy method, which directly measures the difference between the target and the reference.

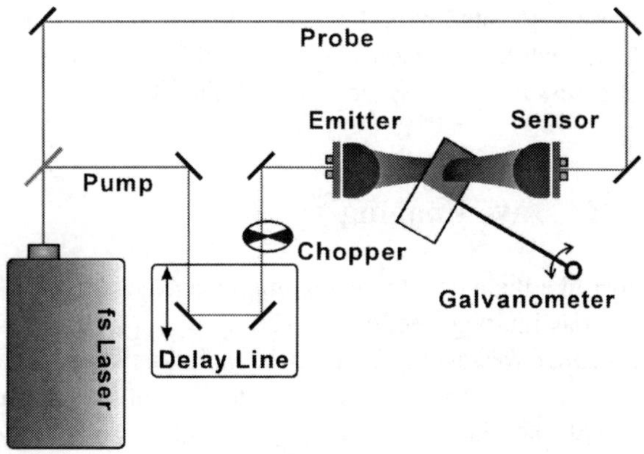

Fig. 3.5 Experimental setup of a THz time-domain differential spectrometer

Figure 3.5 shows the concept of THz time-domain differential spectroscopy, where a double lock-in technique is used. The two lock-in amplifiers are serially connected. One of them uses a higher reference frequency ω_H, with a shorter integration time constant T_H. The output of this lock-in amplifier is used as the input of the other lock-in amplifier, which uses a lower reference frequency ω_L and a longer integration time constant T_L. Figure 3.6 illustrates the process of double lock-in technology. To make the double lock-in technique work, the following relationship must be satisfied,

$$\omega_H >> 1/T_H >> \omega_L >> 1/T_L. \tag{15}$$

In differential spectroscopy measurements, the THz beam alternately passes through the sample and reference with a frequency of ω_L. For instance, a galvanometer can be used to shake the sample in and out of the THz beam. The THz source is modulated, i.e. using an optical copper, with a frequency of ω_H. The first lock-in amplifier with higher reference frequency was used to suppress the background

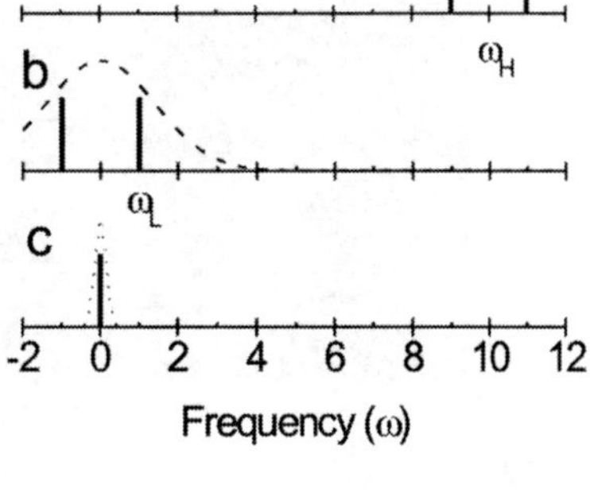

Fig. 3.6 Concept of double lock-in technique. (**a**) Frequency distribution of a signal modified by two frequencies, (**b**) frequency of signal is shifted by $-\omega_H$ after the first lock-in amplifier, and (**c**) frequency of signal is shifted by $-\omega_L$ after the second lock-in amplifier

noise associated with the detection system, i.e. power fluctuation of the probe laser beam, while the second lock-in amplifier with lower reference frequency is used to suppress the noise associated with the THz wave.

THz Wave Imaging

Just like the adjacent bands, microwaves and infrared radiation, THz waves can be used as imaging media in a variety of applications [2]. Terahertz wave imaging is attractive for several reasons: the radiation is non-ionizing and poses very few safety risks, it is capable of submillimeter spatial resolution and significantly, and a number of materials, including paper, plastics, and cardboard are relatively transparent in this frequency band. Figure 3.7 shows the concept of raster scanning THz wave imaging. The THz wave is focused by a lens or parabolic mirror. The target is placed across the THz beam at its focal spot. It is then raster scanned in a plane perpendicular to the THz beam. THz wave transmission through or reflected from each spot of the target is recorded, which forms the THz wave image of the target.

Different from a common optical image or X-ray image, each pixel in a pulsed THz wave image contains the entire THz waveform rather than just the intensity of the beam. Fourier transform of the THz waveform extracts the spectral information of that pixel. Therefore, THz-wave imaging not only identifies the target by its profile but also obtains composite information of the target. Besides the reflection distribution, pulsed THz wave imaging can also profile the target depending on its refractive index distribution, which causes a phase change of the THz pulse. Figure 3.8 shows a THz wave image of the water mark in a 100 dollar bill, which was made by the phase change of the THz pulses.

Fig. 3.7 Concept of raster scanning THz-wave imaging

Fig. 3.8 THz-wave image of a water mark in a $100 bill

Imaging with THz pulses requires scanning in three dimensions, including space (2-D) and temporal (1-D), thus it is usually very time consuming. If spectral information is not required, one can take the THz wave image of a target at a fixed temporal delay, i.e. at the peak of the THz waveform. Only amplitude of the THz field is recorded while scanning the target across the THz beam. A similar image can also be taken using a cw-THz source, by recording the intensity of the transmitted or reflected THz beam. Figure 3.9 shows cw-THz wave transmission images of a tea pot when it is empty and when it is half-full with water. Table 3.3 compares pulsed and cw-THz wave imaging.

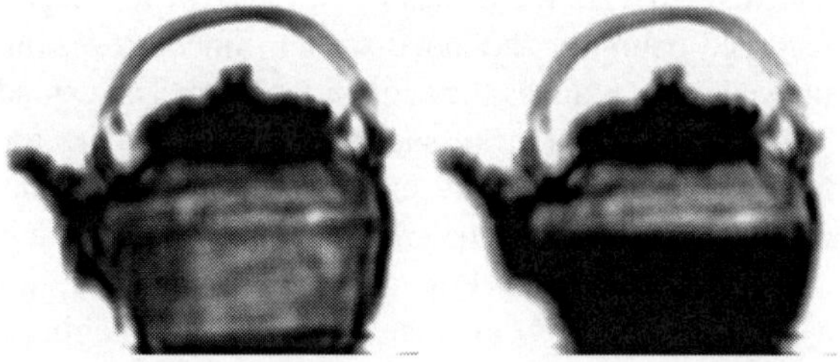

Fig. 3.9 cw THz-wave images of a tea pot. *Left*, an empty tea pot, and *right*, a half full tea pot

Table 3.3 Comparing of cw and pulsed THz wave imaging systems

	cw-THz wave imaging	Pulsed THz wave imaging
Cost	\$50,000–\$150,000	\$200,000–\$1,000,000
System complicity	Low	High
Weight	3 kg	10 kg
Speed	100,000 point/s	< 4,000 point/s
Data complicity	Low	High
Spectral information	No	Yes
Depth information	No	Yes
Refractive index	No	Yes

Besides being time consuming, THz wave imaging also faces other challenges. THz waves are highly reflected by metal surfaces, thus it cannot see through metal containers. THz waves are highly absorbed by water, so they cannot penetrate into a material which contains a lot of water. THz wave imaging cannot be used for medical diagnostics of organs inside the human body except when the endoscope technique is being used. Additionally, since THz waves have longer wavelengths than visible and IR waves, spatial resolution of THz wave imaging is limited to submillimeter in the far-field. Near-field imaging must be used in order to break the diffractive limitation for even higher spatial resolution.

2-D Focal Plane THz Wave Imaging

The Raster scanning method of THz wave imaging utilizes the entire THz radiation generated from the emitter to investigate each pixel of the image, so that it provides a very high signal-to-noise ratio. However, the speed of linearly transporting the target with finite mass back and forth is quite limited. The low frame rate is one of the major technical obstacles blocking THz wave imaging in its approach to real world applications. One can use a similar method that is used in a traditional optical imaging system, THz wave imaging optics, which can be a lens or a concave mirror that directly images the THz wave transmission or reflection from the target onto an extended THz wave sensor. The THz wave image of the target is simultaneously recorded using the extended sensor. Since no scanning is required, 2D THz wave image highly reduces the acquisition time. The extended THz wave sensor could be THz wave detector array, such as pyroelectric detector array, micro-bolometer array, heterodyne detector array, et. al. When using EO sampling to detect THz wave, the extended sensor could be an EO crystal with sufficient aperture.

Figure 3.10 presents the concept of taking THz wave 2-D imaging system using a large aperture EO crystal. A THz wave imaging optic, such as a polyethylene lens, is used for forming the THz wave image of the target, while the target and EO crystal locate the object and image plane, respectively. Thus, the THz wave distribution on the EO crystal carries spatial information of the target. An extended, linearly polarized probe beam, which covers the entire THz wave distribution area, is collinearly propagated with the THz beam through the EO crystal. The spatial distribution of the THz wave is printed onto the probe beam via the EO process. A cross-detection method is used to detect THz wave modulation of the probe beam, where an analyzer with cross-polarization is placed after the EO crystal and the leaking of the probe beam is imaged onto a CCD camera. Through a serial imaging process, the optical image in the CCD camera reflects the THz wave image of the target.

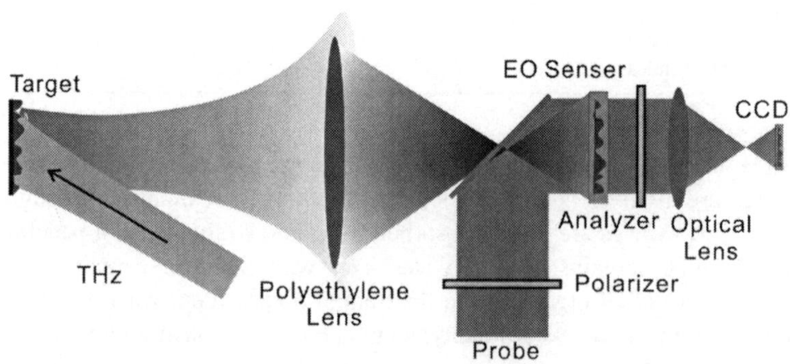

Fig. 3.10 Concept of 2D THz wave imaging using a large aperture EO crystal as extended focal plane detector

The spatial resolution of a far-field imaging system is limited by diffraction of the carrier wave.

$$\Delta = 1.22\lambda \frac{l}{D}, \tag{16}$$

where λ is the wavelength of the carrier wave, l is the distance from the target to the imaging lens, and D is the aperture diameter of the lens. The depth of field plays an important role in an imaging system, which determines a range, within which a target can project a clear image on a fixed imaging plane. Complicated formulas have been developed for photographers to calculate the depth of field of their cameras. The exact formation of depth of field could be different for different lenses or image taken conditions. To get a brief idea without involving detail lens parameters, we can consider an imaging system using an ideal lens. In such a system, a point source which does not locate on the object plane projects an extended spot on the image plane. One can consider a point within the depth of field if its extended image size is smaller then the defined spatial resolution on the image plane. When depth of field is much smaller than the object distance, it can be described as

$$L = \delta Dl/(\delta D' \pm D), \tag{17}$$

Here δD is the required spatial resolution on the target and $\delta D'$ is the required resolution on the imaging plane, while the sign indicates the depth of field at different sides of the target. δD and $\delta D'$ have the following relationship: $\delta D' = \delta D(l'/l)$, where l' is the image distance, which approximately equals to the focal length of the imaging lens for a far field object. A wave with 1 THz frequency has a wavelength of 300 μm, which is much longer than the optical wavelength. Thus, the THz wave image usually has a much lower spatial resolution than an optical image. Using a 40 cm diameter lens to image a target at 10 m away, the spatial resolution is 9.15 mm if the frequency of the carrier wave is 1 THz. If the required spatial resolution equals to the diffraction-limited resolution, then the depth of field is 45.8 cm.

The size of the EO crystal is determined by the target size and focal length of the imaging optics. The object distance in an imaging system is usually much longer than the focal length of the imaging optics. Therefore, the dimension of the EO crystal is defined as the image size.

$$D_S \approx D_T \frac{f}{l}, \tag{18}$$

D_T is the dimension of the target, and f is the focal length of the imaging optics. In the previous example, if the target is a circle with 1 m diameter, and focal length of the imaging lens is 44.7 cm (NA of the lens is 0.5), then the diameter of the EO crystal is 4.47 cm. The thickness of the crystal can be estimated with two-times of the focal depth, which is:

$$L' = \frac{\delta D' f}{D} \approx \frac{\delta D'}{2\text{N.A.}}, \quad (19)$$

where $\delta D'$ denotes the required resolution on the imaging plane, which can be set by the diffraction-limited spatial resolution. N.A. is the numerical aperture of the lens. In the previous example, the maximum thickness of the crystal will be 0.7 mm in order to maintain the spatial resolution. The above discussion does not take into consideration the refractive index of the crystal. Since commonly used EO crystals have fairly large refractive index, the EO crystal could be a few mm thick without affecting the spatial resolution.

Telecentric Beam Scanning THz Wave Imaging

Since the THz wave is diluted onto the extended sensor in a 2-D imaging system, the detection signal-to-noise ratio is usually reduced. A strong THz source is desired in such a 2-D imaging system. An alternative way to perform the THz wave image with high speed is to scan the THz beam with flying or shaking optics rather than the target. Since less mass is associated with scanning the THz beam, a much faster imaging process is expected. Unlike optical imaging, where scattering light is the major information carrier, with THz waves, due to their longer wavelength, the scattering or diffusion are usually less important in the imaging process than transmission and specular reflection. As a result, collection of transmitted or reflected THz waves is essentially important in a beam-scanning THz wave imager. A telecentric beam-scanning technique can be used to ensure a high collection coefficient in the THz wave beam-scanning imaging process.

The concept of a telecentric beam-scanning imager is presented in Fig. 3.11. The collimated THz beam is guided into the imaging system and steered by bending about two orthogonal axes using a pair of shaking mirrors. The output beam from the shaking mirror pair is then guided into a telecentric lens. The telecentric lens could be a single spherical lens, while the shaking mirror locates at one focal spot of the lens and the target locates at its focal plane on the opposite side. THz beam output from the telecentric lens is normal to the focal plane and is focused onto the target. Shaking the mirror pair scans the THz beam across the target. The THz

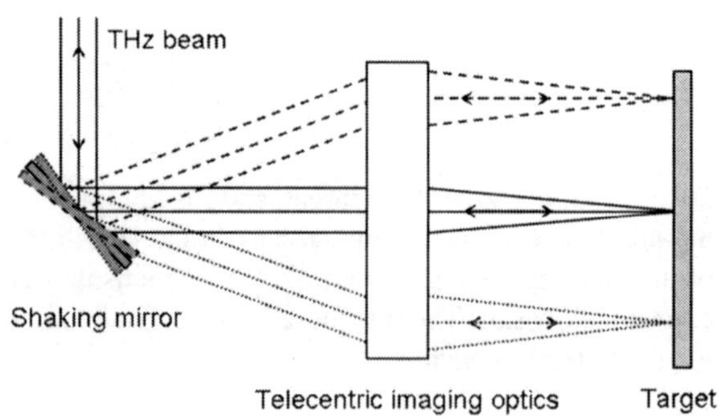

Fig. 3.11 Schematic of a telecentric beam scanning imager

wave reflected by the target is collected by the same lens, and fed back following the same direction. It is picked up by the same shaking mirror pair and counter propagates with the input beam. One can use either a THz wave transceiver, which emits and detects THz radiation, or use a beamsplitter to guide the returning THz beam into the detector.

A common understanding is that 2-D focal plane imaging, which is a parallel process, should be faster than a raster scanning imaging, which is a serial process. This statement is true for passive imaging, as well as active imaging with sufficient dynamic range. In THz-wave imaging, however, once the processing speed is fast enough, the dynamic range becomes a bottleneck due to the low-intensity source and less sensitive detector being used. In a raster scanning imaging process, every pixel uses the entire THz radiation, but shares data acquiring time with others. The dynamic range of the measurement limits how long the THz beam should stay on one pixel of the image. In a 2-D imaging process, each pixel fully uses the data acquisition time, but shares THz power with others, which leads to lower dynamic range for all pixels. If the overall dynamic range of the imager is limited, the speed of a 2-D focal plane imaging process cannot be faster than a raster scanning imaging process.

Time-of-Flight Imaging

The entire THz waveform is recorded for each pixel when pulsed-THz-wave-imaging is taken. If two THz pulses are reflected from two surfaces located at different depths, the reflected THz pulses have different time-delays due to different optical paths. From time-delay, one can retrieve depth information of each pixel, and thus present a topographic profile of the target. This imaging method is called THz wave time-of-flight imaging. Figure 3.12 shows concept of time-of-flight imaging.

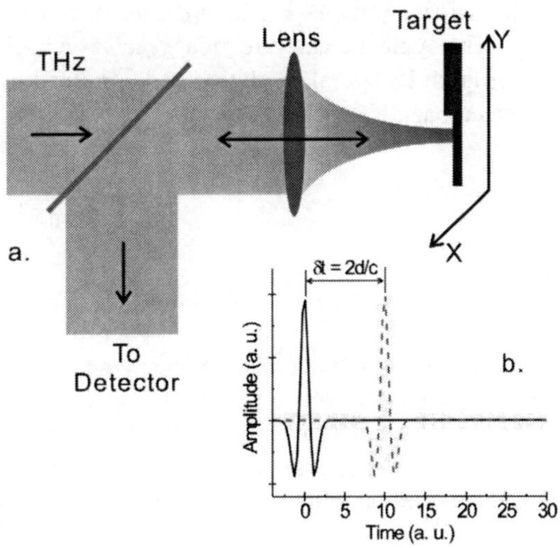

Fig. 3.12 Concept of THz-wave time-of-flight imaging. Inset shows time delay between two THz waveforms

The depth at resolution in THz-wave time-of-flight imaging is limited by temporal resolution in measuring the THz waveform, which is in principle related to the THz wavelength. The actual depth (temporal) resolution one can get experimentally is usually shorter than the THz central wavelength. When THz pulses with a 2 THz central wavelength are used, the smallest depth resolution can be a few microns. Although time-of-flight imaging provides a 3-D topographic profile or layer structure of the target, it is usually not considered a full 3-D imaging technique, since it cannot provide interior information of the target if there is no layer structure presented. Figure 3.13 gives a THz wave time-of-flight image of a spark plug, which shows the surface topographic profile and layer structure inside the spark plug.

Synthetic Aperture and Interference Imaging

Equation (16) gives the spatial resolution of THz wave imaging. To obtain high spatial resolution, especially when imaging a target at a stand-off distance, one needs to use an imaging optic with large aperture. An imaging optic with a large aperture is not only expensive but also not easy to operate and transport due to its bulky size and heavy weight. Synthetic aperture and interference imaging uses either one mobile detector or detector array with discrete spatial distribution to create an image of the target. In these techniques, the spatial resolution of the image is no longer limited by the aperture of each individual detector but is limited by the movement or location of the detector(s). As a result, using the synthetic aperture and interference imaging method one can get high spatial resolution with small imaging elements.

The synthetic aperture and interference imaging technique was originally developed for RF and microwave. Figure 3.14 gives a schematic diagram of the synthetic aperture imaging technique with airborne radar. The pulsed radar carried by an aircraft illuminates a carrier wave onto the ground to one side of the air-craft. Assume that the speed of the air-craft is v, its flying attitude is H, the distance to the detection spot is R, the radar pulse width is T, the aperture is D, and the wavelength of the carrier wave is λ. The aperture of radar is $\Phi = \lambda/D$ viewed from the far-field. The synthetic aperture radar receives a back-scattering signal from objects on the ground. Its lateral resolution is determined by time-delay between radar pulses reflected back from different spots

$$r_g = \frac{Tc}{2\sin\theta}. \tag{20}$$

Here θ is the angle presented in Fig. 3.14. Along the flight direction, the spatial resolution is defined by the optical path difference from an individual spot on the ground to a different location on the plane. Determined by the aperture of the radar, the carrier wave covers a length of $L = \Phi R$ on the ground. From a different view, the radar wave can cover the same spot within a flight distance L. As a result the imaging aperture is extended from the aperture of radar, D, to the flight distance L. Spatial resolution in the flight direction can be estimated using Equation (16)

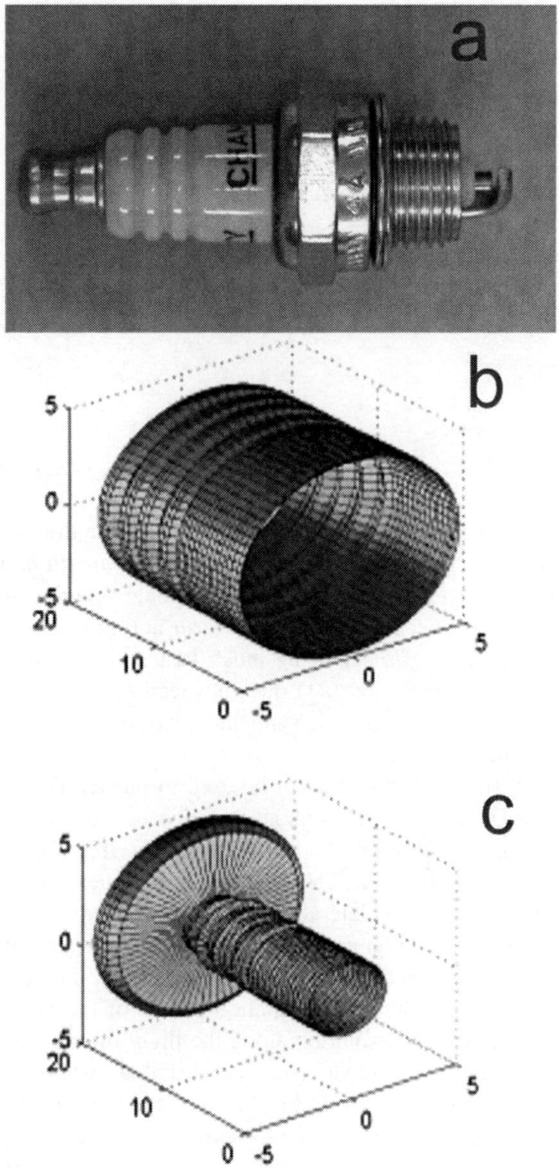

Fig. 3.13 THz-wave time-of-flight image of a spark plug. (**a**) photo of the spark plug, (**b**) THz wave image of the external shell, and (**c**) THz-wave image of the inner metal bar

$$r_a = \lambda \frac{L}{R} = D. \tag{21}$$

Here the factor of 1.22 is dropped. Typically, the spatial resolution at the flight direction is defined as *D/2*. Equation (21) shows that, due to the relationships among λ, D, and Φ, the spatial resolution along the flight direction is not dependent on

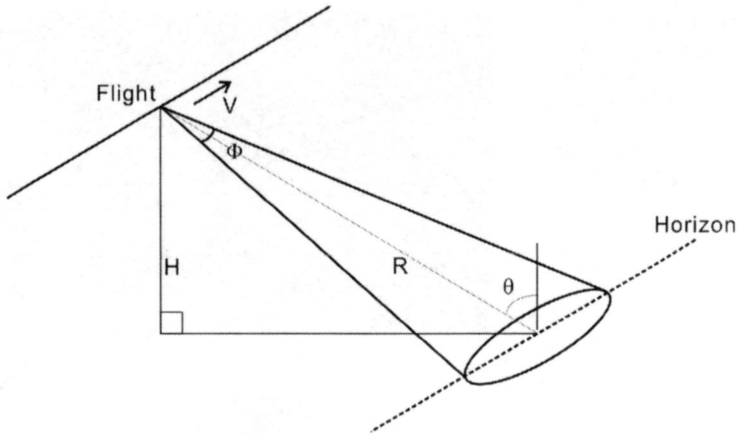

Fig. 3.14 Concept of an airborne synthetic-aperture radar imaging

either the wavelength of the carrier wave or the distance between the target and the radar. In addition, the smaller the radar aperture, the finer the spatial resolution. This is different from a common optical imaging setup, where a larger aperture results in finer resolution. It is worth noting that the above discussion is based on the aperture of radar being much larger than the wavelength of the carrier wave. Therefore, Equation (21) does not lead to the following conclusion: by reducing the size of the radar, the synthetic aperture image can have sub-wavelength spatial resolution.

A similar technique can be used in pulsed THz-wave synthetic-aperture imaging. The pulse width is about one picosecond for the THz pulses, which leads to a lateral resolution of hundreds of microns. In reality, the temporal resolution in the measurement of the THz pulses can be much smaller than the THz pulse width, which results in a lateral resolution on the sub-μm scale. The spatial resolution in the flight direction can be as small as the wavelength level. This is different from common synthetic aperture radar imaging. A common radar pulse width is usually much longer than the oscillation period of the carrier wave. As a result, it gives a better spatial resolution along the flight direction, while the THz wave synthetic aperture image gives a better lateral spatial resolution.

If the relative phase shift between the various sensor locations is recorded, the synthetic aperture imaging system can take an interference image. In an interference imaging setup, any pair of detectors, i and j, form a base line. Each base line corresponds to a point in phase space. It is assumed that all the detectors are distributed within an x-y plane. The coordinate of base line ij in the phase space is

$$\begin{aligned} u_{ij} &= \frac{x_i - x_j}{\lambda} \\ v_{ij} &= \frac{y_i - y_j}{\lambda} \end{aligned}, \qquad (22)$$

where u and v are coordinates in phase space, and x and y are those in the real space. The signal amplitude at that point is a product of the signal amplitudes of those two detectors, which is $A_{ij} = A_i A_j$, while the phase is defined as the phase difference between these two detectors: $\Phi_{ij} = \Phi_i - \Phi_j$. If the total number of detectors is N, then there will be $N(N-1)$ points in the phase space. The target image is extracted from the Fourier transform of the signal in the phase space, which is

$$I(x',y') = \int\int A(u,v)e^{-i\Phi(u,v)}e^{-i2pxu}e^{-i2\pi yv}dudv; \tag{23}$$

x' and y' are coordinate units in the imaging space. Spatial resolution of the interference imaging setup is determined by the length of the longest base line.

References

1. D. Grischkowsky, S. R. Keiding, M. P. van-Exter, and C. Fattinger, "Far-infrared time-domain spectroscopy with terahertz beams of dielectrics and semiconductors," *J. Opt. Society Am. B* **7**, 10, 2006–2015 (1990).
2. B. B. Hu, and M. C. Nuss, "Imaging with terahertz waves," *Opt. Lett.* **20**, 16, 1716–1718 (1995).

Chapter 4
THz Wave Interaction with Materials

To develop technologies utilizing THz waves, one needs to first understand the interaction between THz waves and materials. In this chapter we will discuss the dynamics of THz wave interaction with different kinds of materials.

Dielectric Constant in the THz Band

Wave free-carrier interaction is one of the fundamental wave material interactions. Wave free-electron interaction dominates in THz waves interacting with conductors or semiconductors having high free-carrier density. A THz wave interacting with free carriers can be solved using the classic Drude model. In this model, individual carriers are independent of each other, and no interaction between carriers is considered except collisions. Collision between carriers is considered an instantaneous event, and the mean interval between two collisions involving a same carrier is defined as an average collision time τ, which is independent of the location or velocity of the carrier. Based on these approximations, one has the following motion equation describing a free carrier driven by an electric wave:

$$m^* \frac{d^2x}{dt^2} + \frac{m^*}{\tau}\frac{dx}{dt} - qE = 0, \tag{1a}$$

where m^* denotes effective mass of the carrier, q is the charge of the carrier, and E is the electric field. Under equilibrium condition, the average collision time can be described as

$$\tau = \frac{m^* \mu}{q}. \tag{1b}$$

Here μ is mobility of the carrier. For instance, the effective mass of an electron in undoped silicon is $m^* = 0.19\, m_0$, and its mobility is $\mu = 1{,}400\ \mathrm{cm^2/Vs}$. This results in an average collision time between electrons of 1.5 ps. The electric field induces polarization of the material formed by carrier displacement.

X.-C. Zhang, J. Xu, *Introduction to THz Wave Photonics*,
DOI 10.1007/978-1-4419-0978-7_4,

$$P = (\varepsilon - \varepsilon_\infty)\varepsilon_0 E = Nqx\,, \tag{1c}$$

where ε_∞ is the high frequency relative permittivity of the material, N is free carrier density, ε_0 is permittivity in a vacuum, and ε is the relative permittivity of the material at the frequency of the interacting electromagnetic (EM) wave. The relative permittivity is also called the dielectric constant of the material, and usually the prefix of "relative" is ignored and permittivity is used instead. Using the polarization of the material, the motion Equation (1) can be derived to a polarization equation:

$$\frac{d^2P}{dt^2} + \gamma\frac{dP}{dt} - \frac{Nq^2}{m}E = 0. \tag{2}$$

Here $\gamma = 1/\tau$, denotes the coherent decay factor of the electrons in the material. Any EM wave can be described as the sum of a series of monochromatic waves via Fourier transform. The interaction between wave and material can also be presented as the sum of material interacts with the series of monochromatic waves. Each monochromatic wave and its induced polarization are simple harmonic oscillations, with formations of $E = E_0 e^{-i\omega t}$, and $P = \chi\varepsilon_0 E_0{}^{-i\omega t}$, where χ is the electric susceptibility of the material and defined as $\varepsilon = \varepsilon_\infty + \chi$. Equation (2) becomes

$$\chi(\omega^2 + i\gamma\omega) + \frac{Nq^2}{\varepsilon_0 m} = 0. \tag{3}$$

Solving Equation (3) results in the complex permittivity of the material at certain frequency, $\tilde{\varepsilon} \equiv \varepsilon_1 + i\varepsilon_2$

$$\begin{aligned}\varepsilon(\omega) &= \varepsilon_\infty - \frac{\varepsilon_\infty\omega_p^2}{\omega^2 + i\omega\gamma} \\ &= \varepsilon_\infty\left(1 - \frac{\omega_p^2}{\omega^2+\gamma^2} + i\frac{\omega_p^2\gamma}{\omega(\omega^2+\gamma^2)}\right),\end{aligned}$$

with

$$\omega_p = \sqrt{\frac{Ne^2}{m^*\varepsilon_\infty\varepsilon_0}}\,, \tag{5}$$

which is called the plasma oscillation frequency (POF) of the material. The POF is proportional to the square root of the free-carrier density in the material. For instance, metals have very high electron density, so that they have high POF located in the UV band. The free-carrier density in semiconductors varies with different materials and environmental conditions, such as doping, temperature, et al. Therefore, semiconductors have a variety of POFs. Crystalline silicon has a high frequency permittivity of $\varepsilon_\infty = 11.7$. When the free-electron density is 6×10^{13} cm^{-3}, its POF is 0.047 THz. For intrinsic silicon, whose free-electron density is only in the

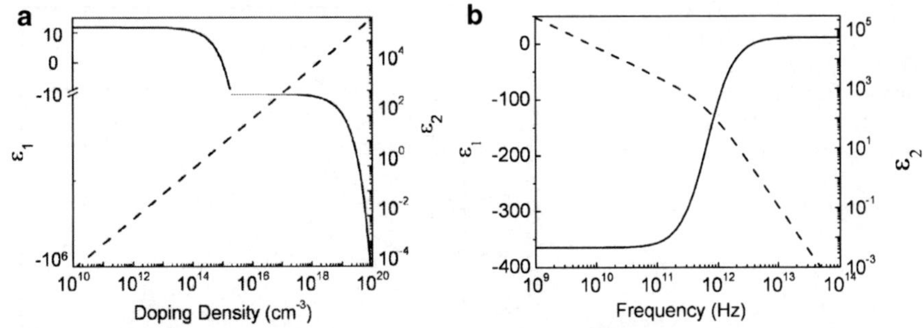

Fig. 4.1 (**a**) Real (*solid*, ε_1) and imaginary (*dashed*, ε_2) part of permittivity for n-type silicon with different doping density at 1 THz. (**b**) Real (*solid*, ε_1) and imaginary (*dashed*, ε_2) part of permittivity for n-type silicon with 10^{16} cm^{-3} doping density at different frequency. Both calculated by Equation (4)

order of 10^{10} cm^{-3}, ω_p is less than 1/100 of THz. As a result, a THz wave is transparent in intrinsic silicon. Figure 4.1a shows the permittivity of silicon at 1 THz with different doping densities. The change in effective mass and mobility of electrons due to doping density is not considered in the calculation. For low doping density, THz wave frequency is much higher than the POF of silicon. Its permittivity is almost a real value, which is approximately equal to the high frequency permittivity. As the level of doping increases, a silicon crystal has a higher POF. Until ω_p^2 is comparable to $\omega^2 + \gamma^2$, the real part of the dielectric constant becomes a much smaller value than the high frequency limit. When the doping density becomes even higher and ω_p^2 becomes larger than $\omega^2 + \gamma^2$, the real part of the dielectric constant becomes a negative value and its imaginary part cannot be considered negligible anymore. The permittivity strongly presents its complex nature and silicon becomes lost to the THz wave. Figure 4.1b shows permittivity of silicon with 10^{16} cm^{-3} electron density at different frequencies, which is calculated under the same approximations as in Fig. 4.1a. When the wave frequency is much higher than the POF of the material, the dielectric constant is real and positive and the material is transparent. On the other hand when wave frequency is lower than the POF of the material, the dielectric constant shows more complex behavior with the negative real part, and the material is opaque.

Carriers in semiconductors, i.e., GaAs crystals, have a dynamic equilibrium without disturbance from the outside. Although there are free carriers in such a material due to doping, no macro carrier motion is presented. If the crystal is excited with femtosecond laser pulses, the photo-induced free carriers are accelerated, driven by the semiconductor surface field. This instant photo-carrier emits THz wave radiation. At the same time, motion of the free electron may also induce plasma oscillation in the semiconductor crystal. Separation of electrons and holes generates an instant field in the semiconductor and may exist as those "cold" electrons form plasma oscillation too. Both types of plasma oscillation may generate THz waves. Figure 4.2 shows THz pulses generated from laser-excited "cold" plasma oscillation

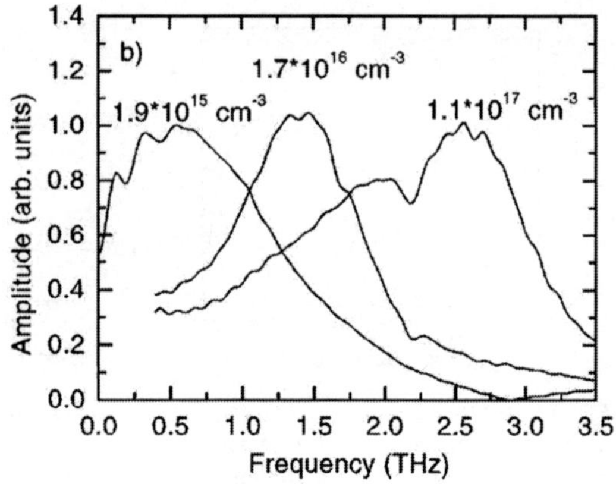

Fig. 4.2 Spectra of THz pulses generated from n-type GaAs wafer with different doping density. Excitation density was smaller than doping density for all cases (Courtesy of Dr. Kersting)

[1]. The central frequency of the THz wave is proportional to the square root of the doping density in the GaAs crystal.

Refractive Index in the THz Band

Considering a THz wave as an EM wave, the corresponding property of material is its permittivity. Considering a THz wave as an optical wave, the most important property of material is its refractive index $\tilde{n} \equiv n + i\kappa$. Propagation of a monochromatic wave through material is described as

$$E_T = E_0 e^{\text{inkl}} e^{-\kappa \text{kl}}, \tag{6}$$

where l is the propagation distance through that material. The decay of amplitude and delay of phase caused by propagation through the material can be directly extracted from the imaginary and real parts of its refractive index. According to electrodynamics, the refractive index of material can be derived from its relative permittivity and permeability as $\tilde{n}^2 = \tilde{\varepsilon}\tilde{\mu}$. If the material is not a ferromagnetic material, its relative permeability $\mu \approx 1$. In this case the refractive index of material is $\tilde{n} = \sqrt{\tilde{\varepsilon}}$. If the average collision time is much longer than the oscillation period of the electromagnetic wave, then $\gamma \approx 0$. The complex refractive index of material is

$$\tilde{n} = n_\infty \sqrt{1 - \frac{\omega_p^2}{\omega^2}}. \tag{7}$$

Here $n_\infty \equiv \sqrt{\varepsilon_\infty}$ is defined as the high frequency refractive index of the material. If the frequency of the wave is lower than the POF of the material, then the refractive index of material is a pure imaginary value. An EM wave decays when propagating

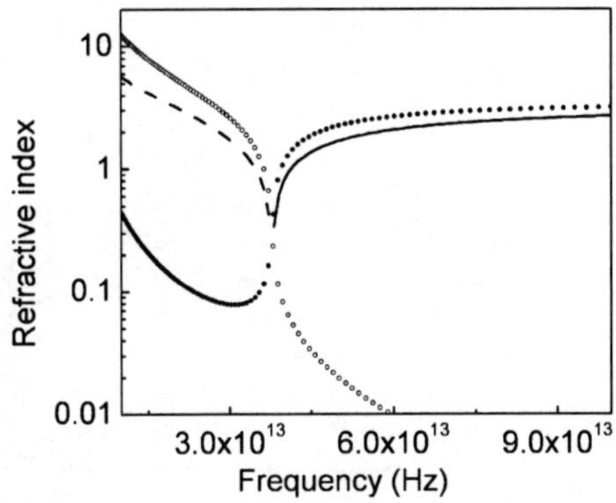

Fig. 4.3 Refractive index of n-type silicon with 10^{18} cm^{-3} doping density as a function of frequency. *Solid* and *dashed curves* are real and imaginary part of refractive index calculated by Equation (7). *Solid* and *open dots* are real and imaginary part of refractive index calculated by Equation (4)

in such a material. When $\omega_p < \omega$, the refractive index is a real value and no attenuation occurs for the EM wave in propagation. Since a THz wave has low frequency, $\omega >> \gamma$ may not be always satisfied. As a result, in reality, the refractive index of material is a departure from the equation presented in Equation (7). Figure 4.3 compares refractive index of silicon with 10^{18} cm^{-3} free-electron density, calculated by Equations (4) and (7), respectively.

Characterize Carrier Properties Using THz Wave Spectroscopy

Carrier properties, such as density, effective mass, mobility, et al., affect material interaction with THz waves. One can use THz wave spectroscopy to characterize carrier properties in material, especially semiconductor and superconductor, et al. One important characteristic of semiconductor material is its impedance, especially the impedance at carrier wave frequencies for high speed semiconductor devices. The operation frequency of a semiconductor device is well above GHz and climbs still to an even higher frequency. The high frequency response of a semiconductor material is essentially important. THz spectroscopy can be used to evaluate the high frequency response of semiconductor materials. The complex conductance of material and its permittivity has the following relationship:

$$\tilde{\varepsilon} = \varepsilon_\infty + i\frac{\tilde{\sigma}}{\omega\varepsilon_0}. \tag{8}$$

Combining Equation (8) with Equation (4) one has

$$\tilde{\sigma} = \varepsilon_0\varepsilon_\infty\frac{\omega_p^2}{-i\omega + \gamma}. \tag{9}$$

Once the complex permittivity is measured using THz wave spectroscopy, one can calculate its complex conductivity as a function of frequency based on

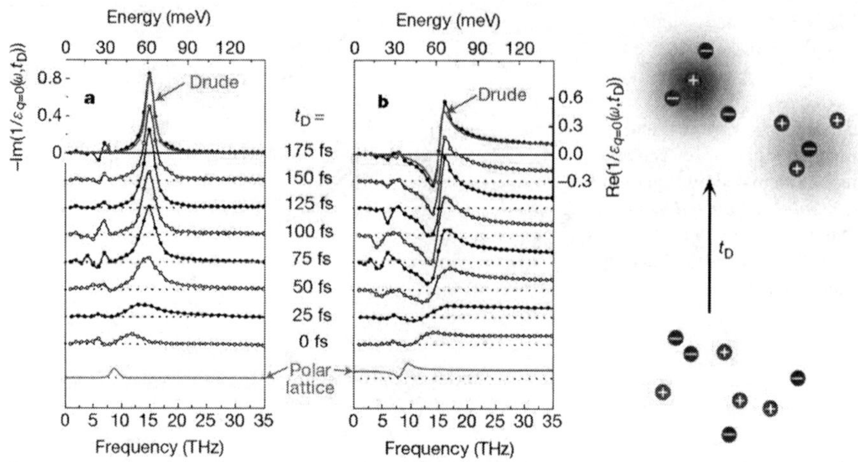

Fig. 4.4 Temporal evolution of Coulomb screening process and plasmon scattering. *Right figure* gives concept of excited free carriers' evolution (Courtesy of Dr. Leitenstorfer)

Equation (9). For example, using an optical pump THz wave probe spectroscopy one can measure the Coulomb screening process after the free electron has been excited with laser pulses. Figure 4.4 gives the transmission spectrum of THz pulses with different time delay after optical excitation [2]. The THz transmission spectrum and Coulomb screening process can be simulated using the classic Drude model. The absorption peak indicates photo carrier and phonon interaction. THz wave interaction with phonons will be discussed later in the chapter.

It is worth to notice that all of the above discussions are based on the classic Drude model. This model can be used only when free electron approximation is satisfied. If this approximation is not satisfied one needs to modify this model or use quantum electrodynamics for rigid calculation in order to solve the problem.

THz Wave Interaction with Resonant States

Different from free carrier, carriers, which are bonded by a potential barrier, present discrete energy levels according to the quantum mechanism. Those discrete energy states play essential roles in the wave material interaction. Each energy state involved in the interaction can be considered as a simple harmonic oscillator with a frequency of $\omega_0 = E_0/\hbar$, where E_0 is energy of that state and $\hbar$ is the Planck constant. When the resonant energy state is involved in wave material interaction, Equation (1) can be modified to

$$\frac{d^2x}{dt^2} + \gamma\frac{dx}{dt} + \omega_0^2 x = \frac{q}{m^*}E. \tag{10}$$

And Equation (4) becomes

$$\varepsilon = \varepsilon_\infty + \frac{\varepsilon_\infty \omega_p^2}{\omega_0^2 - \omega^2 - i\gamma\omega}. \tag{11}$$

Refractive index of material is

$$n^2 = \varepsilon_\infty \left(\frac{\omega_p^2}{\omega_0^2 - \omega^2 - i\gamma\omega} + 1 \right). \tag{12}$$

If the wave frequency is far from the resonant frequency, the real and imaginary part of refractive index can be estimated as

$$\begin{aligned} n &= n_\infty \left(1 + \frac{\omega_p^2(\omega_0^2 - \omega^2)}{2[(\omega_0^2 - \omega^2)^2 + \gamma^2\omega^2]} \right), \\ \kappa &= n_\infty \frac{\omega_p^2 \omega\gamma}{2[(\omega_0^2 - \omega^2)^2 + \gamma^2\omega^2]}. \end{aligned} \tag{13}$$

For those waves whose frequency is much lower than resonant frequency, the real part of its refractive index is

$$n \approx n_\infty \left(1 + \frac{\omega_p^2}{2\omega_0^2} \right). \tag{14}$$

It is a constant value and is independent of the wave frequency, but determined by the resonant frequency and the POF of the material. If the wave frequency is much higher than resonant frequency, the real part of the refractive index is

$$n \approx n_\infty \left[1 - \frac{\omega_p^2}{2(\omega^2 + \gamma^2)} \right], \tag{15}$$

which is independent of the resonant frequency.

The most interesting part of EM wave interaction with a material having resonant energy structures occurs when the EM wave frequency is close to the resonant frequency, i.e., $\delta\omega \equiv (\omega_0 - \omega) << \omega_0$. In this case refractive index of the material is

$$\begin{aligned} n &= n_\infty \left(1 + \frac{\omega_p^2 \frac{\delta\omega}{4\omega_0}}{\delta\omega^2 + (\gamma/2)^2} \right), \\ \kappa &= n_\infty \frac{\omega_p^2 \gamma / 8\omega_0}{\delta\omega^2 + (\gamma/2)^2}. \end{aligned} \tag{16}$$

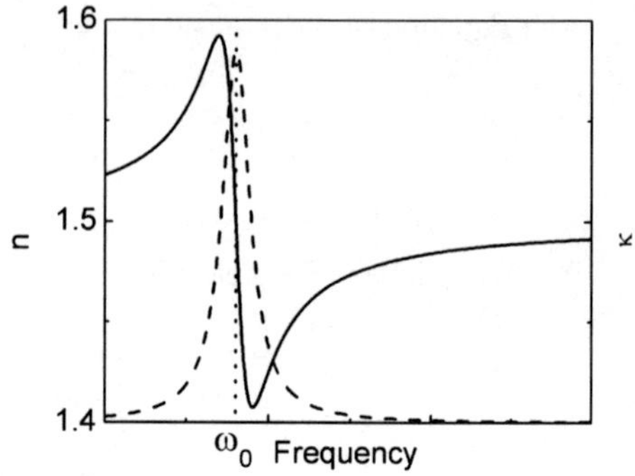

Fig. 4.5 Real (*solid*) and imaginary (*dashed*) parts of refractive index calculated by Equation (16). The *dotted line* indicates location of resonant frequency

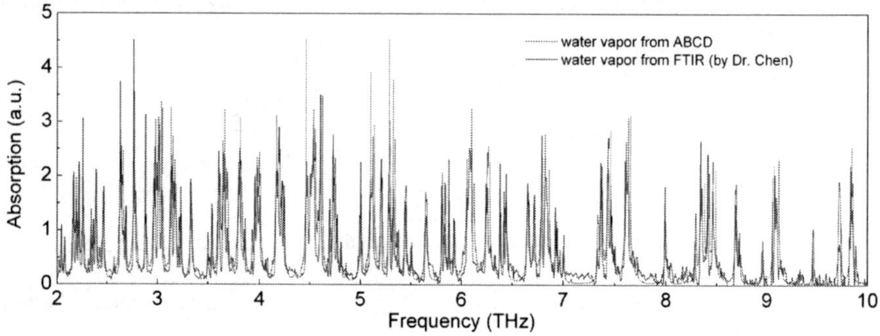

Fig. 4.6 Absorption spectrum of water vapor (in 2–10 THz)

Figure 4.5 shows refractive index of material as a function of EM wave frequency. The imaginary part of the refractive index presents a peak at the resonant frequency, which indicates the absorption peak of the material according to Equation (6). When THz wave spectroscopy is measured, one can identify those resonant energy levels through the absorption peaks. Figure 4.6 gives water vapor absorption spectroscopy from 2 to 10 THz, measured via THz air-breakdown-coherent-detection (ABCD) system and FTIR respectively. Absorption lines in this spectrum reflect vibration and rotation transitions of water molecules.

THz Wave Reflection Spectroscopy and Phonon Oscillation

Energy structure is not only shown in THz wave absorption spectroscopy, but also presented in its reflection spectrum. According to the Fresnel principle, the EM wave reflected from the surface of a medium is

$$\begin{aligned} \tilde{r}_{//} &= \frac{\tilde{n}\cos\theta_i - \cos\theta_t}{\tilde{n}\cos\theta_i + \cos\theta_t}, \\ \tilde{r}_{\perp} &= \frac{\cos\theta_i - \tilde{n}\cos\theta_t}{\cos\theta_i + \tilde{n}\cos\theta_t}. \end{aligned} \tag{17}$$

Here $\tilde{r}_{//}$ and $\tilde{r}_{\perp}$ are used to denote reflection of p and s polarization wave, respectively. θ_i and θ_t are incident angle and transmission angle. According to Equation (17), one can obtain the complex refractive index as a function of frequency from reflective spectroscopy, and thus obtain carrier properties and energy structures of the material. Figure 4.7 shows an experimental setup used to measure phonon resonance in semiconductor material using THz wave reflection spectroscopy.

Fig. 4.7 Experimental setup of THz wave ABCD reflection spectroscopy

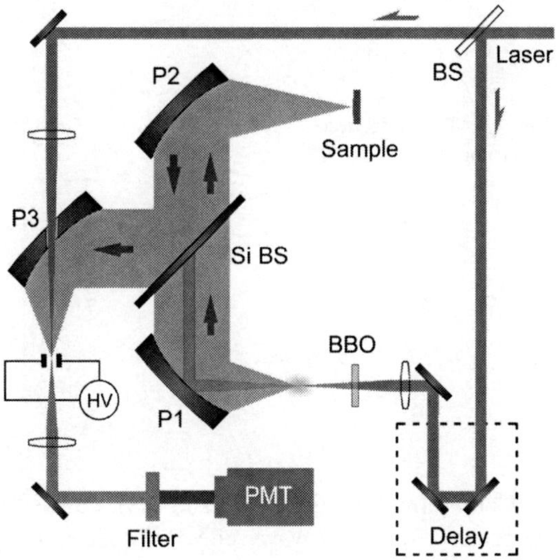

In a crystalline structure, all cells are periodically distributed in space and vibration of each cell can be coupled into collective vibration modes and presented as phonons. If each cell of the crystal contains more than one atom, the phonon can be further defined into an acoustic phonon and optical phonon. According to the vibration direction related to the direction of propagation, a phonon can also be defined as a transverse phonon, where the vibration is perpendicular to its propagation and a longitudinal phonon, where propagation travels along with the vibration direction. Optical phonons in a crystal usually contribute to interaction with THz waves. For instance, the first transverse optical (*TO*) phonon and the first longitudinal optical (*LO*) phonon of a GaAs crystal are 8.1 and 8.8 THz, respectively. A GaAs crystal shows strong absorption to THz waves whose frequency is close to its phonon energy. The crystal also shows strong reflection for THz waves between its *TO* and *LO* phonon, and this energy region is called the Reststrahlen band. The complex permittivity and refractive index of the crystal around its Reststrahlen band is

$$\tilde{\varepsilon}(\omega) = (n + i\kappa)^2 = \varepsilon_\infty \left(1 + \frac{\omega_{LO}^2 - \omega_{TO}^2}{\omega_{TO}^2 - \omega^2 - i\gamma\omega}\right). \tag{18}$$

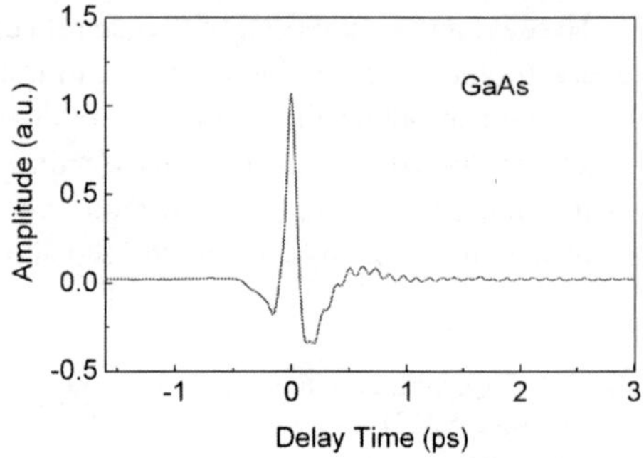

Fig. 4.8 The measured THz waveform in THz wave reflection spectroscopy. GaAs crystal was the sample

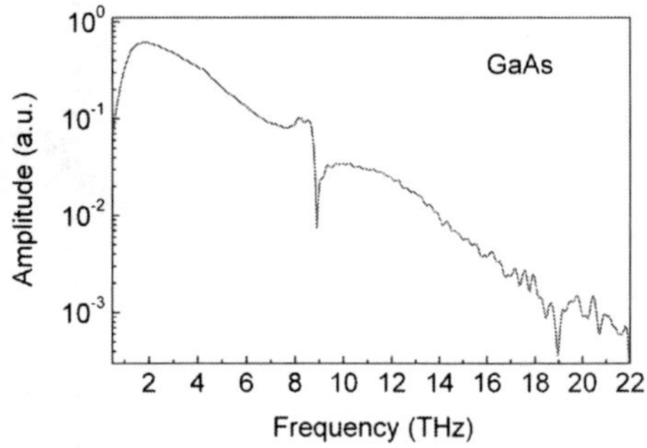

Fig. 4.9 THz reflection spectrum of GaAs crystal

Combining Equations (17) and (18), the Reststrahlen band of the crystal can be characterized in THz wave reflection spectroscopy. Figure 4.8 shows the waveform of THz pulses reflected from the GaAs crystal. After reflection, the THz waveform shows strong oscillation. Fourier transform of the waveform in Fig. 4.8 gives the reflection spectrum, which is presented in Fig. 4.9. There is a reflection peak present between the *TO* and *LO* phonon of the GaAs crystal.

Interaction Between Phonon and Free Carriers – THz Wave Emission Spectroscopy

The phonon properties of a semiconductor crystal can also be characterized through THz wave emission spectroscopy, which is generated from photon-induced free carriers and interacts with optical phonons in the crystal. In a semiconductor material, which has low symmetry in its cell structure, the laser pulse excitation gives a shock to the atoms in the crystal cell, and displaces it from a position of equilibrium. The atoms vibrate around the equilibrium position. Under homogenous carrier distribution approximation, the interaction between plasma and phonon is

$$
\begin{aligned}
&\frac{\partial^2}{\partial t^2}P + \gamma_e \frac{\partial}{\partial t}P + \omega_{P}^2 p = \omega_P^2 (E^{\text{ext}} - 4\pi \gamma_{12} W) \\
&\frac{\partial^2}{\partial t^2}W + \gamma_{Ph} \frac{\partial}{\partial t}W + \omega_{LO}^2 W = \frac{\gamma_{12}}{\varepsilon_\infty}(E^{\text{ext}} - 4\pi P),
\end{aligned}
\tag{19}
$$

P is the polarization and W is the normalized atom displacement. γ_e, γ_{Ph}, and $\gamma_{12} = \omega_{TO}\sqrt{(\varepsilon_{DC} - \varepsilon_\infty)/4\pi}$ are the decay factor of electrons, phonons and electron–phonon interaction, respectively. ε_{DC} is the low frequency permittivity of the semiconductor material. Equation (19) can be simplified under the following approximations. First of all, a low-doped semiconductor has a very low POF, so the plasma–phonon interaction can be ignored. Secondly, since the laser pulse width is much shorter than the plasma oscillation period, the plasma as a function of time can be ignored. Additionally, the electron hole recombination time is much longer than the plasma oscillation period; subsequently, the free-carrier density can be treated as a constant. Equation (19) can be solved to

$$P(t) = \frac{\omega_P^2 E^{\text{ext}}}{\varepsilon_\infty}[1 - I(\omega_P)\cos(\omega_P t)]\theta(t)$$

$$W(t) = \left[-\frac{\gamma_{12}E^{\text{ext}}}{\varepsilon_\infty \omega_{LO}^2} + W_0 I(\omega_P)\frac{\omega_{LO}^2}{\omega_{LO}^2 - \omega_P^2}\cos(\omega_P t) - \right.$$
$$\left. - W_0 I(\omega_{LO})\frac{\omega_P^2}{\omega_{LO}^2 - \omega_P^2}\cos(\omega_{LO}t)\right]\theta(t)$$

$$E(t) = E^{ext} - \theta(t)E^{ext}\left[1 + I(\omega_P)\cos(\omega_P t) + I(\omega_{LO})\frac{\omega_P^2}{\omega_{LO}^2 - \omega_P^2}\frac{\varepsilon_{DC} - \varepsilon_\infty}{\varepsilon_{DC}}\cos(\omega_{LO}t)\right]. \tag{20}$$

The above discussion is based on carrier homogenous distribution approximation. In reality, the photo-induced carrier is not uniformly distributed. This non-homogenous distribution leads to a broadening of plasma phonon interaction, thus its contribution to THz wave emission is not significant. Only the phonon oscillation, which is not affected by carrier distribution, contributes to THz wave emission. The coherent phonons generate a macro dipole oscillation. This dipole oscillation emits an EM wave with a frequency equal to the *LO* phonon.

$$E_{\text{rad}}(r,t + r/c) = \frac{\sin\phi V d^2 E}{\varepsilon_\infty c^2 r dt^2}, \tag{21}$$

where ϕ is the incident angle, V is the excitation volume, and r is the distance between the sample and the detector.

Figure 4.10 shows a setup of THz wave emission spectroscopy. The THz wave emitter is the sample that needs to be evaluated. To obtain high frequency information, a very short laser pulse (12 fs) is used as the excitation source and a 20 μm thick, (110) orientation ZnTe crystal is used to detect THz pulses. According to the discussion in Chapter 2, this detection system has a high frequency response. Figure 4.11 gives the THz pulse waveform (a) and emission spectrum (b) emitted from an undoped GaAs crystal. The phonon-oscillation-induced dipole oscillation is directly reflected in the time-domain oscillation of the THz field. The Fourier transform of the THz waveform gives its emission spectrum. An emission peak located at 8.8 THz is indicated by the *LO* oscillation of the GaAs crystal.

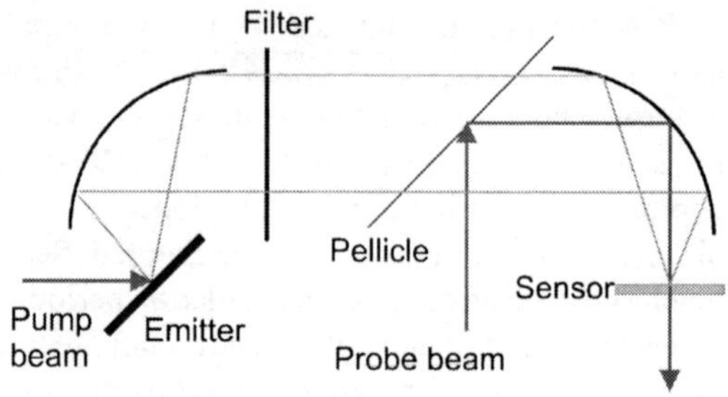

Fig. 4.10 Experimental setup of THz wave emission spectroscopy

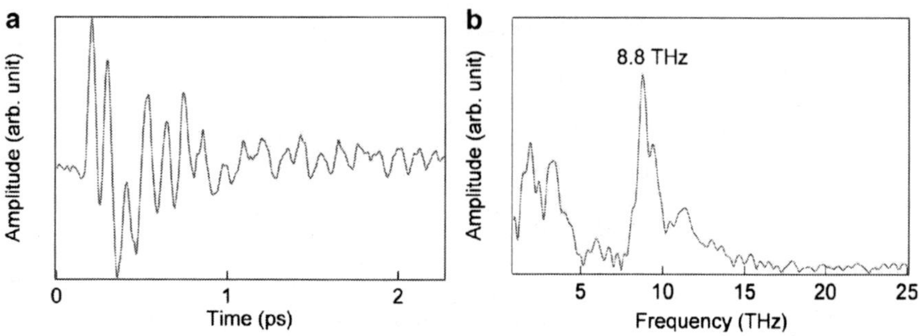

Fig. 4.11 (**a**) Waveform of THz pulses generated from GaAs crystal, and (**b**) THz wave emission spectrum

THz Wave Propagating in Free Space

So far we discussed THz wave interaction with material as an EM wave, now we will discuss THz wave propagation and how different kinds of materials affect THz wave propagation. The simplest case is THz wave propagation in the free space. Except in some extreme cases, when discussing THz wave propagating in free space, the THz field can be approximately treated as a scalar value. Its dynamic is governed by the following Maxwell equation:

$$\Delta U - \frac{1}{c^2}\frac{\partial^2}{\partial t^2}U = 0. \tag{22}$$

Kirhoff and Sommerfeld gave the integral solution of this equation for a monochromatic wave, as

$$U(P_0) = \frac{1}{i\lambda}\iint\limits_{\Sigma} U(P_1)\frac{\exp\left(ikr_{01}\right)}{r_{01}}\cos\left(n,r_{01}\right)ds. \tag{23}$$

This equation gives the diffraction properties of a monochromatic wave. Here P_0 is the field point, P_1 is the source point, $\boldsymbol{r}_{01}$ denotes the distance between the field point and the source point, $\boldsymbol{n}$ is normal of the local emission area. This equation integrates the entire emission area. If radiation has a temporal profile, it can be decomposed into individual monochromatic waves using Fourier transform. Apply Equation (23) to each monochromatic wave and sum them together, one has

$$u(P_0,t) = \iint\limits_{\Sigma} \frac{\cos(n,r_{01})}{2\pi c r_{01}} \frac{\partial}{\partial t} u\left(P_1, t - \frac{r_{01}}{c}\right) ds. \tag{24}$$

Equation (24) indicates that the electric field at the field point is determined by the time derivative of the source electric field. To understand this in a straight forward manner, it is the change of the electric field and not the electric field itself that emits EM waves. When applying this equation to an optical beam, however, the derivative effect is not usually pronounced. For example, you would not expect a light source to look brighter if you turned it on quicker. A normal optical pulse, even an ultrashort laser pulse, can be considered as a monochromatic wave being modulated with a temporal profile $u(t) = a(t)\exp(-i\omega_0 t)$. Its time derivative is $\frac{d}{dt}u(t) = \left[\frac{d}{dt}a(t) - i\omega_0 a(t)\right]\exp(-i\omega_0 t)$. Since for most of the optical pulse, its pulse width is much wider than its oscillation period, and therefore $\omega_0 >> d[a(t)]/dt$ is always true. As a result, the differential item is not significant. Only when laser pulse width is only a few fs, which only contains a few periods of oscillation in a laser pulse, the differential effect can be observed from the pulse shape and phase shifting during propagation.

A typical THz pulse contains a half to a few oscillations, and its bandwidth is even beyond one octave. Consequently, the differential effect is clearly observable during propagation of THz pulses. For EM waves, whose dynamics are governed by Equation (23), if both paraxial ($r^2 >> d^2$) and far field ($r >> d^2/\lambda$) conditions are satisfied, where d is the distance from the field point to the optical source, its diffraction is named the Fraunhofer diffraction. Equation (25) gives single slit Fraunhofer diffraction of THz pulses

$$u(\theta,t) = C \int_{-\infty}^{+\infty} U(\omega) \frac{\mathrm{Sin}\left(\frac{\omega a}{2c}\mathrm{Sin}(\theta)\right)}{\mathrm{Sin}(\theta)} \exp(ikz)\exp(-i\omega t)\, d\omega . \tag{25}$$

According to Equation (25), ω and θ cannot be separated in the integration. As a result, diffraction of the THz pulse does not only change its temporal waveform, but also affects the spatial distribution in propagation.

A THz wave excited using laser pulses can be considered to have a Gaussian spatial distribution. A THz waveform evolves when the THz beam propagates with a Gaussian format. If we consider the waist of the Gaussian beam as the reference point, the waveform modification beside the waist with a distance much larger than the Rayleigh length, is just like the Fraunhofer case. The waveform is the temporal differential of the waveform at the waist. Figure 4.12 shows the temporal waveform of THz pulse evolution from both sides of a Gaussian waist. One can use

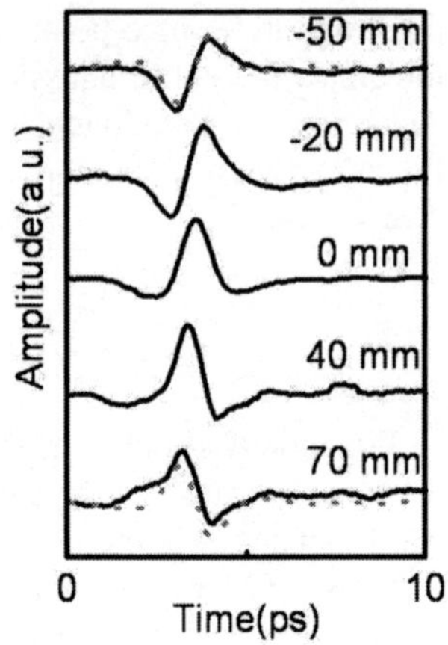

Fig. 4.12 Evolution of THz waveform during propagation

THz waveform propagation to solve a differential calculation, and this calculation is made in light speed. Additionally, Equation (25) indicates that the temporal and spatial functions are tangled together in THz pulse propagation. This means a pulsed wave and monochromatic wave have different spatial distribution in diffraction. In general, an ultra broadband pulse has a smaller angular distribution in propagation.

The Waveguide Propagation of a THz Wave

Unlike propagation in the free space, when the EM wave is confined in a limited volume, such as propagation within a cavity with metal walls, its propagation properties are much different. Propagation of EM waves confined within a limited space is called the waveguide propagation. Waveguide propagation is widely used in applications, such as telecommunications. The common waveguides include parallel plate waveguide, stripe waveguide and optical fiber, et al. Figure 4.13 shows a simple parallel plate waveguide, consisting of two parallel plates with infinity area composite with ideal metal. For EM waves, those two ideal metal plates form mirrors, which confine the EM wave propagation in between those two plates. The EM wave propagating inside a waveguide must be self-consistent, and therefore all components can be coherently constructed during propagation. As a result, an EM with a certain wavelength can only propagate in the waveguide if it is incident with certain angles. Those discreet, propagating incident angles are called the modes in the waveguide. For the parallel plate waveguide, the propagating angle is

$$\sin\theta_m = m\frac{\lambda}{2d}, \tag{26}$$

where m can be any natural number. Equation (26) indicates that, if the wavelength $\lambda > 2d$, then despite what number of m is selected, there is no incident angle to fulfill Equation (26). $\lambda = 2d$ is called the cutoff wavelength of this waveguide,

Fig. 4.13 Propagation of EM wave in parallel plate waveguide

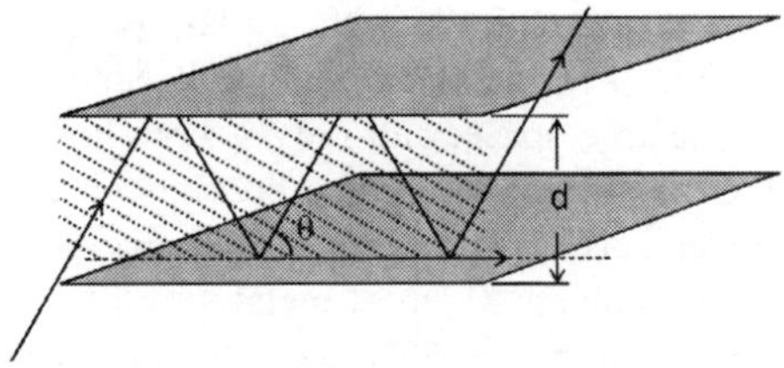

which indicates the longest wave which can propagate through that waveguide. When an EM wave propagates within a waveguide, the components of its wave vector perpendicular and parallel to the plate are

$$k_{\perp} = m\frac{\pi}{d}$$
$$k_{//} = \sqrt{\omega^2/c^2 - m^2\pi^2/d^2}. \tag{27}$$

The propagation group velocity is $v_P = d\omega/dk_{//}$. According to Equations (26) and (27), group velocity of the EM wave in the waveguide is

$$v_P = c\frac{k_{//}}{k_0} = c\cos\theta_m. \tag{28}$$

Equation (28) shows that, even though there is no medium presented, the wave propagation in the waveguide has chromatic dispersion. It also has mode dispersion, which indicates different propagation speed for different modes even though the same color of light is propagated. The discussion above is based on the simplest condition, the parallel plate waveguide; guiding mode in strip waveguide, cylindrical waveguide or if there is media in the waveguide will be more complicated.

The most important feature for a waveguide is its loss. Low frequency waves, such as microwaves usually propagate inside metal waveguides. A metal waveguide is not very suitable for waves with higher frequency since no real metal can be considered as an ideal metal for EM waves with high frequency. This leads to a high extinction ratio in propagation. Fortunately, there are dielectric materials, which are very transparent for optical waves. For instance, fused silica has well below 1 dB/Km attenuation for near IR waves in several transmission windows. This allows an optical wave to propagate inside the dielectric waveguide, particularly the optical fibers. THz waves have higher frequency than microwaves, thus a metal waveguide presents higher loss for THz waves. On the other hand, a dielectric material which has super low loss for THz waves has still not been found. Plastics, such as polyethylene, Teflon, et al., are transparent for THz waves, however, their extinction ratio for THz waves is still in cm^{-1} scale. As a result, it is difficult to use such a material to make a long THz waveguide. To produce a THz wave guide, one needs to have THz wave interaction with the guiding material as small as possible.

Interacting with the surface plasmon of a metal wire, a THz wave can bond with the metal wire and travel along it. Since a major portion of a THz wave propagates in the free space, and only a very limited portion interacts with the metal wire, the metal wire waveguide presents very low attenuation and almost no dispersion under 1 THz [3]. Figure 4.14 shows waveform and spectrum of THz pulses after guiding along 20 cm of metal wire [4]. A similar technique can also be applied to holey core plastic fiber and sub-wavelength fibers, while majority of the THz wave propagates within free space, and allows for a small portion of the wave interacting with media to bond to the wave. For a pulsed THz wave, dispersion is also very important due to its broad band. Figure 4.15 shows waveform and spectrum of THz pulses after propagating through a parallel plate waveguide [5]. Comparing the guided pulse with the original pulse, the waveform and spectrum are similar. This indicates that the parallel plate waveguide presents low dispersion for THz pulses. Moreover, the high resolution of waveguide spectroscopy is applied to the study of biological molecules [6]. An ordered polycrystalline film on a metal waveguide plate was made, which can significantly reduce the inhomogeneous broadening associated with THz vibration mode. Then the incorporation of the metal plate into a single transverse electromagnetic mode parallel-plate waveguide makes the film interrogated by a THz beam with a high sensitivity. Figure 4.15c shows the corresponding amplitude spectra for the THz pulses transmitted through waveguide with deoxycytidine film. Recently, time-resolved THz spectroscopy in a parallel plate waveguide was reported [7]. They apply a novel parallel plate waveguide where one of the metallic plates is replaced by a transparent conducting oxide. The absorption coefficient of the transparent waveguide is reduced in their work compared to previously by a factor of 3 to approximately 4 cm^{-1} at 05 THz. They propose a useful tool for time-resolved studies of photoexcitations in thin films with low absorption in the THz range.

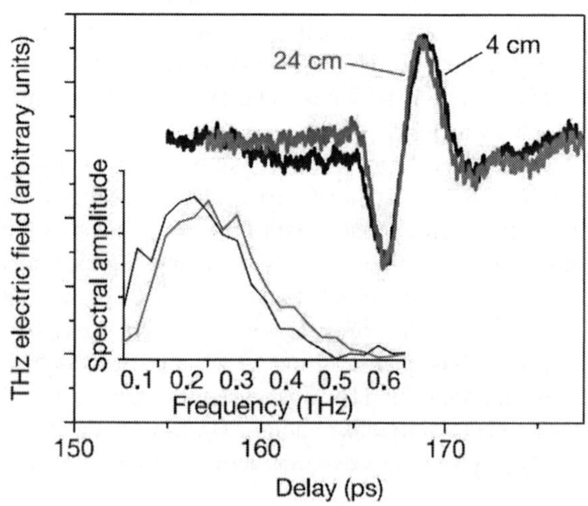

Fig. 4.14 Waveforms and spectra of THz pulses after propagation along different length of metal wire (Courtesy of Dr. Mittleman)

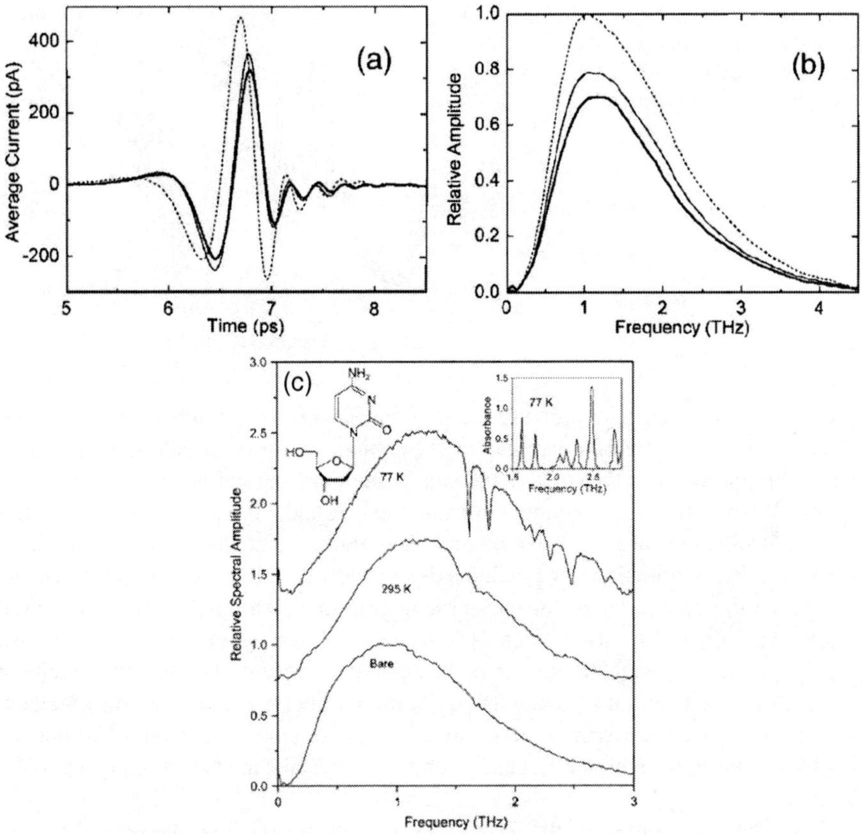

Fig. 4.15 (**a**) Original THz waveform (*dashed curve*), and waveforms of THz pulse propagation through 12.6 mm (*Thinner solid curve*) and 24.4 mm (*Thicker solid curve*) parallel plate waveguide made by copper plates. (**b**) spectra of THz pulses in a [5]. (**c**) corresponding amplitude spectra for the THz pulses transmitted through bare waveguide and waveguide with deoxycytidine film. (*Inset*) Absorbance spectrum at 77 K. Spectrum at 295 K is normalized to unity (Courtesy of Dr. Greschkowsky)

As for the THz quantum-cascade laser (QCL) – it is based on a chirped superlattice design with a novel surface-plasma waveguide first demonstrated by Köhler et al. [8]. Since that initial breakthrough, major developments have taken place in both the multiple-quantum-well gain medium and waveguide. The use of a metal-metal ridge waveguide, similar in form to a microstrip transmission line, has been successfully used to provide a high-confinement, low-loss cavity for terahertz lasers [9].

THz Wave and Photonic Crystal

Like traditional crystal interaction with electrons, a photonic crystal manipulates photons in a similar way. A photonic crystal is made by media with a periodically distributed refractive index. The periodic distributed refractive index

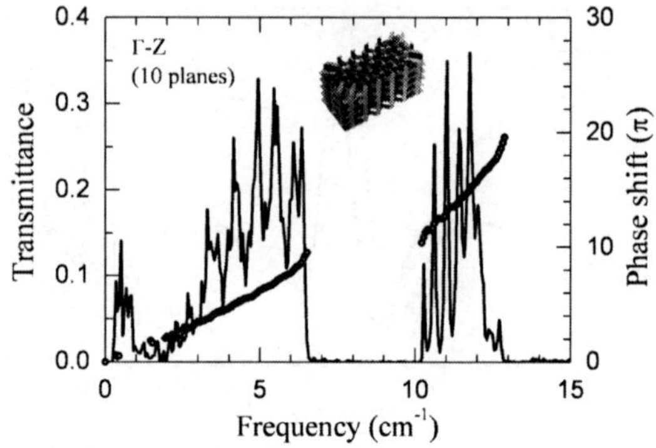

Fig. 4.16 THz wave transmission spectrum of a photonic crystal (*inset* shows the structure) (Courtesy of Dr. Tani)

provides a band and gap structure for photons, called the photonic band and gap. Photons located at the photonic gap of the crystal cannot propagate in this crystal, while the photons located in the band can. A photonic crystal has interesting properties. When designing a photonic crystal, one can make it act as a band pass filter with controllable central wavelength and bandwidth. It can also be used as a mirror with broadband reflection. By playing a defect trick, the photonic crystal can be use to confine the optical beam for only propagation in a desired location and desired mode, which could be used as an optical cavity or waveguide, etc. One can also use the photonic crystal structure to make a photonic crystal fiber, which has unique properties which cannot be made using traditional fibers. Those properties include, single mode propagation in a large mode field, anonymous dispersion for normal dispersive wave, ultra high NA, et al. A photonic crystal can also be scaled up to the THz band.

Since the wavelength of the THz wave is much longer than the optical waves, the techniques used to make a photonic crystal in the THz band are usually different than the optical band. Mechanical stacking rather than photolithography is generally used to make a photonic crystal in the THz band. Plastic is the most used material for THz waves, just as glass is for optical waves. The refractive index of plastic is around 1.5. When a material with high refractive index is required, silicon or other high resistivity semiconductors or ceramic materials are typically used. The major applications of a THz photonic crystal include filter and modulator. Figure 4.16 shows a THz wave photonic crystal and its band structure [10]. Figure 4.17 shows a THz wave photonic crystal fiber made by polyethylene [11]. The spectrum of THz pulses propagating through 2 cm of such a fiber is presented.

Surface Plasmon and Metamaterial

To creatively utilize EM waves, researchers often design artificial materials instead of using a natural material to manipulate EM waves. Waveguides and photonic crystals can be considered as artificial materials. Recently, an artificial material fabricated on a 2D conductive layer has drawn attention, as surface plasmon and as a metamaterial.

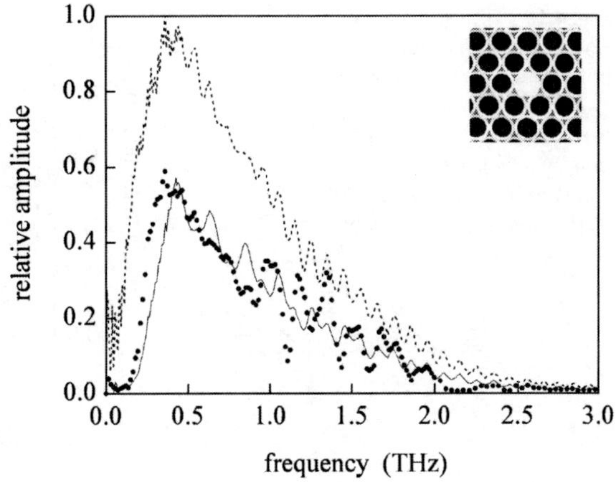

Fig. 4.17 THz wave photonic crystal fiber (*inset*) and spectrum of THz pulse after propagation through 2 cm fiber. *Dots* are experimental data and *solid curve* gives calculation result. *Dashed curve* indicates the original THz spectrum (Courtesy of Dr. Han)

The investigation of EM wave interaction with periodic conductive structure began with the study of a metal grating polarizer for infrared radiation. A metallic grating polarizer consists of repeating conductor (metal)/dielectric strips with a period shorter than the wavelength of the EM wave. If the EM wave is incident onto the grating with its polarization perpendicular to the strips, the grating is transparent for the incident wave. However, if its polarization is parallel to the strips, then the grating reflects the EM wave. As a result, such a grating acts as a polarizer for IR waves. The interaction between EM waves and the metallic grating can be calculated through an impedance matching model between the grating and EM wave. If the period of grating is d, and the width of the metal strip is a, then the transmission of the grating for s and p polarization waves are [12]

$$t_{\perp} = \frac{4(X_0/Z_0)_{\perp}^2}{4(X_0/Z_0)_{\perp}^2 + 1}$$
$$t_{//} = \frac{4(X_0/Z_0)_{//}^2}{4(X_0/Z_0)_{//}^2 + 1}. \tag{29}$$

Here Z_0 denotes the free space impedance, and X_0 is the incident coupling impedance of the grating. For an EM wave whose polarization is perpendicular and parallel to the grating, there is

$$\left(\frac{Z_0}{X_0}\right)_{\perp} = \frac{4d}{\lambda}\left\{\ln\left[\csc\frac{\pi(d-a)}{2d}\right] + \frac{Q_2\cos^4[\pi(d-a)/2d]}{1+Q_2\sin^4[\pi(d-a)/2d]}\right.$$
$$\left. + \frac{1}{16}\left(\frac{d}{\lambda}\right)^2\left[1-3\sin^2\frac{\pi(d-a)}{2d}\right]^2\cos^4\frac{\pi(d-a)}{2d}\right\}, \tag{30a}$$

and

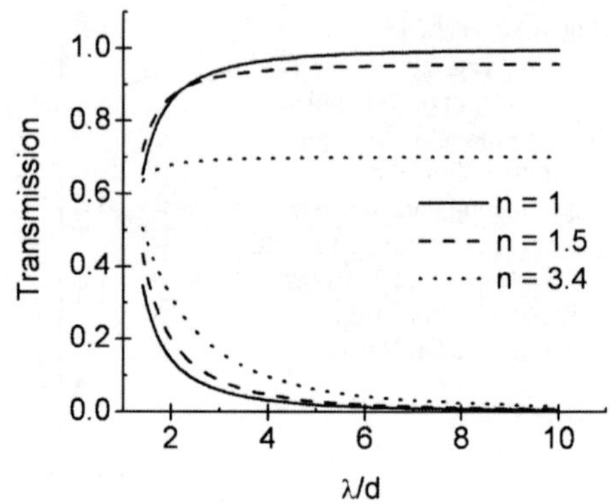

Fig. 4.18 Transmittance of EM waves with different polarization as a function of λ/d. Refractive index of dielectric medium is 1 (air, *dotted curve*), 1.5 (plastic, *solid curve*), and 3.4 (silicon, *dashed curve*), respectively

$$\left(\frac{X_0}{Z_0}\right)_{//} = \frac{d}{\lambda}\left[\ln\left(\csc\frac{\pi a}{2d}\right) + \frac{Q_2\cos^4(\pi a/2d)}{1 + Q_2\sin^4(\pi a/2d)} + \frac{1}{16}\left(\frac{d}{\lambda}\right)^2\left(1 - 3\sin^2\frac{\pi a}{2d}\right)^2\cos^4\frac{\pi a}{2d}\right], \tag{30b}$$

where

$$Q_2 = \left[1 - (d/\lambda)^2\right]^{-1/2} - 1. \tag{30c}$$

Transmission of metallic grating and its polarization extinction ratio (PER) for EM waves with different wavelength can be calculated through Equations (29) and (30). Figure 4.18 gives transmission and PER of grating as a function of λ/d. Metallic grating with a period much smaller than the EM wave wavelength leads to high PER.

Metals have high free-electron density, and thus have POF in the UV band. Besides of bulk material, plasma oscillation can also happen on the boundary between the metal and dielectric material, called surface plasmon. The properties of surface plasmon are much different than bulk plasmon in terms of frequency and dispersion. The wave vector of surface plasmon generated in a metal film coated on a dielectric material is

$$k_{sp} = k_0\left[\frac{\varepsilon_d}{(\varepsilon_{rs} + \varepsilon_d)^2 + \varepsilon_{is}^2}\right]^{1/2}\left[\frac{\varepsilon_e^2 + (\varepsilon_e^4 + \varepsilon_d^2\varepsilon_{is}^2)^{1/2}}{2}\right]^{1/2}, \tag{31}$$

where k_0 is wave number of the EM wave in the free space, ε_d denotes dielectric constant of the dielectric material, ε_{rs} and ε_{is} denote real and imaginary parts of metal permittivity, respectively, and $\varepsilon_e^2 = \varepsilon_{rs}^2 + \varepsilon_{is}^2 + \varepsilon_d \varepsilon_{rs}$. Under most circumstances, permittivity of the metal is much larger than that of the dielectric material. Thus Equation (31) can be simplified to

$$k_{sp} = k_0 \sqrt{\varepsilon_d}. \tag{32}$$

Surface plasmon oscillation propagates along the surface, and decays in a direction perpendicular to the surface. And, the dispersion property of surface plasmon oscillation is not matched with the EM wave in the free space. Spectral techniques need to be applied in order to provide sufficient coupling between surface plasmon oscillation and the free space EM wave. Such techniques include using a prism, grating or utilizing the total internal reflection process.

One method to couple surface plasmon oscillation with the free space EM wave is to make periodic holes on the metal film. Assume the period of holes on the metal film is L, then the wave vector of surface plasmon, which is able to couple with the free space EM wave is described as

$$\vec{k}_{sp} = \vec{k}_{//} + m\vec{K}_x + n\vec{K}_y. \tag{33}$$

Here $\vec{k}_{//}$ denotes the wave vector of the EM wave parallel to the boundary, m and n can be any natural numbers, and $K_x = K_y = 2\pi/L$ are the wave number of the periodic structure in x and y direction, respectively. For a normal incident EM wave, there is$k_{//} = 0$. Combining Equations (32) and (33), there is the wavelength of an EM wave suitable to couple with the surface plasmon oscillation:

$$\lambda = \frac{L\sqrt{\varepsilon}}{\sqrt{n^2 + m^2}}. \tag{34}$$

Coupling with surface plasmon helps EM wave transmission through a metal film with periodic holes. According to the EM wave's diffraction principle, transmission is very low through sub-wavelength holes on metal film. However for periodic holes on metal film, if the EM wave has a wavelength that satisfies Equation (34), its transmission will be highly enhanced due to coupling with surface plasmon in the metal film. The transmission is much higher than the sum of all single-hole transmission and even higher than the ratio between the sum of the hole area and the entire area of the sample. Figure 4.19 gives the spectrum of THz pulses transmitted though the hole's array on metal film [13]. The resonating coupling results in transmission peaks. Equation (34) indicates that the resonance wavelength is related to the dielectric constant of the dielectric material. As a result, the changing of permittivity could change the resonance spectrum. This allows the use of surface plasmon oscillation for spectroscopy measurement. If the dielectric material has a strong absorption feature, its dielectric constant could widely vary at the wavelength close to the absorption feature. In this case, the strong variation of permittivity could generate a resonance coupling on the absorption feature. This phenomenon is shown in Fig. 4.20.

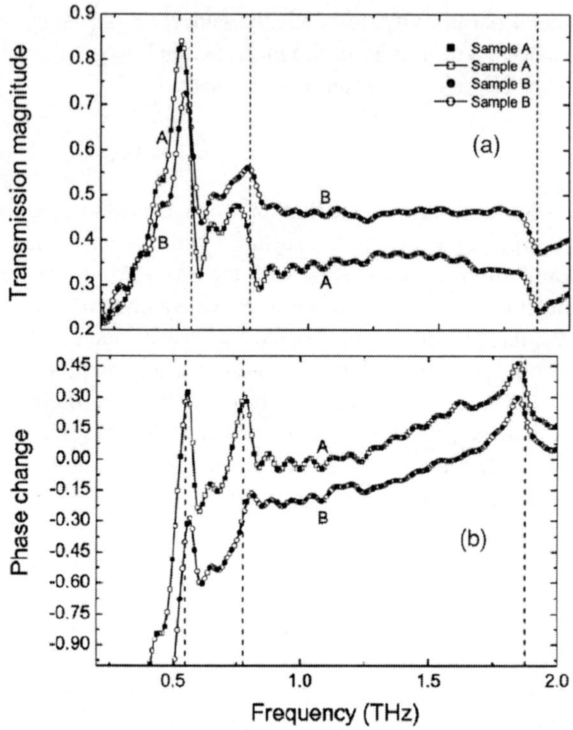

Fig. 4.19 Transmission spectrum of metal film with periodic holes. The period of holes array is 160 μm. (**a**) THz transmission spectroscopy, (**b**) the phase shift. Sample A has 80 μm by 100 μm *rectangular holes* and sample B has 100 μm diameter *round holes* (Courtesy of Dr. Grischkowsky)

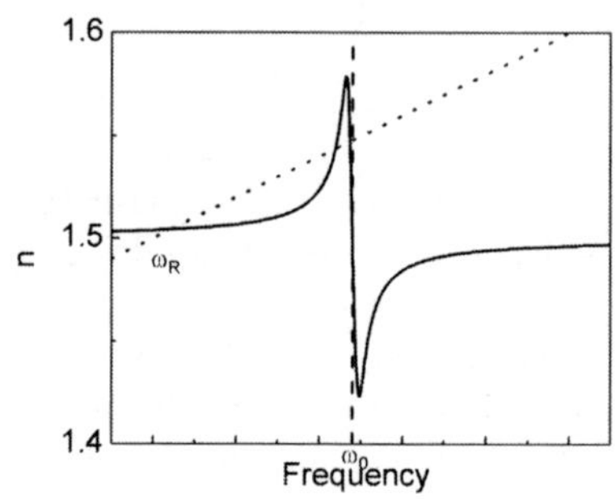

Fig. 4.20 Medium refractive index determined surface plasmon oscillation frequency. *Solid curve* indicates refractive index of the medium, *dashed curve* shows dispersion of the surface plasmon oscillation determined by the periodic structure

If the material, which has negative permittivity also has negative permeability, it presents a negative refractive index for the EM wave. This material is called the left-handed material, which indicates that it does not obey the "right-handed rule" which a natural material does. A left-handed material has unique features such as a super lens which "amplifies" the evanescent wave, and cloaking which "bends" the passing EM wave surrounding it. The former could be used in imaging with spatial resolution better than the diffraction limit. The latter could be used to make an "invisible" container. According to Equation (4), a material, such as metal, whose

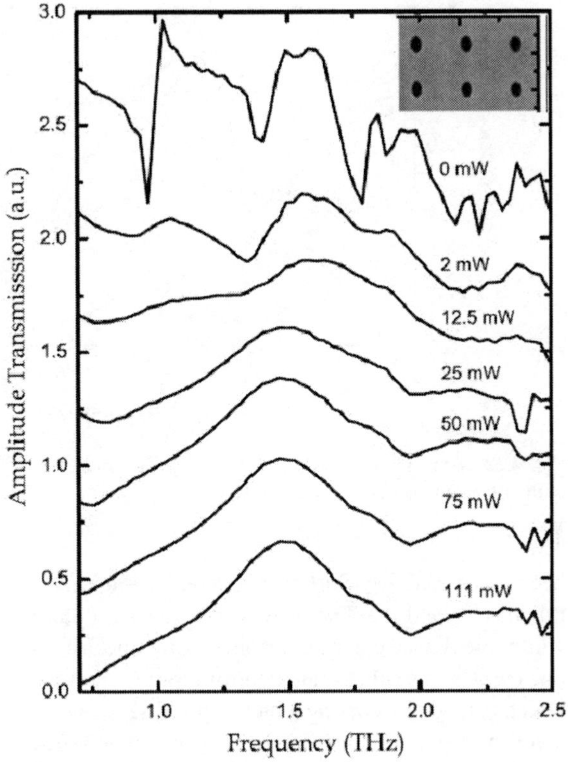

Fig. 4.21 Transmission spectra of the surface plasmon device with different optical excitation power. *Inset* gives structure of the device

POF frequency is much higher than EM wave frequency, will have negative permittivity. However, a natural material does not have both negative permittivity and permeability. One method to make a left-handed material is to format a microcircuit structure which couples with the EM wave and excite carrier motion. Material with negative permittivity and (or) permeability for certain EM wave may not be readily exist in nature. While those properties may be conferred to an artificial material with engineered structure, which is so called metamaterial. One of the most famous meta material is the split-ring resonator.

Study of surface plasmon oscillation and metamaterials in the THz band is not only for pure scientific interest but also enables development of promising devices used to manipulating THz waves [14]. Surface plasmon oscillation and metamaterials show resonating structures, which lead to high modification for THz waves at those frequencies. These devices could be used to modulate THz waves with high modulation depth. This is especially useful for dynamic modulation of THz waves. Figure 4.21 shows the THz wave as modified by a surface plasmon device [15]. This device is made by thin intrinsic silicon wafer with periodic holes. Since silicon is a dielectric material, the device is a photonic crystal for THz waves without excitation. Excited with a laser beam generating free carriers in the silicon wafer makes it

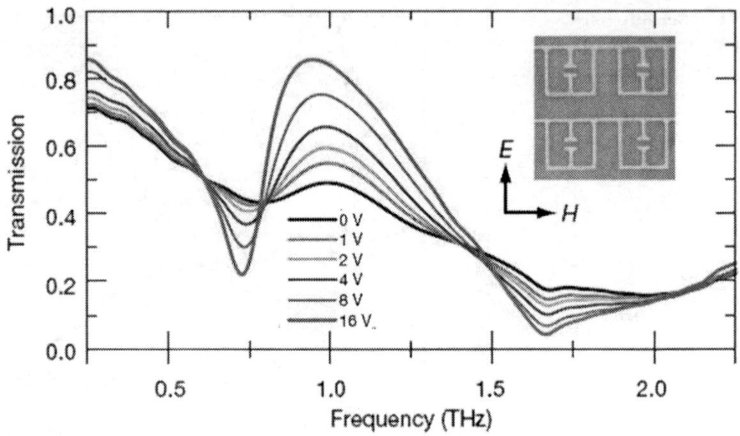

Fig. 4.22 Transmission spectra of the active THz metamaterial device (*inset* gives the structure) with different bias (Courtesy of Dr. Chen)

become conductive. Surface plasmon oscillation dominates the interaction between this device and the THz waves. Using an excitation laser pulse with sub ps pulse width, the device can be controlled with ultrafast speed. Besides optical controlling, one can also use electronic controlling. Figure 4.22 shows a THz metamateral [16], which is made by coating a metal split ring resonator array on top of an n-type GaAs layer. Applying voltage between the metal and the semiconductor controls electron density in the n-GaAs around the split. As a result, resonation is switched on/off using biased voltage.

References

1. R. Kersting, K. Unterrainer, G. Strasser, H. F. Kauffmann, and E. Gornik, "Few-cycle THz emission from cold plasma oscillations," *Phys. Rev. Lett.* **79**, 3038 (1997).
2. R. Huber, F. Tauser, A. Brodschelm, M. Bichler, G. Absteiter, A. Leitenstorfer, "How many-particle interactions develop after ultrafast excitation of an electron-hole plasma," *Nature* **414**, 286 (2001).
3. J. A. Deibel, K. Wang, M. D. Escarra, and D. M. Mittleman, "Enhanced coupling of terahertz radiation to cylindrical wire waveguides," *Opt. Lett.* **14**, 279 (2006).
4. K. Wang, and D. M. Mittleman, "Metal wires for terahertz wave guiding," *Nature* **432**, 376 (2004).
5. R. Mendis, and D. Grischkowsky, "Undistorted guided-wave propagation of subpicosecond terahertz pulses," *Opt. Lett.* **26**, 846 (2001).
6. N. Laman, S. S. Harsha, D. Grischkowsky, and J. S. Melinger, "High-resolution waveguide THz spectroscopy of biological molecules," *Biophys. J.* **94**, 1010 (2008).
7. D. G. Cooke, and P. U. Jepsen, "Time-resolved THz spectroscopy in a parallel plate waveguide," *Phys. Status Solidi A* **206**, 5, 997 (2009).
8. R. Köhler, A. Tredicucci, F. Beltram, H. E. Beere, E. H. Linfield, A. G. Davies, D. A. Ritchie, R. C. Iotti, and F. Rossi, "Terahertz semiconductor-heterostructure laser," *Nature* **417**, 156 (2002).

9. B. S. Williams, S. Kumar, H. Callebaut, Q. Hu, and J. L. Reno, "Terahertz quantum-cascade laser at $\lambda \approx 100$ μm using metal waveguide for mode confinement," *Appl. Phys. Lett.* **83**, 2124 (2003).
10. T. Aoki, M. W. Tekeda, J. W. Haus, Z. Yuan, M. Tani, K. Sakai, N. Kawai, and K. Inoue, "Terahertz time-domain study of a pseudo-simple-cubic photonic lattice," *Phys. Rev. B* **64**, 045106 (2001).
11. H. Han, H. Park, M. Cho, and J. Kim, "Terahertz pulse propagation in a plastic photonic crystal fiber," *Appl. Phys. Lett.* **80**, 2634 (2002).
12. J. P. Auton, "Infrared transmission polarizers by photolithography," *Appl. Opt.* **6**, 1023 (1967).
13. D. Qu, D. Grischkowsky, and W. Zhang, "Terahertz transmission properties of thin, subwavelength metallic hole arrays," *Opt. Lett.* **29**, 896 (2004).
14. H. Chen, W. J. Padilla, M. J. Cich, A. K. Azad, R. D. Averitt, and A. J. Taylor, "A metamaterial solid-state terahertz phase modulator," *Nat. Photonics* **3**, 148 (2009).
15. W. Zhang, A. K. Azad, J. Han, J. Xu, J. Chen, and X.-C. Zhang, "Direct observation of a transition of a surface plasmon resonance from a photonic crystal effect," *Phys. Rev. Lett.* **98**, 183901 (2007).
16. H. Chen, W. J. Padilla, J. M. O. Zide, A. C. Gossard, A. J. Taylor, R. D. Averitt, "Active terahertz metamaterial devices," *Nature* **444**, 597 (2006).

Chapter 5
THz Air Photonics

THz wave spectroscopy and imaging technologies are promising in security inspection applications. However, the following hurdles prevent THz technologies to be used in in situ applications, especially when standoff detection is required. First of all, the attenuation of THz waves in the atmosphere is higher than 100 dB/km, so it was previously thought impossible to perform long distance broadband THz wave sensing and spectroscopy, due to severe water vapor attenuation. Secondly, pulsed THz wave emitters using either real or virtual photocurrents, saturate when high excitation intensities are used. Further increase of the excitation power may even cause damage to the emitter. The saturation and damage of THz wave emitter limits the strength of the THz fields that can be generated from such emitters. Additionally, although pulsed THz wave generation and detection systems provide broadband spectral coverage, the spectrum does not generally cover the entire terahertz band continuously. Semiconductors or nonlinear crystals usually have phonon modes in the THz band. Absorption and dispersion due to photons result in dark areas in the measured THz spectrum. Finally, the reflection of THz waves, by both surfaces of the emitter or sensor, generates interference patterns in the THz spectrum. Confronted by those hurdles, using ambient air as the THz wave emitter and sensor becomes more and more interesting. By using ambient air as THz wave emitter and sensor, one can generate and detect THz waves close to the sample. Sending an optical beam instead a of THz wave, benefits long-distance standoff detection due to the relatively low attenuation experienced in the atmosphere. Since air or other gases are easily replaceable, damage is not a concern even if a strong laser field is used to generate the THz pulses. As a result, it is preferable in the generation of high intensity THz pulses. Finally, dry air has neither phonon bands nor boundary reflection surfaces, and thus provides continuous coverage along the entire bandwidth.

THz Wave Generation in Ambient Air

THz waves can be generated in air via several different mechanisms. When a high-intensity laser pulse ionizes a gas, a THz transient would be formed through a

X.-C. Zhang, J. Xu, *Introduction to THz Wave Photonics*,
DOI 10.1007/978-1-4419-0978-7_5, © Springer Science+Business Media, LLC 2010

coherent plasma oscillation driven by either the ponderomotive force or by transition radiation from accelerated electron bunches driven by wakefield acceleration. The latter provided high-energy electrons that would emit transition radiation in the THz range when impinging on a sharp dielectric constant gradient. Electrons driven with ponderomotive force induce an oscillation primarily along the direction of the pulse propagation vector, and thus radiates in a conical pattern, while electrons driven by the wakefield provided THz radiation in the forward direction. The above experiments were carried out with low-repetition rate (10 Hz), high-energy (many mJ) laser systems, which were required in order to provide the high ionization probabilities and ponderomotive potentials required for their respective effects. Due to the highly nonlinear nature of the effects, they are inefficient at low intensities, and so work with more common regenerative amplifiers with kHz repetition rates required alternate methods. Such lasers are powerful enough to ionize the target gas, but using them efficiently requires a more direct method of accelerating the electrons. Applying of a DC bias to the plasma accelerates the electrons as they are ionized, resulting in a rapidly increasing current and the radiation of a THz pulse. In this case, the direction of the photocurrent is determined by the direction of the bias field, and so it can be set to be orthogonal to the pulse propagation direction, leading to coherent build-up of the THz pulse amplitude along the plasma. Doing this led to an order of magnitude enhancement of the THz pulse amplitude over what was generated through ponderomotive acceleration alone, and was limited mainly by the strength of the applied bias, which was constricted by electrical breakdown of the gas as the field approached 30 kV/cm in the atmosphere.

An alternative method generates a strong THz pulse by combining the fundamental laser pulse with a pulse at its second harmonic frequency, originally described as four-wave rectification, a third-order nonlinear process based on four-wave mixing. The actual physical details are quite different from four-wave mixing, as will be discussed later, but it provides a convenient framework for describing several experimental results. This framework is similar to second-order (three-wave) optical rectification, where two photons near the fundamental laser frequency (ω and $\omega+\Delta\omega$) are coupled, producing their difference frequency $\Omega = (\omega + \Delta\omega) - \omega$, with the frequency offset $\Delta\omega$ provided by the bandwidth of the optical pulse. Such a second-order process does not occur in a centrosymmetric medium such as a gas, so one is forced to move on to third-order effects. To perform four-wave rectification, one must couple three photons to produce a nearly DC output, which precludes the possibility of all photons having approximately the same energy. If one photon has approximately twice the energy as the other two, the difference between its energy ($2\omega+\Delta\omega$) and the sum of the energies of two fundamental photons ($\omega+\omega$) will indeed produce the desired difference frequency. THz wave generation through four-wave mixing is described in the following equation:

$$(2\omega + \Omega_{\mathrm{THz}}) - \omega - \omega - \Omega_{\mathrm{THz}} = 0. \tag{1}$$

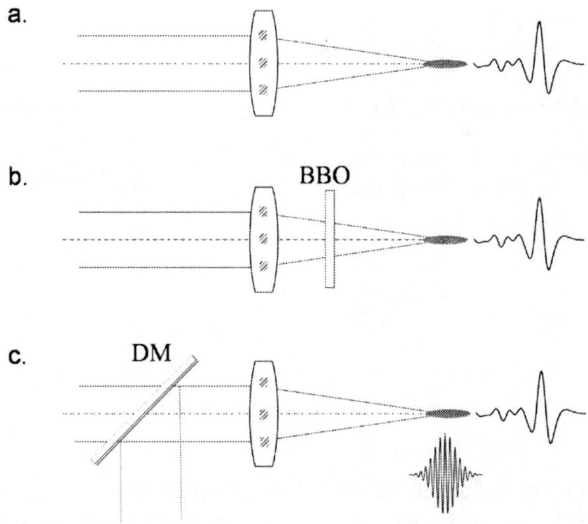

Fig. 5.1 Concepts of generating a THz wave in gas. (**a**) Generating THz wave using a single (semi-monochromatic) laser beam. (**b**) Generating THz wave using fundamental laser beam combining with its second harmonic. (**c**) Independent controlling of delay between the fundamental beam and second harmonic beam can control phase shift in THz wave generation

Figure 5.1 shows the concept of THz pulse generation in air through four-wave mixing processes. To have sufficient efficiency, a femtosecond laser amplifier with sub mJ pulse energy is usually used.

The concept is presented in Fig. 5.1b. A BBO crystal is placed in front of the laser focal spot. Second harmonic frequency is generated in the crystal and both fundamental and second harmonic lasers are focused at the same focal spot. The frequency mixing process is

$$\Omega_{\mathrm{THz}} = (2\omega + \Omega_{\mathrm{THz}}) - \omega - \omega. \tag{2}$$

Figure 5.2 shows THz waveforms generated through these two processes respectively. The THz pulse generated by one excitation laser beam has a similar waveform to those generated by a combination of different color laser beams. However, the THz field generated via the latter process is orders stronger than the former.

Air has a very low third order nonlinear coefficient, $\chi^{(3)} = 1.68 \times 10^{-25}(m/V)^2$ at 20°C. As a result, THz wave generation through non-ionized air is very weak. To have high generation coefficient, one need to use a strong laser to ionize air in order to benefit from nonperturbative effects. Figure 5.3 shows THz field as a function of the fundamental laser intensity (a) and intensity of the second harmonic laser (b). Figure 5.3 indicates that above the ionization threshold, the generated THz field is proportional to intensity of the fundamental laser and is also proportional to the square root of the second harmonic laser. This relationship consists of Equation (2).

$$E_{\mathrm{THz}} \propto \chi^{(3)} \sqrt{I_{2\omega}}\, I_{\omega}. \tag{3}$$

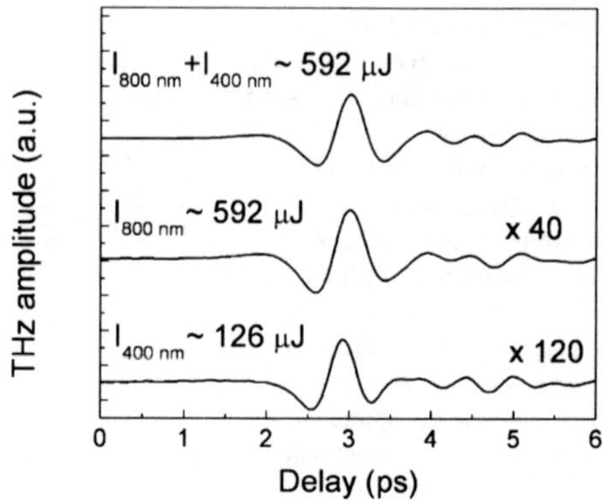

Fig. 5.2 THz waveforms generated in air. From the top to the bottom, THz pulse generated by combining of 800 nm and 400 nm laser beams, and by 800 nm laser pulse, 400 nm laser pulse, respectively

In Fig. 5.3a, the THz field has a much lower value when the fundamental power is less than 55 mW. Higher excitation intensity gives much larger slope. This indicates that there is an excitation power threshold for the generation of THz waves. This threshold is the threshold for significant ionization of the air. No excitation threshold exists in Fig. 5.3b since the fundamental frequency itself has sufficient power to ionize the air even when the second harmonic frequency power is low. When the method shown in Fig. 5.1b is used to generate THz waves, the THz field can be as high as 100 KV/cm. However due to dispersion between the fundamental laser and the second harmonic laser, the phase shift between these two beams varies as they propagate. The phase shift variation affects the THz wave generation coefficient. Using the concept shown in Fig. 5.1c one can independently control the phase shift between the fundamental pulse and the second harmonic pulse, as well as their power and polarization. Figure 5.4 shows the amplitude of THz pulse generated in air as a function of the delay between the fundamental pulse and the second harmonic pulse. The results show that THz field reaches its maximum when

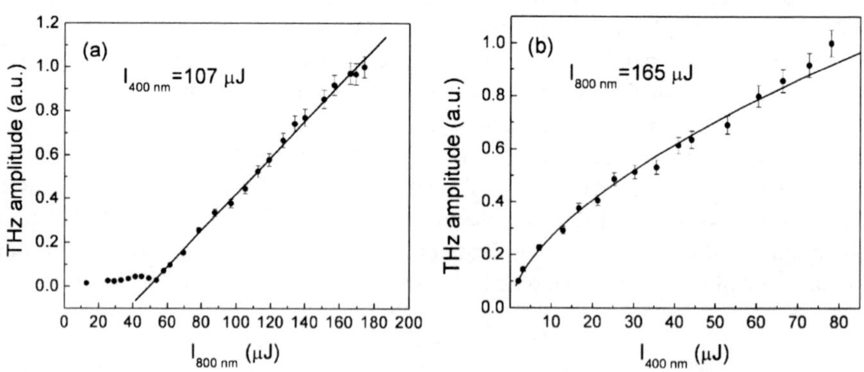

Fig. 5.3 THz field as a function of laser intensities of 800 nm laser beam (**a**) and 400 nm laser beam (**b**) in THz wave generation by combination of 800 and 400 nm fs laser beams

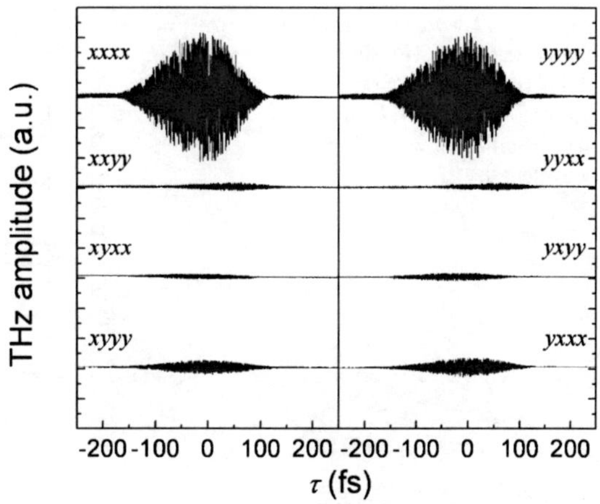

Fig. 5.4 Amplitude of THz pulse as a function of time delay between two laser pulses (800 nm wavelength and 400 nm wavelength). x and y denote coordinate system for polarization of fundamental laser, second harmonic laser, and THz wave

the polarizations of both excitation lasers are parallel to each other. In this case, the THz wave has the same polarization with the excitation laser beams. The THz field oscillates with time delay between the fundamental beam and the second harmonic beam. The strongest field is achieved when the maxima of both laser pulses overlap in time. The field decays with separation between the pulses. Figure 5.5 gives a zoomed in view of the THz field oscillation. The equation describing the THz field as a function of phase shift between those two excitation laser beams is

$$E_{\mathrm{THz}}(t) \propto \chi^{(3)}\, E_{2\omega}(t)\, E^{*}_{\omega}(t)\, E^{*}_{\omega}(t) \cos(\varphi), \tag{4}$$

where the phase shift $\varphi = k_{2\omega}\,\Delta l$ gives the phase change between two excitation lasers. According to Fig. 5.5, the polarity of THz field reverses when the delay between two excitation lasers changes by a half cycle of the second harmonic wave. This is confirmed by the THz waveforms presented in Fig. 5.6 where the time delay

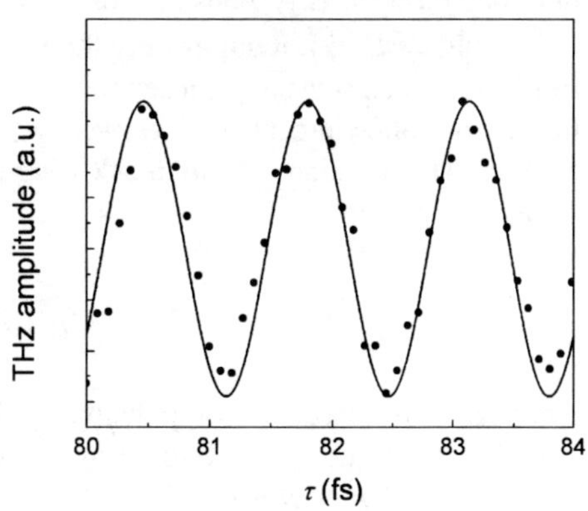

Fig. 5.5 Zoom in of THz pulse amplitude evolution with time delay. *Dots* show experimental data and *solid curve* is calculated result

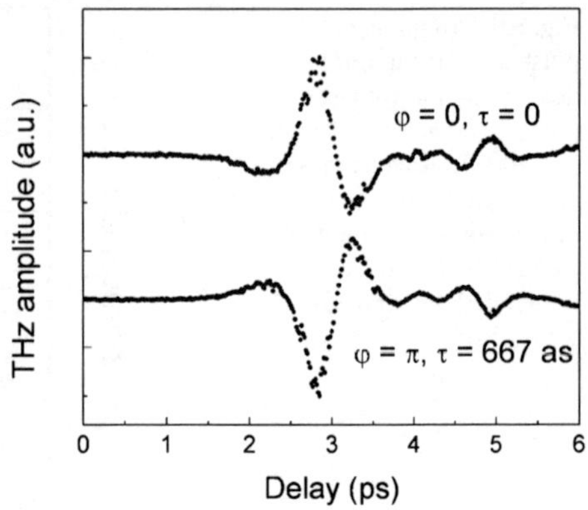

Fig. 5.6 THz waveforms of two THz pulses with time delay (in Fig. 5.4) of 667 attosecond (*half circle* of laser oscillation with 400 nm wavelength)

difference are 667 attosecond between those two waveforms, which equals one half cycle of the 400 nm pulse used in the experiment.

However, the four-wave mixing framework cannot explain all of the observed phenomena in the generation process. Specifically, an intensity threshold coincident with the threshold for ionization of the gas can be observed, in contrast to the simple power-law intensity dependence predicted by four-wave mixing. The ionization process definitely plays an essential role in THz wave generation. The ionization enhanced four-wave mixing can be explained by the unidirectional motion of the electrons during ionization by the two-color field, effectively forming a transient current similar to the electrically biased case, the current is formed due to the average velocity of the electrons. The gas ionization and THz wave generation process can be solved through a quantum mechanical approach based on numerically solving the time-dependent Schrödinger equation (TDSE). This allows for a non-perturbative simulation that includes both the bound and ionized states and transitions between them. Because the bound states are included in the simulation, the effects of perturbative nonlinear optics are faithfully reproduced as well. In the case of the single active electron approximation (where it is assumed that only one electron is responsible for the observed interaction), solving the TDSE, $i\hbar {}^{\partial}/_{\partial t} |\Psi\rangle = H |\Psi\rangle$, requires only the Hamiltonian operator H and the initial electron state, which is usually chosen to be an eigenstate of H in the absence of external fields. The following calculations were performed in the velocity gauge [1]. In this case, the Hamiltonian is written as (using atomic units and assuming a spherically symmetric potential)

$$H = \frac{1}{2} (p + A)^2 + V(r) \tag{5}$$

where A is the laser vector potential, p is the kinetic momentum operator and V is the atomic potential. Exploiting the Coulomb gauge, the expression simplifies to $H = \frac{1}{2}p^2 + p \bullet A + V(r)$. In the case of a linearly polarized laser field, the system of coordinates may be rotated such that the field points along the z-axis, and the interaction term $p \bullet A$ becomes $-iA\frac{\partial}{\partial z}$. When the wavefunction is expanded over spherical harmonics, this operator introduces no coupling between states with different values of the z-axis projection of the angular momentum (m), and the three-dimensional problem reduces to two dimensions, with the wavefunction represented as a series of partial waves $\varphi_\ell(r)$, with the complete wavefunction $\Psi(r,\theta,\phi) = \sum_\ell \varphi_\ell(r) Y_\ell^m(\theta,\phi)$. Performing the simulation amounts to solving the TDSE using a propagation scheme based on the Crank-Nicolson method. This provides the electron wave function at each time step in the simulation, which leads naturally to visualizations to aid understanding of the underlying effects. The square of the wavefunction modulus provides the electron density distribution, which when converted to Cartesian coordinates can show how the ionization process takes place. In Fig. 5.7, a series of these images is presented for various points in the ionization of an argon atom by a 50 fs, 800 nm pulse (with a $\sin^2$ envelope) with a peak electric field of 200 MV/cm, combined with a 400 nm pulse of the same duration and 20 MV/cm peak field, with the relative phase between them set to $23\pi/12$.

One can see in Fig. 5.7 that the evolution of the density distribution is strongly asymmetric, with a relatively large proportion of the density propagating down the page. In contrast, when the phase between the fundamental and second harmonic waves is shifted by $\pi/2$, the distribution becomes markedly more symmetric, as

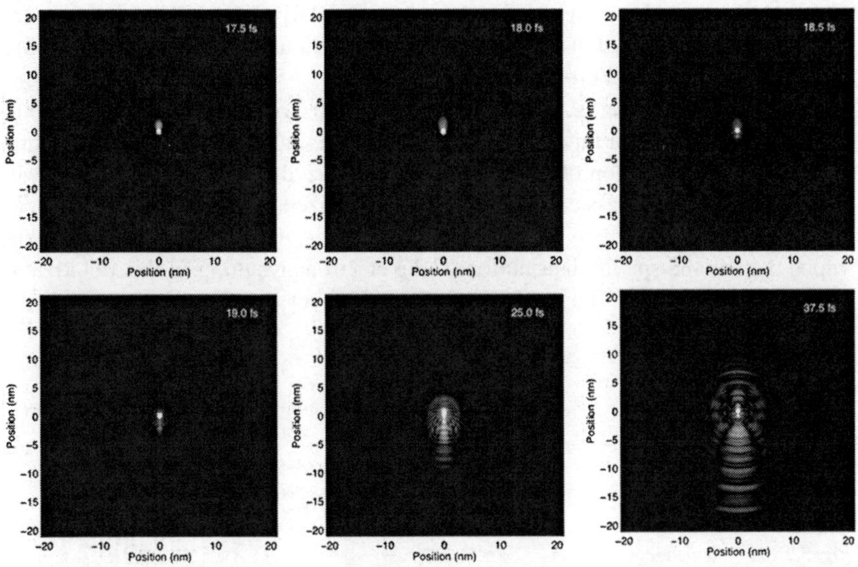

Fig. 5.7 Evolution of the electron density distribution in Cartesian coordinates for a 50 fs, 800 + 400 nm pulse with relative phase $23\pi/12$ and 200 MV/cm 800 nm field amplitude. The time scale used in the simulation starts at 0 fs at the beginning of the optical pulse, with the electric field envelope reaching its maximum at 25 fs and returning to zero at 50 fs. Scaled logarithmically

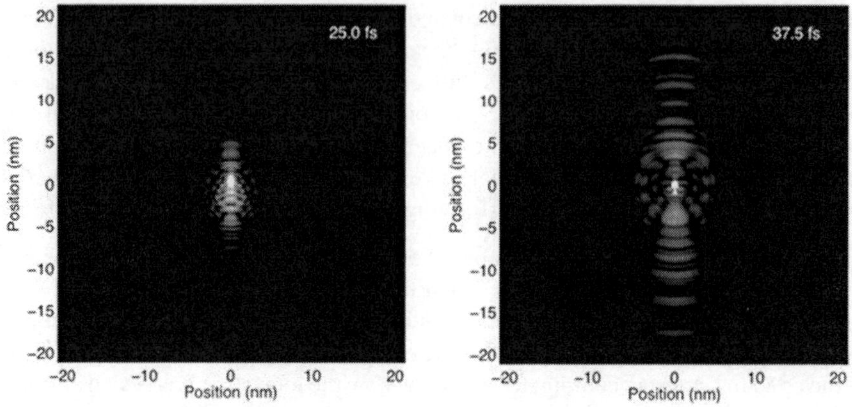

Fig. 5.8 Evolution of the electron density distribution under similar conditions as Fig. 5.1, but with the relative phase shifted by π/2

shown in Fig. 5.8. When comparing Figs. 5.7 and 5.8, one can clearly see that the final state of the asymmetric ionization process will exhibit a larger electrostatic polarization than the symmetric case. The time evolution of the polarization is the source of the emitted radiation, and so this is one of the key observables measured over the course of the simulation. In atomic units, the polarization P is simply determined by the expectation value of the electron along the desired axis. For a laser with polarization along the z-axis, the relevant polarization will be $P_z = -\langle z \rangle$, which can be calculated at each time step in the simulation. Plots of P_z for three different 400 nm phases are shown in Fig. 5.9.

The effect of the intense, bichromatic pulse is thus to induce a polarization in the atom that increases as a function of time, consistent with the semiclassical pictures. However, the distribution of energy among the ionized electrons differs from what would be calculated classically. This can be visualized by taking a time-dependent slice of the electron density distribution (in this case, the density along the z-axis), which depicts the spatial distribution of the electrons resulting in this polarization as a function of time. This is shown in Fig. 5.10 for the phase $\phi = 11\pi/12$. The

Fig. 5.9 Time dependent polarization for argon atoms subjected to 800 + 400 nm optical pulses with various values of the relative phase of the two frequencies

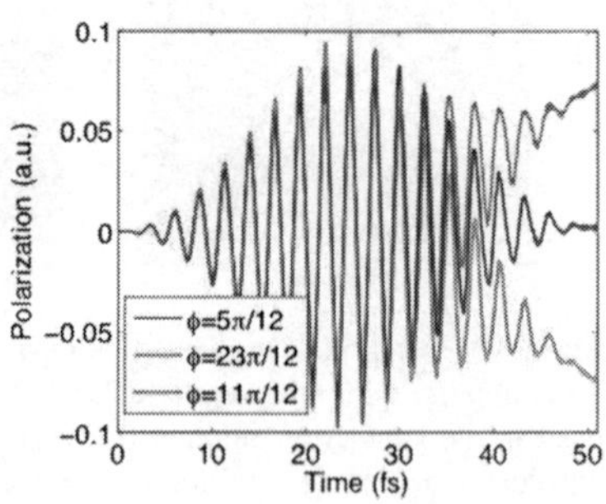

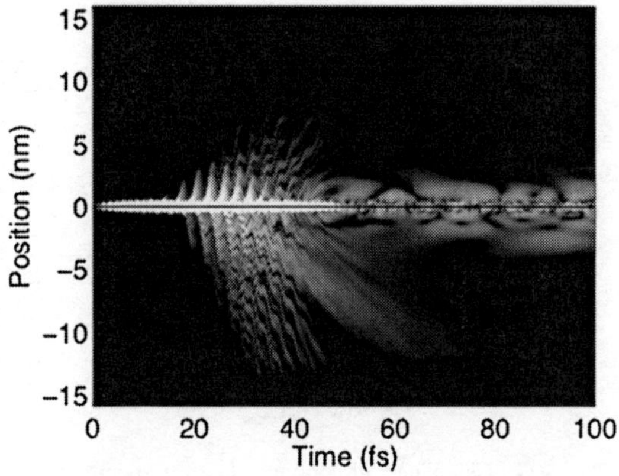

Fig. 5.10 *Z*-axis electron density distribution as a function of time for the case where $\phi = 11\pi/12$

density distribution is not continuous; there are allowed paths with high density, and disallowed ones with low probability. The source of this effect can be understood by viewing the electron energy spectrum, which can be calculated through the energy window method [2]. Applying such a window to the wavefunction returns only the portion of the wavefunction that lies within a specified bandwidth centered around the chosen energy, Ψ_n. By integrating the probability density in such a windowed wave function for each energy, the energy spectrum can be constructed from the constants $c_n^2 = \langle \Psi_n \,|\, \Psi_n \rangle$. This can be taken a step further by weighting the partial waves by the values

$$\upsilon_\ell^{(\mathrm{pos})} = \sqrt{\frac{2\ell+1}{4\pi}\frac{(\ell-m)!}{(\ell+m)!}}, \tag{6}$$

$$\upsilon_\ell^{(\mathrm{neg})} = (-1)^\ell\, \upsilon_\ell^{(\mathrm{neg})}, \tag{7}$$

which, when applied to the windowed wavefunction prior to probability integration, return a value proportional to the probability of the electron moving in the positive or negative z direction, respectively. The directional electron spectra corresponding to the polarizations shown in Fig. 5.4 are presented in Fig. 5.11.

The structure in the time-dependent density plot in Fig. 5.10 can now be understood – the electrons do not possess a continuous spectrum of energy, but instead are localized around values given by $N\hbar\omega - U_P$, where N is an integer and Up is the ponderomotive energy. The electrons are thus grouped in wave packets with center frequencies separated by the photon energy. As time passes, the electrons propagate away from the parent atom with velocities proportional to the square root of the energy. In terms of the THz radiation expected from a single atom, the energy distribution may not seem important – the observed THz signal is related to the average polarization, and in principle all that matters is how the wave packets are accelerated as a function of time, not their final state. However, the THz radiation

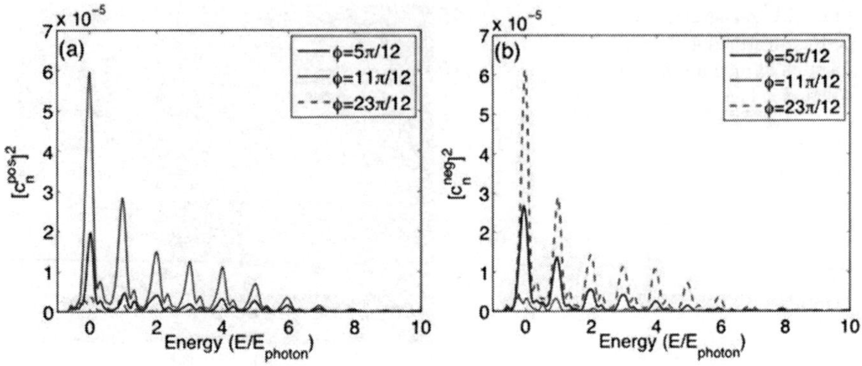

Fig. 5.11 Calculated directional electron spectra for the three phases shown in Fig. 5.9. (**a**) Electrons moving in the positive z direction. (**b**) Electrons moving in the negative z direction

observed experimentally has not been from isolated atoms, but rather in a collection of atoms forming an ionizing gas. In this case, it is not sufficient to consider only the laser-atom interaction since other factors can influence the evolution of the polarization.

One possibility is that as the electron propagates into the surrounding medium, it is scattered by a collision with another atom. This results in a change in its velocity, and thus bremsstrahlung. Usually, bremsstrahlung is incoherent radiation, but in the case of the first collision of the electron after it leaves the atom, the expectation value of its velocity is aligned with the laser polarization axis, with direction determined by ϕ. Thus, after collision, the direction of the velocity change along the laser polarization direction will be anti-parallel to its original propagation direction. This determines the phase of the emitted bremsstrahlung, and so for collisions that take place within one half of a cycle of the THz wave, the collisionally-induced radiation can build coherently. In the case where the collision rate is increasing as a function of time due to the dispersion of the electron wave packets, the time-dependent amplitude c_n of each wave packet can be described by

$$c_n(t) = c_n(0)\exp\left(-at^2/2\right), \tag{8}$$

$$a = \pi v_\perp v_{||} r_A \rho, \tag{9}$$

where $c_n(0)$ is the amplitude obtained via the energy spectrum, $v_\perp$ is the speed of dispersion of the electron wave packet, $v_{||}$ is its translational velocity away from the atom, r_A is the scattering radius of the surrounding gas atoms, and ρ is the gas density. This results in the postionization coherent polarization due to each wave packet $|\varphi_n\rangle$ taking the form

$$P_n(t) = -\langle\varphi_n|\,\hat{z}\,|\varphi_n\rangle = -c_n^2(0)v_{||}t\exp\left(-at^2\right). \tag{10}$$

The full polarization can then be constructed from the polarization at the time of ionization obtained from the quantum mechanical simulation combined with the sum of the bremsstrahlung contributions:

$$P(t) = \sum_n \left(P_n^{(\mathrm{pos})}(t) - P_n^{(\mathrm{neg})}(t)\right). \tag{11}$$

The spectral contribution of this "echo" signal to the measured THz wave can be approximated by performing a Fourier transform on its third derivative:

$$E(\omega) \propto \int_{-\infty}^{\infty} dte^{i\omega t}\frac{d^3}{dt^3}P(t) = -\frac{\sqrt{\pi}v_{||}\omega^4}{2a^{3/2}}e^{-\omega^2/(4a)}. \tag{12}$$

The full THz generation process can be described in two steps: first, asymmetric ionization, followed by disruption of the original trajectories by the surrounding gas or plasma, resulting in a coherent echo. The full process is pictured in Fig. 5.12.

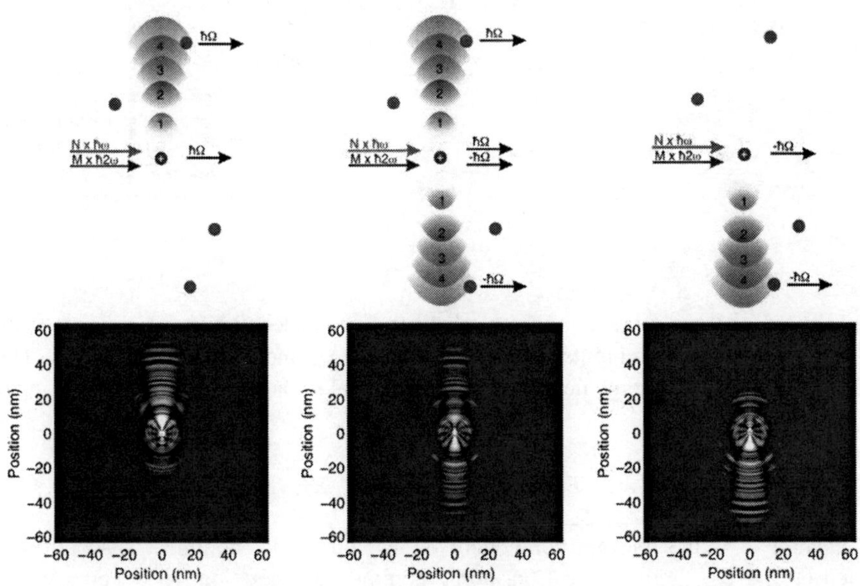

Fig. 5.12 Diagram of the THz emission process, for three different second harmonic phases (−π/12, 5π/12 and 11π/12, from right to left). When the wave packets (numbered by approximate energy in photons) propagate in the positive *z* direction, (*left*), the asymmetric ionization emits THz radiation (ħΩ). When a second beam is released in the opposite direction (*center*), there is cancellation of the radiation emitted by the two beams. In the case where the dominant beam is in the negative direction, radiation with phase opposite that of the left-hand case is released (−ħΩ)

Detection of THz Waves in Gases

Since a nonlinear optical process can be used to generate THz waves in gases, the question arises as to whether detection can be performed in a similar manner. Just as the generation of THz waves in a centrosymmetric medium requires an odd number of input photons, the symmetry requirements also dictate that an odd number of input photons are coupled in detection. The concept of detection of a THz wave in gas is presented in Fig. 5.13. This is accomplished through a four-wave mixing process, where two input photons are at the fundamental laser frequency, and one is the THz photon. This produces an output near the second harmonic. Thus, detection is the phenomenological inverse of generation: whereas in the generation process, fundamental and second harmonic light are mixed to produce THz radiation, in detection THz radiation is mixed nonlinearly with the fundamental laser light to produce a second harmonic signal. However, the physical details are quite different.

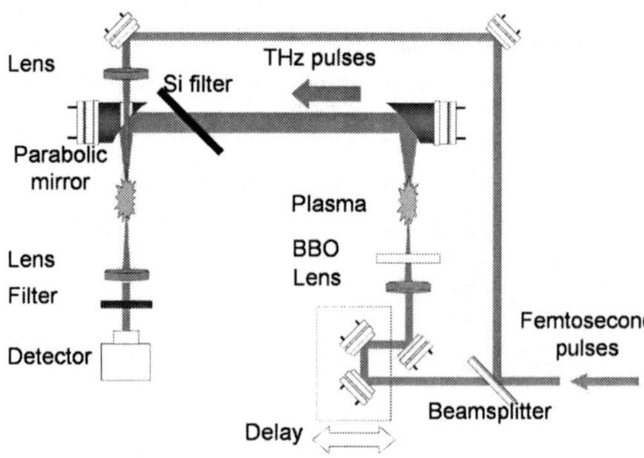

Fig. 5.13 Experimental setup of using air to generate and detect THz wave

The four-wave mixing processes that produce a second harmonic signal from fundamental and THz inputs are $2\omega = \omega + \omega + \Omega_{\mathrm{THz}}$ and $2\omega = \omega + \omega - \Omega_{\mathrm{THz}}$. The emitted second harmonic field will be proportional to the product of the three input fields:

$$E_{2\omega} \propto P_{2\omega} = \chi^{(3)} E_{\omega} E_{\omega} E_{\mathrm{THz}}, \tag{13}$$

where $\chi^{(3)}$ is the relevant component of the third-order nonlinear susceptibility tensor and $P_{2\omega}$ is the second harmonic nonlinear polarization. However, when a measurement is performed, it is the power of the second harmonic that is measured, not the electric field. As a result, the measured quantity is proportional to intensity, which itself is proportional to the square of the electric field.

$$I_{2\omega} \propto |E_{2\omega}|^2 \propto \left(\chi^{(3)} I_{\omega}\right)^2 E_{\mathrm{THz}}^2. \tag{14}$$

This has the unfortunate consequence that the measured signal is proportional to the square of the THz field, resulting in a loss of phase information. However, this equation only applies in cases where there are no other sources of coherent second harmonic radiation. When THz waves were measured through second harmonic generation in solids and gases, there was a background second harmonic signal, which will here be designated as $E_{LO}^{2\omega}$, that resulted in homodyne detection of the THz field [3]. The background second harmonic signal is contributed from the white light generated from the laser induced air plasma through self-modulation and self-steeping. Thus, the resulting second harmonic intensity then became (assuming identical beam structures for both second harmonic fields)

$$I_{2\omega} \propto |E_{2\omega}|^2 \propto \left(\chi^{(3)} I_\omega\right)^2 E_{\mathrm{THz}}^2 + 2\left(\chi^{(3)} I_\omega\right) E_{\mathrm{THz}} E_{\mathrm{LO}}^{2\omega} + \left(E_{\mathrm{LO}}^{2\omega}\right)^2, \tag{15}$$

which contains a cross term with a linear dependence on the THz field. In the event that $E_{\mathrm{LO}}^{2\omega}$ is much larger than the field of the THz-induced second harmonic, that is, the laser induced air plasma density is high, the cross term will dominate the E_{THz}^2 term, resulting in quasi-coherent detection of the THz wave. Figure 5.14 shows the typical THz waveforms measured with air sensor at three different probe power. The detection is coherent only when $E_{\mathrm{LO}}^{2\omega}$ dominates. Figure 5.15a and b give the relationships between the signal second harmonic beam intensity and the optical beam and THz beam with homodyne technique, respectively.

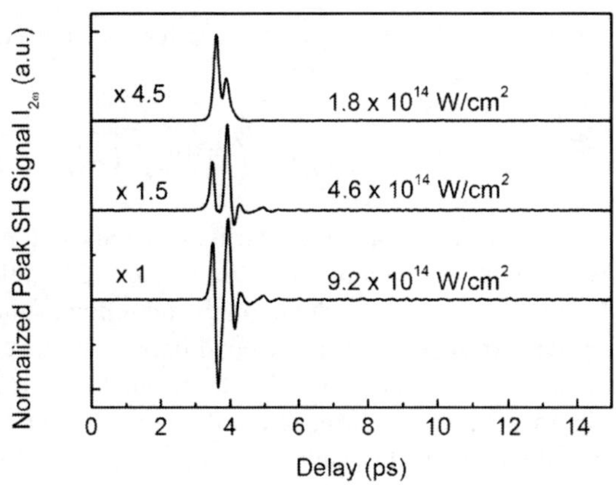

Fig. 5.14 Typical time-resolved SH waveforms (*solid lines*) measured with a gas sensor at three different estimated probe intensities: 1.8×10^{14} W/cm^2 (*upper*), 4.6×10^{14} W/cm^2 (*middle*), and 9.2×10^{14} W/cm^2 (*lower*), respectively. The waveform offsets are shifted for clarity

The homodyne technique partially solves the problem of coherent THz detection through the four-wave mixing process. However, it has significant downsides: it is only coherent within a certain range of THz field values, and will result in distortion of the waveform if the field is too high. Additionally, the requirement that E_{LO} be much larger than the signal dictates the presence of a large background signal, which will produce difficulties in obtaining an adequate dynamic range for time-domain spectroscopy. It would be better if the intrinsic limit on THz field strength could be

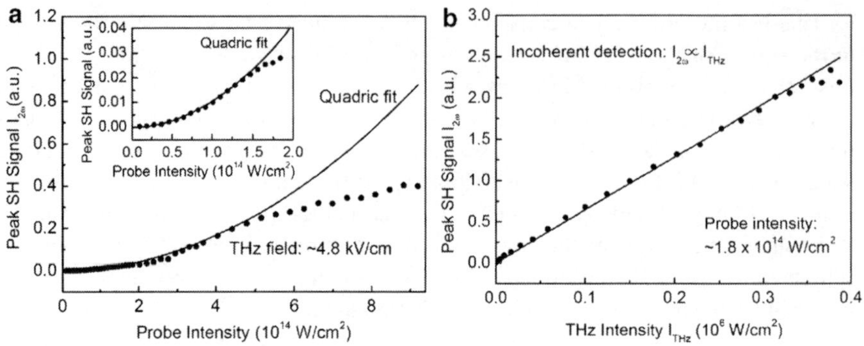

Fig. 5.15 Relationship between the signal intensity and the probe laser beam intensity (**a**) and THz field (**b**) in the homodyne detection process

lifted, and coherent detection be guaranteed. This can be done through a heterodyne technique [4].

Much like the THz-field induced second harmonic, a second harmonic signal can also be produced using a DC electric field as one of the inputs. Assuming that the nonlinear susceptibility for the two processes is the same and that all beams are plane waves, this gives the expression

$$E_{2\omega} \propto \chi^{(3)} E_\omega E_\omega \left(E_{\mathrm{THz}} + E_{\mathrm{DC}}\right), \tag{16}$$

which again has a coherent cross term in the equation for the second harmonic intensity:

$$I_{2\omega} \propto |E_{2\omega}|^2 \propto \left(\chi^{(3)} I_\omega\right)^2 \left(E_{\mathrm{THz}}^2 + 2E_{\mathrm{THz}}E_{\mathrm{DC}} + E_{\mathrm{DC}}^2\right). \tag{17}$$

Unlike the case of a second harmonic local oscillator generated by surface effects or white light generation, the phase of the field induced second harmonic may easily be controlled. Simply changing the direction of the electric field results in a π shift of the carrier phase of the second harmonic pulse. This will switch the sign of the cross term, but leave the other terms unchanged. Thus, by employing a modulated electrical field synchronized with the repetition of the laser pulses, the cross term will be modulated at the modulation frequency. As a result of this, applying a modulated bias results in a heterodyne process that allows only the coherent term of the measurement to be detected.

$$I_{2\omega} \propto \left(\chi^{(3)} I_\omega\right)^2 E_{\mathrm{THz}}E_{\mathrm{DC}}. \tag{18}$$

Accordingly, there are no requirements placed on the relative amplitudes of the THz and bias fields to achieve coherent detection, allowing a large THz field to be measured against a small background, improving the possible dynamic range. THz waveform detected through the homodyne process and its spectrum are presented

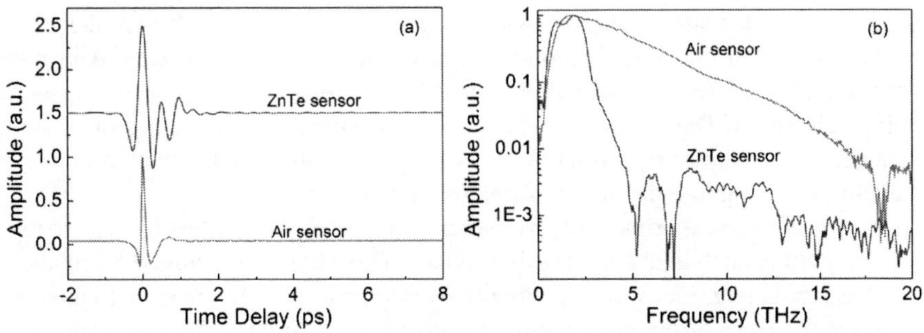

Fig. 5.16 (**a**) THz waveforms detected using air and ZnTe crystal, respectively. (**b**) Spectra of THz pulses detected using air and ZnTe crystal, respectively

in Fig. 5.16a and b. THz waveform and spectrum detected through an EO process using a ZnTe crystal are used as comparisons.

And also, Equation (18) provide several approaches to enhance the second harmonic intensity by simply increasing probe pulse power and electrical field. Figure 5.17 shows the dependence of second harmonic intensity on probe pulse energy and applied DC electrical field with different gases in heterodyne technique, respectively. The optical or electrical breakdown limits the intensity of second harmonic for each circumstance.

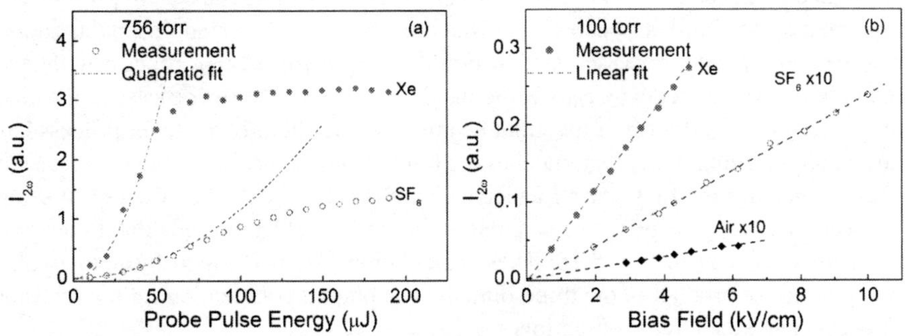

Fig. 5.17 The second harmonic intensity versus probe pulse power (**a**) and DC field strength (**b**) with different gases

As in the case of THz wave generation in gases through fundamental and second harmonic light, THz detection in gases was initially assigned to a four wave mixing process. In terms of the underlying physics, this is an invocation of perturbation theory. Unlike the case of generation, perturbation theory is quite successful in explaining all observed features of the detection process, as will be detailed below. The same quantum mechanical treatment that was given to the generation process may also be applied to detection. However, in this case, no threshold appears. Instead, the dependence on optical field strength is quadratic, consistent with four wave mixing theory, as shown in Fig. 5.17a. In the measurements and quantum mechanical calculations, the behavior of the detection process below the intensity threshold for tunnel ionization is consistent with four-wave mixing. Above the threshold, the TDSE solution and measured data no longer converge with four-wave mixing, however. The measurement falls below the quadratic fit, suggesting that the main effect of the plasma formation is intensity clamping, and that there is no major benefit of moving into the non-perturbative regime.

As a result, the experimentally useful range can be well described by perturbation theory and the analytical solutions it provides. This allows for detailed calculations of phase matching, focusing and other macroscopic effects. In order to do this, one requires the form of the polarization, which for the two possible processes $2\omega = \omega + \omega + \Omega_{\mathrm{THz}}$ and $2\omega = \omega + \omega - \Omega_{\mathrm{THz}}$, is given by

$$\begin{aligned} P_{+} &= \chi^{(3)} E_{\omega}^{2} E_{\mathrm{THz}} + c.c. \\ P_{-} &= \chi^{(3)} E_{\omega}^{2} E_{\mathrm{THz}}^{*} + c.c. \end{aligned} \tag{19}$$

where E_{THz}^{*} denotes the complex conjugate. The difference between these forms of the polarization is only in the phase; in terms of amplitude, they are identical. For a nonlinear effect, the phase is critical, however, since it determines how the pulse amplitude builds up as it propagates. In a low-dispersion gas such as air, the phase matching is not dominated by the difference in refractive index between the three different wavelengths involved, but by the Gouy phase. This phase term can be separated from each electric field term as a factor of $\exp\left[i \arctan\left(z/z_R\right)\right]$ if they are assumed to be Gaussian beams with Raleigh length z_R. The resulting polarization for the two separate processes will differ: the $2\omega = \omega + \omega + \Omega_{\mathrm{THz}}$ process will have a phase of $\exp\left[i3 \arctan\left(z/z_R\right)\right]$ while the $2\omega = \omega + \omega - \Omega_{\mathrm{THz}}$ process retains the $\exp\left[i \arctan\left(z/z_R\right)\right]$ term of the input beams. As a result, only the latter process has a spatially-varying phase that matches the fundamental Gaussian mode, whereas the former matches the first order Laguerre Gauss (LG1) mode. The effect of this can be seen in a Huygens principle calculation of the far-field amplitude due to the two processes, as seen in Fig. 5.18. These calculations are performed by numerically integrating the product of the three-dimensional phased polarization for the relevant processes and the Green's function.

Figure 5.18 shows that the output beam of the $2\omega = \omega + \omega - \Omega_{\mathrm{THz}}$ process is single mode and of far greater amplitude than the $2\omega = \omega + \omega + \Omega_{\mathrm{THz}}$ process. Additionally it matches the $2\omega = \omega + \omega + \mathrm{dc}$ beam, which is important for heterodyne detection where the two must have similar phase profiles for the beating between them to be easily measurable through a power measurement.

As a result, the following analysis will focus on the $2\omega = \omega + \omega - \Omega_{\mathrm{THz}}$ process, and use a Gaussian trial solution of the second harmonic output beam with the same Rayleigh length as the fundamental input beam and beam waist smaller by $\sqrt{2}$. This yields the terahertz field induced second harmonic amplitude

$$E_{2\omega}^{(\mathrm{THz})} = \left(\frac{i 8 \pi^{2} \omega \chi^{(3)}}{n c} E_{\omega} E_{\omega} E_{\mathrm{THz}} \right) \frac{z_R z_T}{z_R + z_T} e^{z_T \Delta k}, \tag{20}$$

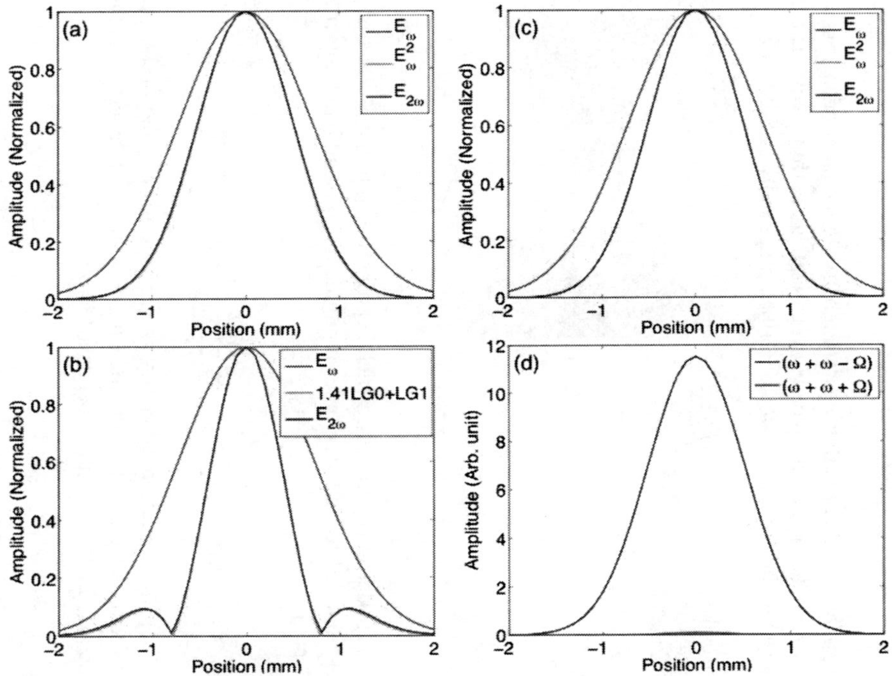

Fig. 5.18 Huygens principle calculations of the THz detection process. (**a**) Input beam (E_ω) and output beam ($E_{2\omega}$) for the $2\omega = \omega + \omega - \Omega_{\mathrm{THz}}$ process. The square of the input beam (E_ω^2) is shown for reference. (**b**) Input beam (E_ω) and output beam ($E_{2\omega}$) for the $2\omega = \omega + \omega + \Omega_{\mathrm{THz}}$ process. A fit of the output beam to a sum of the zero order (LG0) and first order (LG1) Laguerre Gauss modes is shown. (**c**) Input beam (E_ω) and output beam ($E_{2\omega}$) for the $2\omega = \omega + \omega + \mathrm{dc}$ process. The square of the input beam (E_ω^2) is shown for reference. (**d**) Comparison of the amplitudes of the two THz detection processes

where n is the index of refraction of the gas, z_R is the Rayleigh length of the probe beam, z_T is the Rayleigh length of the THz beam and $\Delta k = 2k_\omega - k_{2\omega} - k_\Omega$ is the phase mismatch between the fundamental, second harmonic and THz beams (respectively) and is assumed to be negative. The DC field induced second harmonic can be calculated in a similar manner, using the spatially dependent bias field function [5], which yields

$$E_{2\omega}^{(DC)} = \left(\frac{i8\pi^2\omega\chi^{(3)}V}{nc\cosh^{-1}(l/2a)}E_\omega E_\omega E_{\mathrm{THz}}\right)\frac{z_R}{d+2z_R}e^{d\Delta k/2}, \tag{21}$$

where V is the applied voltage, l is the separation between the electrodes, a is their radius, and $d^2 = l^2 - 4a^2$ is the distance between equivalent thin electrodes. The parameters involved in these equations are described graphically in Fig. 5.19.

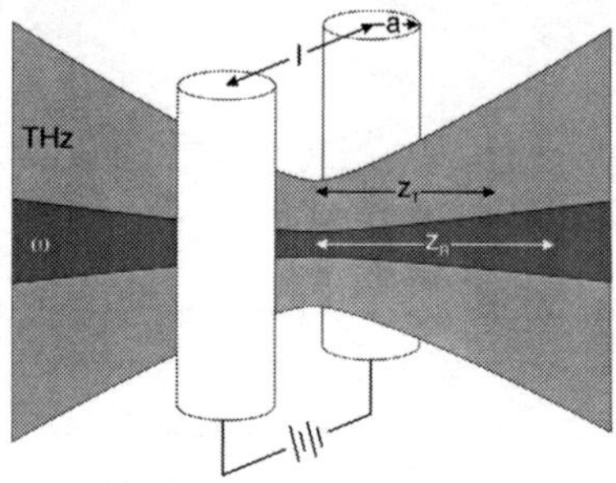

Fig. 5.19 Diagram of the parameters contained in the equations describing the THz detection process

The coherent THz detection process, which is proportional to the product of the DC and THz field induced second harmonic amplitudes at constant input power can be expressed as the intensity

$$I_{2\omega} \propto (\chi^{(3)}I_\omega)^2 E_{\mathrm{THz}}E_{\mathrm{DC}}\frac{z_R^2 z_T}{(z_R+z_T)(2z_R+d)}e^{(z_T+d/2)\Delta k}. \tag{22}$$

From Equation (22), the detection efficiency is sensitive to how the probe beam and terahertz are focused. Considering the focusing condition of terahertz beam and optical beam, and also make pressure dependence explicit,

$$I_{2\omega} \propto \frac{\chi_0^{(3)}p^2\sqrt{z_T}}{(z_R+z_T)\,(2z_R+d)}e^{-(z_T+d/2)dk\,p}, \tag{23}$$

where

$$dk = -\Delta k/p = -2\pi\frac{n_{800}^{1\mathrm{atm}} - n_{400}^{1\mathrm{atm}}}{400\,\mathrm{nm}}, \tag{24}$$

and $\chi_0^{(3)}$ is the value of $\chi^{(3)}$ at one atmosphere of pressure, the pressure p is expressed in atmospheres and the approximations $k_\Omega << k_{2\omega}$ and $n \approx 1$ have been made. The above equations allow for the coherent THz detection process to be understood analytically, including effects related to the medium and focusing of the optical and THz beams.

From Equation (23), the optimum pressure of a certain gas for THz detection can be derived:

$$p_{\rm opt} = \frac{2}{(z_T + d/2)dk}. \tag{25}$$

The optimal pressure for a certain gas sensor is related to the Rayleigh length of the THz beam, the geometry of the electrodes and optical dispersion. The signal at optimum pressure is then

$$I_{2\omega} \propto \frac{4\sqrt{z_T}}{(z_R + z_T)(2z_R + d)(z_T + d/2)^2}\left(\frac{\chi_0^{(3)}}{dk}\right)^2. \tag{26}$$

According to above equation, all of the terms related to the gas are confined to the ratio of $\chi_0^{(3)}$ to dk, and so it is possible to introduce a figure of merit (*FOM*) to characterize the sensitivity of gases:

$$FOM = \chi_0^{(3)}/dk \tag{27}$$

Figure 5.20 shows the relative signal obtained from xenon as a function of pressure together with a fit from Equation (26). In Fig. 5.20, the coherent terahertz signal is varied by changing the pressure and terahertz focusing conditions for the representative gases xenon, which possess sufficient dispersion for phase matching effects to be visible. The results in Fig. 5.20 show that the detection process is sensitive to not only the phase matching of the various wavelengths of radiation involved, but also to how the terahertz beam is focused. However, the dependence is predictable and can be well understood through the analysis above.

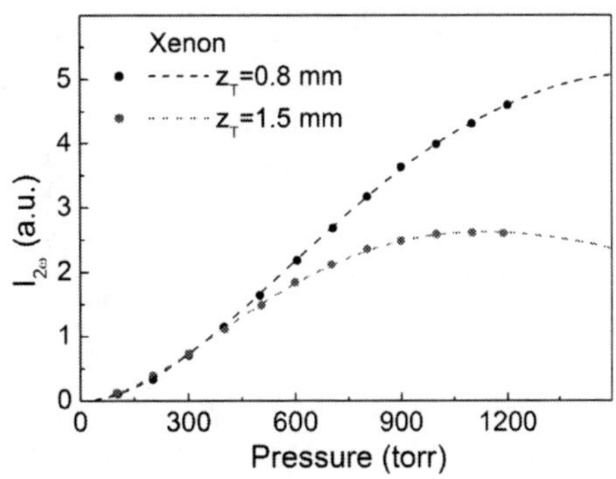

Fig. 5.20 Measured pressure dependence of terahertz detection in xenon. The Rayleigh length of the terahertz z_T is altered by the insertion of an aperture in the beam path

Optimization of THz Systems Based on Gas Photonics

Understanding the physical underpinnings of the generation and detection processes allows for improved design of systems based on the effects. The first consideration is the gas used for generation or detection. In each case, the efficiency of THz or second harmonic generation can be altered by an order of magnitude through the choice of species.

Since the THz wave is generated while the gas is in the process of being ionized through a tunneling process, the most relevant parameter for the characterization of a gas as a THz source is its ionization potential. Atoms or molecules with lower ionization potentials will ionize at a higher rate, and thus smaller pulse energies will enable higher THz fields. The study of pulsed THz radiation generated from five noble gases provides the quantitative relationship between the THz wave amplitude generated from each gas versus its ionization potential (IP) (Fig. 5.21a) [6]. The noble gases are chemically inert and structurally simple, hence no vibrational structures and photo-induced fragmentation occur. They are suitable for use as a test-bed system for the THz generation mechanism. Experimental results reveal that terahertz generation efficiency of these noble gases increases with decreasing ionization potential. Xe has the lowest ionization potential among noble gases, and it provides more than twice the field strength than that from dry nitrogen. However, the gases with lower ionization potential tend to have relative large dispersion, which will result in walk-off between ω and 2ω beam. And also, a dense plasma will lead to losses of THz generation efficiency through phase mismatching effects. A phase compensator is one of effective approaches to solve the above problems.

In the detection process, the signal-to-noise ratio is often limited by the dark noise of the photomultiplier tube used to measure the second harmonic light. In this case, increased second harmonic amplitude will improve the dynamic range. Since the coherent heterodyne signal is proportional to $\left(\chi^{(3)}\right)^2$, using a different gas is one clear method to improve the signal amplitude (although one should take care that

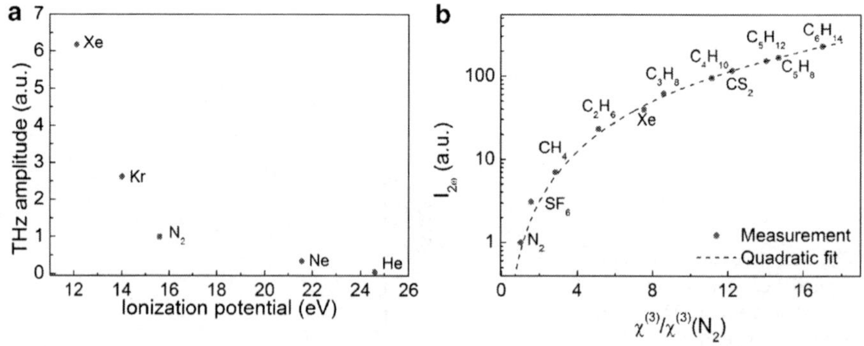

Fig. 5.21 (**a**) The generated terahertz field amplitude of noble gases versus ionization potential at 100 torr. (**b**) The detected second harmonic intensity versus third-order nonlinear susceptibility

the insertion losses of the gas cell do not negate the benefits of changing the gas). Figure 5.21b shows the detected second harmonic signal as a function of normalized third order nonlinear susceptibility $\chi^{(3)}$ together with quadratic fits (dashed line). With n-Butane gases, two orders enhancement of second harmonic intensity is observed compared to nitrogen sensor [7].

Since the detection process occurs in the perturbative regime, its power dependence is not as dramatic as generation, although it does have a nonlinear dependence on intensity, which results in increased efficiency at higher probe intensities. However, in the measurements this increase in signal at pulse energies above the ionization threshold is not strong enough to compensate for the defocusing caused by the plasma as shown in Fig. 5.17a. Since the formation of a plasma also results in the creation of numerous sources of background noise, the best dynamic range is achieved when the intensity is slightly below the ionization threshold. Knowing this, one should partition the laser energy such that the probe pulse does not contain more energy than necessary (typically less than 120 μJ for an 80 fs pulse).

Since the interaction length associated with detection in gases is long compared to the thickness of the nonlinear crystals typically used for THz photonics, the effects of phase matching become critical in determining the spectrum and dynamic range of the system. In this case, the phase matching is affected by how the THz beam is focused in addition to the dispersion properties of the gas. Using Equation (26), the optimal focusing conditions for the THz beam can be determined. The probe beam should be focused as tightly as possible without generating a plasma, which will allow for its Rayleigh length z_R to be known. Using this value, the optimal Rayleigh length for the THz beam becomes

$$z_T^{\text{opt}} = \frac{\sqrt{4p^2 dk^2 z_R^2 + 12p\, dk\, z_R + 1} - 2p\, dk\, z_R - 1}{4p\, dk}. \tag{28}$$

A typical value of the phase mismatch dk in the atmosphere (p=1) is 120 m^{-1}.

In the heterodyne detection technique, one also needs to consider the production of the local oscillator. Due to the effects of the Gouy phase, the polarization that produces the DC field induced second harmonic has twice the phase anomaly of the input beams, and thus coherent build-up of the local oscillator intensity only occurs in a confined region around the focus. As a result, the process is more efficient with a spatially-confined electric field. This can be obtained using thin electrodes and placing them as close together as is possible without blocking the THz or optical beams. Close placement of the electrodes has the additional advantage of requiring a smaller bias in order to achieve comparable electric field strengths (which are limited to ~30 kV/cm due to the static breakdown of the atmosphere).

Careful control of the above parameters allows for the creation of THz systems with broad bandwidths (limited by the optical pulse duration) and dynamic ranges high enough for high-quality spectroscopic data to be obtained.

THz Wave Air Break-Down Coherent Detection (ABCD) System

Unlike solid state emitters and sensors (such as semiconductors or electro-optic crystals) which are commonly used in THz time-domain spectroscopy, gases have no phonon resonances or echoes due to the multiple THz or optical reflections, enabling broadband THz spectroscopy free of instrumental artifacts. Using ionizing gases as the emitter and sensor media solves both of these issues, as they exhibit much lower dispersion than solids and are continuously renewable. Those unique features make the THz wave air break-down coherent detection system an ideal spectrometer.

A system design that has been implemented is shown in Fig. 5.22. A laser pulse from a Ti:Sapphire regenerative amplifier is focused through a 100 μm thick type-I beta barium borate (BBO) crystal to generate its second harmonic beam. The fundamental and its second harmonic beams generate a THz wave at the ionizing plasma spot. After focusing, the radiated THz wave is collected by a 90° off-axis parabolic mirror and focused again by another parabolic mirror. The probe beam is sent through a time delay stage and then focused by a lens through a hole in second parabolic mirror. Thus, the THz wave propagates collinearly with the probe beam and is focused at the same spot. A 500 Hz electric bias field is applied to the detection region, supplying a second harmonic local oscillator for coherent detection via a photomultiplier detector. The second harmonic signal is passed through a pair of 400 nm bandpass filters and detected by a photomultiplier tube (PMT).

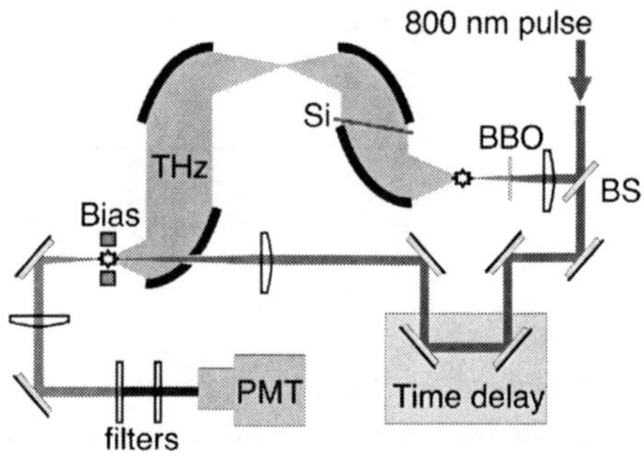

Fig. 5.22 Schematic setup of heterodyne THz ABCD system

Since the generation and detection of the THz waves occurs in a gas, the spectrum of the THz time-domain spectrometer is almost solely limited by the properties of the laser pulse. Figure 5.23a and b show the measured THz waveform and its spectrum, respectively. Unlike the case of electro-optic sampling in a crystal such as ZnTe or GaP, there are no crystal phonon modes to introduce dispersion or absorption into the detection region, and there are no optical or THz reflections from the solid emitter and sensor interfaces. As a result, the detected spectrum can be continuous and cover the full bandwidth of the input laser pulse. Due to the use of a high-resistivity silicon wafer in the THz beam path to remove the residual optical beam, several features appear near 18.5 THz, consistent with Si two-phonon and

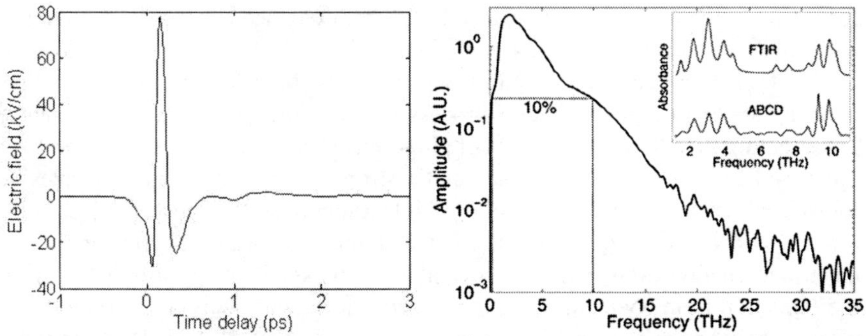

Fig. 5.23 (**a**) Typical THz waveform with a 80 kV/cm THz field, using dry nitrogen as the THz wave emitter and sensor gas. As a comparison, the standard air dielectric breakdown field is about 30 kV/cm. (**b**) The widest spectrum (linear plot 10% bandwidth: <0.1 to > 10 THz, tail reaches 20 THz). Inset: Spectra of 2-2′ Biphenol taken using both the time-domain ABCD system and an FTIR spectrometer. The spectra are offset and baselines removed for clarity

carbon impurity absorption. Figure 5.23b shows that the measured 10% bandwidth spans from 0.3 to 10 THz. The broad spectral range and heterodyne detection capability allow spectroscopic measurement across the full THz range. The inset plot in Fig. 5.23b compares the spectra of 2-2′ Biphenol taken by the THz time-domain ABCD system and one measured by a traditional FTIR spectrometer (Bruker). All the major spectral features are reproduced within the overlapping spectral range of the two methods.

THz Radiation Enhanced-Emission-of-Fluorescence (THz-REEF) from Gas Plasma

The interaction between electromagnetic (EM) waves and laser-induced gas plasma has been extensively studied in most of the spectral regions. Electric field measurements and plasma dynamics characterization in gas DC and RF discharge were demonstrated by various schemes of laser-induced fluorescence spectroscopy (LIF). However, the study in the THz region (0.1–10 THz) has been a challenge in the past due to the lack of strong, tabletop THz sources. A short THz pulse is a promising tool for a time-resolved investigation of plasma dynamics without the limitation of the detector response time. Low energy of the THz photon promises in-situ, non-invasive plasma characterization. For standard laboratory, laser-induced plasma with electron densities of 10^{14}–10^{19} cm^{-3}, the plasma frequency lies in the range of 90 GHz to 28 THz. The large frequency span of THz pulses developed recently encompasses a wide range of plasma densities and allows for the study of resonant interaction between THz wave and plasma. Recent major technical advances in developing intensive THz sources [8–10], have provided us with new opportunities for the investigation of plasma inverse-bremsstrahlung heating and electron impact molecular excitation by THz waves.

The transition between ground electronic states, high Rydberg states, and fluorescing states in laser excitation process of gas molecules is very important to investigating the ultrafast dynamics of photo-electrons in gas plasma. In the photo-ionization process of gas molecules under strong laser field, some molecules are ionized while there are some others in high Rydberg states which are just below ionized states. The energy difference between them is very small compared to the electronic energy levels, which is several eV. This small energy difference makes the transition from Rydberg states to ionized states more likely through molecule-electron collision in plasma. This means that gas molecules, being in Rydberg states, can gain energy from its inelastic collision with other surrounding energetic electrons and then become ionized. External electric field, which can accelerate the photoelectrons, can increase the possibility of this transition and increase the gas ionization rate. The fluorescence emission of the gas plasma is therefore expected to be enhanced under external field because more molecules are ionized and the population of the fluorescing states is increased by electron collisional excitation. In this sense, the THz wave is an ideal probe tool to study this process since THz photon energy is too low to directly ionize gas molecules and THz pulse cycle is on the order of ps which is long enough to significantly change electron momentum.

In all, the plasma-THz wave interaction leads to the increased electron temperature and enhanced fluorescence emission which would be dependent on the THz field. It is very intriguing to uncover the mechanism of the THz field induced electron collisional excitation in different gas molecules and its dependence on gas density and plasma temperature. Furthermore, fast plasma channel formation, which is usually on the order of the laser pulse duration, can provide high temporal resolution for the study of this process, and can also be utilized for measurement of the amplitude and phase of THz radiation. This provides a promising method for remote THz sensing which has been a great challenge for a long time.

Under the influence of the THz radiation, the electron dynamics in laser-induced plasma are determined by the laser photo-ionization process, gas density, and by the amplitude and phase of the THz field. The intense illumination of an ultrashort laser pulse releases free electrons from air molecules via a multi-photon ionization (MPI) or tunneling photon ionization (TI) process. The initial free electrons density $n_e(0)$ and electron temperature $T_e(0)$ after ionization depends on the laser pulse intensity and molecule ionization potential. After the passage of the optical pulse, the electron motion, in the presence of a THz field $\vec{E}_{\mathrm{THz}}(t)$, can be described semiclassically [8]

$$\frac{d\vec{v}(t)}{dt} + \frac{\vec{v}(t)}{\tau} = -\frac{e}{m}\vec{E}_{\mathrm{THz}}(t), \tag{29}$$

where $\vec{v}(t)$ is the electron velocity, τ is the electron collision relaxation time and m is the electron mass. $\vec{v}(t)\backslash\tau$ is the damping term which is accountable for the energy transfer from electrons to molecules/ions via collisions. During the THz cycle, the electrons velocity is increased or decreased depending on the transient direction of the THz field and electron velocity. But the average electron kinetic energy is increased because the initial electrons' velocity distribution $\rho(\vec{v}(0))$ is symmetric after single wavelength laser pulse ionization, i.e. $\rho(\vec{v}(0)) = \rho(-\vec{v}(0))$. After an electron is heated by the THz radiation, electron-impact-excitation promotes a fraction of the gas species into upper electronic states that decay and emit light in time scale of ns. Therefore, studying the subsequent molecular fluorescence emission provides information of electron temperature and population of excited molecular states in the presence of the THz radiation.

The total fluorescence emission $FL(t_d) = \int FL_\lambda(t_d)d\lambda = FL_b + \Delta FL(t_d)$ can be considered as a function of the time delay t_d between the peak of the THz pulse and the peak of the laser pulse. Here $t_d>0$ is defined as when the THz pulse is ahead of the laser pulse. FL_b, the background plasma fluorescence emission without the THz field, is directly from the laser pulse excitation. $\Delta FL(t_d)$ is the change of the fluorescence by the THz pulse. The amount of the fluorescence emission enhanced by the THz pulse would take the form

$$\Delta FL \propto Cn_e(\beta_{ei},t_d)\sum_{i=1}^{\infty}\langle \Delta E_i(\tau,t_d)\rangle. \tag{30}$$

Here C describes the electron-impact-excitation efficiency constant. $n_e(\beta_{ei},t_d)$ is the electron density. β_{ei} is the electron-ion recombination rate. $\langle \Delta E_i(\tau,t_d)\rangle$ is the average energy transferred from one electron to ion/molecules during ith collision at t_i. To simplify the calculation, most of the electron kinetic energy gained from THz field between neighboring collisions is assumed to be transferred to the molecules/ions in inelastic collision. Therefore, due to $\rho(\vec{v}) = \rho(-\vec{v})$, $\langle \Delta E_i(\tau,t_d)\rangle$ can be reduced to $m\Delta\vec{v}_i^2/2$ where $\Delta\vec{v}_i = \vec{v}(t_i) - \vec{v}(t_i - \tau) = \int_{t_i-\tau}^{t_i} e\vec{E}_{\mathrm{THz}}(t)dt/m$ is the velocity change by the THz field between neighboring collisions at $t_i-\tau$ and t_i. Under gas pressure P, the electron collision relaxation time $\tau(P)$ is $\tau(P) = \tau_0 P_0/P$ where τ_0 is the electron collision relaxation time at atmosphere pressure P_0 and τ_0 is a few hundreds of femtosecond at ambient pressure.

In one extreme case when pressure is very low and thus $\tau(P)$ is much longer than THz pulse duration $\tau_{\mathrm{THz}}\sim$ 1 ps so that first electron collision happens long after THz pulse passes, $\sum_{i=1}^{\infty}\Delta E_i(\tau,t_d)$ can be approximated with only one time energy transfer

$$\sum_{i=1}^{\infty}\Delta E_i(\tau >> \tau_{\mathrm{THz}},t_d) = \frac{m\Delta\vec{v}_1^{\,2}}{2} = \frac{e^2}{2m}\left(\int_{-\infty}^{+\infty}\vec{E}_{\mathrm{THz}}(t)\cdot H(t-t_d-t_\varphi)dt\right)^2, \tag{31}$$

where t=0 is defined by $\vec{E}_{\mathrm{THz}}(0) = \vec{E}_{\mathrm{peak}}$. The step function $H(t-t_d-t_\varphi)$ represents the fast formation of the plasma channel within the laser pulse duration. t_φ is the phase delay caused by the plasma formation dynamics at the early stage. Consequently ΔFL has the form

$$\lim_{\tau>>\tau_{\mathrm{THz}}}\Delta FL \propto Cn_e(\beta_{ei},t_d)\frac{e^2}{2m}\left(\int_{t_d+t_\varphi}^{\infty}\vec{E}_{\mathrm{THz}}(t)dt\right)^2 \propto \vec{A}^2(t_d+t_\varphi). \tag{32}$$

Therefore, at low pressure the ΔFL is proportional to the square of the vector potential of the THz pulse at $t_d + t_\varphi$. In another extreme case when pressure is very high and $\tau(P) << \tau_{\mathrm{THz}}$ so that $\vec{E}_{\mathrm{THz}}(t)$ can be considered constant between neighboring collisions, $\sum_{i=1}^{\infty} \Delta E_i(\tau, t_d)$ can be approximated as

$$\begin{aligned} \sum_{i=1}^{\infty} \Delta E_i(\tau << \tau_{\mathrm{THz}}, t_d) &= \sum_{i=1}^{\infty} \frac{m\Delta \vec{v}_i^{\,2}}{2} \\ &= \frac{e^2}{2m} \int_{-\infty}^{+\infty} \tau \vec{E}_{\mathrm{THz}}^2(t) \cdot H(t - t_d - t_\varphi) dt \end{aligned} \quad . \tag{33}$$

Therefore ΔFL is proportional to the energy of the THz pulse at high pressure

$$\lim_{\tau << \tau_{\mathrm{THz}}} \Delta FL \propto C n_e(\beta_{ei}, t_d) \frac{e^2 \tau}{2m} \int_{t_d + t_\varphi}^{+\infty} \vec{E}_{\mathrm{THz}}^2(t) dt \tag{34}$$

In both cases of Equations (32) and (34), the ΔFL carries the amplitude and phase information of the THz time-domain waveform.

The influence of the THz pulse on the plasma can be measured in a THz-REEF experiment whose schematic is shown in Fig. 5.24a. A broadband, single-cycle of free space THz radiation with linear polarization and a peak field of 100 kV/cm, generated from a Lithium Niobate prism using the tilted pulse front scheme, is focused into a plasma region which is formed by focusing a 100 μJ femtosecond laser pulse with a center wavelength of 800 nm. The laser pulses propagate collinearly with the THz pulses. The plasma fluorescence spectrum is measured by a monochromator and a photo multiplier tube (PMT). To give a parallel reference for the plasma fluorescence measurement, the time-domain waveform of the THz pulse was measured using standard electro-optic (EO) sampling in a 200 μm thick <110> GaP crystal. The influence of the THz field on the laser-induced plasma emission spectrum is experimentally investigated in the spectral range of 320–400 nm which contains the strongest fluorescence emission of nitrogen gas. Figure 5.24b shows that fluorescence emission from both nitrogen molecules and ions are enhanced by the same factor as THz field is increased. Figure 5.24c shows the measured THz field dependence of the total emission FL at the strongest line 357 nm and the quadratical fit. The fact that the ΔFL at all wavelengths is proportional to the THz intensity, agrees with Equation (34) when high pressure approximation is satisfied at ambient pressure. In the inset of Fig. 5.24c the angular pattern of fluorescence shows an isotropic emission profile. These observations indicate that the THz pulse enhances rather than quenches plasma fluorescence emission and further imply that photo-ionized electrons gain kinetic energy from the THz field, like electron inverse-bremsstrahlung heating by the THz pulse, and transfer kinetic energy to air molecules/ions via collisions.

Utilizing the linear dependence of ΔFL enhanced fluorescence on the THz intensity at ambient pressure in Equation (34), REEF could be used for incoherent THz detection with the temporal resolution determined by the ionizing pulse envelope.

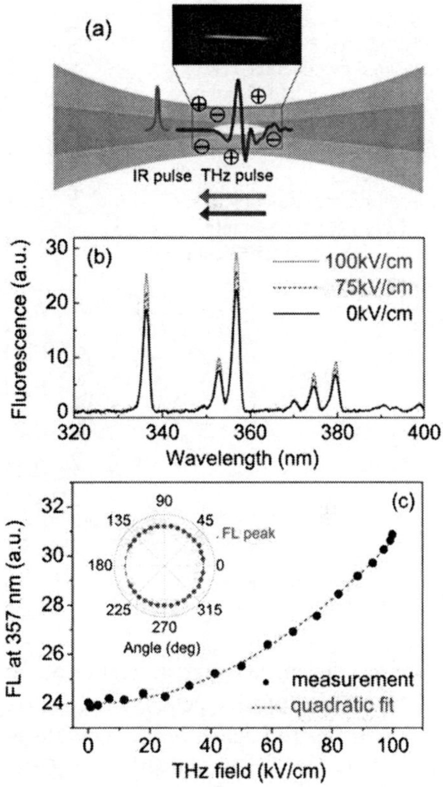

Fig. 5.24 (**a**) Schematics of the interaction between the THz wave and laser-induced plasma. (**b**) The measured fluorescence spectra in the range of 320 and 400 nm versus THz field. (**c**) The measured quadratic THz field dependence of 357 nm fluorescence emission line. *Inset*: The isotropic emission pattern of THz-REEF

Furthermore the coherent detection using REEF is also applicable if an external 20 kV/cm bias parallel with $E_{\mathrm{THz}}(t)$ is applied on the plasma as a local oscillator E_{LO}. The resulting ΔFL is

$$\Delta FL \propto \tau \int_{-\infty}^{+\infty} (\vec{E}_{\mathrm{THz}}(t) + \vec{E}_{\mathrm{LO}})^2 \cdot H(t - t_d - t_\varphi)dt. \tag{35}$$

If the bias is modulated at half of the THz pulse repetition rate, the only term being modulated is the cross term $2\vec{E}_{\mathrm{LO}}\vec{E}_{\mathrm{THz}}(t)$ inside the integral. Therefore

$$\Delta FL \propto \tau \int_{t_d+t_\varphi}^{+\infty} 2\vec{E}_{\mathrm{LO}}\vec{E}_{\mathrm{THz}}(t)dt \propto \vec{E}_{LO}\vec{A}(t_d + t_\varphi). \tag{36}$$

In this manner, the THz waveform can be retrieved from the derivative of the vector potential of the THz pulse $\vec{A}(t_d)$, with a phase delay t_φ. Figure 5.25a shows measured vector potential $\vec{A}(t_d)$ and good agreement between the THz waveform calculated from $d\vec{A}(t_d)/dt_d$ in THz-REEF and that measured by EO sampling. Similar to other THz wave detection methods, THz-REEF can also be used in THz wave

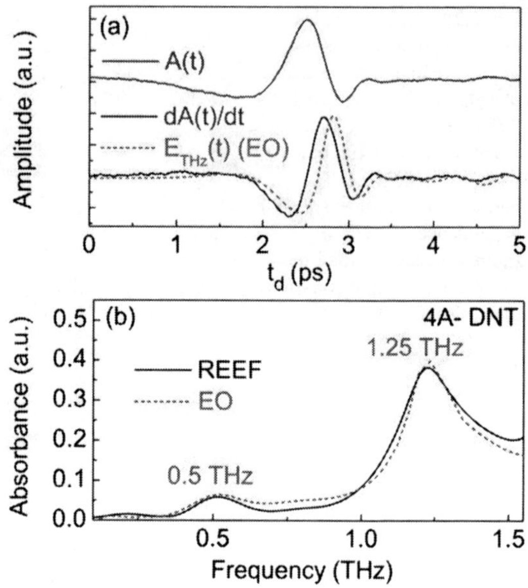

Fig. 5.25 (**a**) Vector potential $\vec{A}(t_d)$ of THz pulse measured by coherent THz detection using REEF and the $d\vec{A}(t_d)/dt_d$ compared to the THz waveform measured by EO detection. (**b**) The THz absorption spectroscopy of 4A-DNT explosive sample measured by THz-REEF and EO sampling

spectroscopy measurement. Figure 5.25b shows the resolved absorption features of 4A-DNT explosive pellet sample at 0.5 and 1.25 THz by REEF compared with results using EO detection. It is worth noting that one of the fundamental differences between detection using THz-REEF and other THz detection methods is that the THz-REEF process is not instantaneous unlike other nonlinear optical methods, such as four-wave-mixing or Pockel effects, but it is determined by the nature of fast plasma formation and long-lived air plasma compared to THz pulse duration. This provides a new method for omni-directional broadband coherent THz wave detection which could be potentially extended to other spectral regions. Furthermore, owing to its omni-directional emission, THz-REEF provides a promising tool for standoff THz detection, which has abundant applications in global environmental monitoring and homeland security.

References

1. H. G. Muller, "An efficient propagation scheme for the time-dependent Schrödinger equation in the velocity gauge," *Laser Phys.*, **9**, 138 (1999).
2. K. J. Schafer, and K. C. Kulander, "Energy analysis of time-dependent wave functions: Application to above-threshold ionization," *Phys. Rev. A* **42**, 5794 (1990).
3. J. Dai, X. Xie, and X.-C. Zhang, "Detection of broadband terahertz waves with a laser-induced plasma in gases," *Phys. Rev. Lett.* **97**, 103903 (2006).
4. N. Karpowicz, J. Dai, X. Lu, Y. Chen, M. Yamaguchi, H. Zhao, X.-C. Zhang, L. Zhang, C. Zhang, M. Price-Gallagher, C. Fletcher, O. Mamer, A. Lesimple, and K. Johnson, "Coherent heterodyne time-domain spectrometry covering the entire 'terahertz gap'," *Appl. Phys. Lett.* **92**, 011131 (2008).

5. R. S. Finn, and J. F. Ward, "DC-induced optical second harmonic generation in the inert gases," *Phys. Rev. Lett.* **26**, 285 (1971).
6. Y. Chen, M. Yamaguchi, M. Wang, and X.-C. Zhang, "Terahertz pulse generation from noble gases," *Appl. Phys. Lett.* **91**, 251116 (2007).
7. X. Lu, N. Karpowicz, Y. Chen, and X.-C. Zhang, "Systematic study of broadband terahertz gas sensor," *Appl. Phys. Lett.* **93**, 261106 (2008).
8. D. J. Cook, and R. M. Hochstrasser, "Intense terahertz pulses by four-wave rectification in air," *Opt. Lett.* **25**, 1210 (2000).
9. K. L. Yeh, M. C. Hoffmann, J. Hebling, and K. A. Nelson, "Generation of 10 μJ ultrashort terahertz pulses by optical rectification," *Appl. Phys. Lett.* **90**, 171121 (2007).
10. T. Bartel et al., "Generation of single-cycle THz transients with high electric-field amplitudes," *Opt. Lett.* **30**, 2805 (2005).

Chapter 6
THz Wave 3D Imaging and Tomography

THz waves are transparent to most of the dry dielectric materials. This property makes THz wave a promising candidate for nondestructive evaluation of the internal structures of targets. THz wave time-of-flight imaging method is one of the techniques to extract the information about the layered structures of a target. If there is no layer structure within the target, or if the interesting features are not located on those layer structures, one needs to use tomographic imaging techniques to extract those interesting information [1].

T-Ray Computerized Tomography

Initially, Computerized tomography (CT) technique was developed for X-ray imaging, and currently is used in medical diagnostics and other nondestructive evaluation applications. Figure 6.1 illustrates the concept of CT. Collimated X-ray beam transmits through the target and the transmitted intensity is recorded. The transmission of the X-ray beam through different paths defined by lateral position and angles is obtained by rotating and laterally shifting the target. For each beam path, the intensity of transmitted beam is determined by the integral absorption along the optical path. If the absorption distribution of the target is defined as $f(x,y,z)$, then the total absorption along the path with rotation angle θ, height z and lateral displacement l is described as

$$p(\theta,l,z) = \int_{L(\theta,l,z)} f(x,y,z)dl \equiv \Re(f(x,y,z)), \tag{1a}$$

where $L(\theta,l,z)$ denotes the beam path through the target. The transmission T and absorption p of the beam path are defined as

$$T = \exp\left[-\int_{L(\theta,l,z)} f(x,y,z)ds\right] \equiv \exp[-p(\theta,l,z)]. \tag{1b}$$

X.-C. Zhang, J. Xu, *Introduction to THz Wave Photonics*,
DOI 10.1007/978-1-4419-0978-7_6,

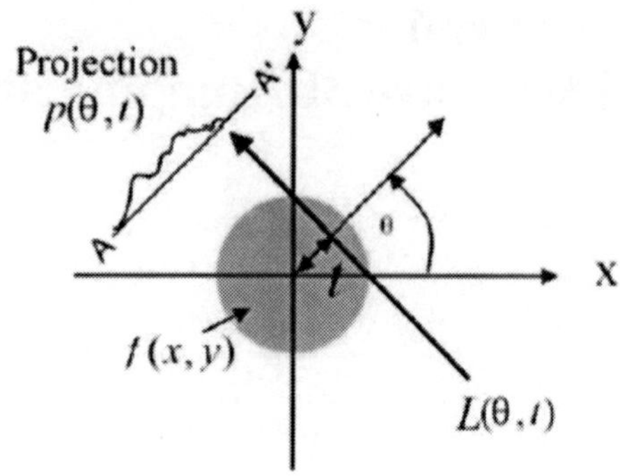

Fig. 6.1 Concept of computerized tomography (CT). Intensity of carrier wave beam is recorded after transmission through the target with certain lateral shifts and rotation angles. Inverse Radon transform of the recorded signal gives 3D information of the target

Equation (1) is known as the Radon transform [2]. The Fourier transform of Equation (1) gives

$$\begin{aligned} P(\theta,v,z) &= \int p(\theta,l,z)\exp(-ivl)dl \\ &= \iint \exp[-i(\eta x - \zeta y)]f(x,y,z)dxdy. \end{aligned} \tag{2}$$

Here v is lateral spatial frequency, which indicates the spatial resolution of the CT imaging process. $\eta = v\sin(\theta)$ and $\xi = v\cos(\theta)$ are spatial frequency components along the x and y direction, respectively. Equation (2) indicates that, Fourier transform of the Radon transform is equivalent to 2D Fourier transform of the absorption distribution.

CT image directly measures transmission of the carrier wave corresponding to each beam path. To get the 3D structure of the target, Equations (1) and (2) are used to calculate the spatial distribution of the absorption, $f(x,y,z)$, in the target

$$\begin{aligned} f(x,y,z) &= \iint P(\theta,v,z)\exp\left[2\pi iv(x\cos\theta + y\sin\theta)\right] v dv d\theta \\ &= \iiint P(\theta,l,z)\exp(-ivl)\exp\left[2\pi iv(x\cos\theta + y\sin\theta)\right] v dl dv d\theta. \end{aligned} \tag{3}$$

This equation is called the reverse Radon transform. Reverse Radon transform of the beam path absorptions at height z gives cross section image of the target at the height z. The above process is repeated for all different heights to obtain the cross section images at the different heights. These different cross section images provide the 3D structure of the target.

Unlike X-ray imaging, which provides only the intensity information, the THz wave imaging provides both the intensity and the phase information. The THz CT enables one to obtain the absorption and the refractive index information of the target. An experimental setup of THz wave CT system is presented in Fig 6.2. To ensure the spatial resolution, THz beam is focused onto the target using either an off-axis parabolic mirror or a lens. The focal spot size determines the spatial resolution of the CT image, while its depth of focus should be at least as long as lateral dimensions of the target. Therefore, the focusing conditions of THz beam needs to

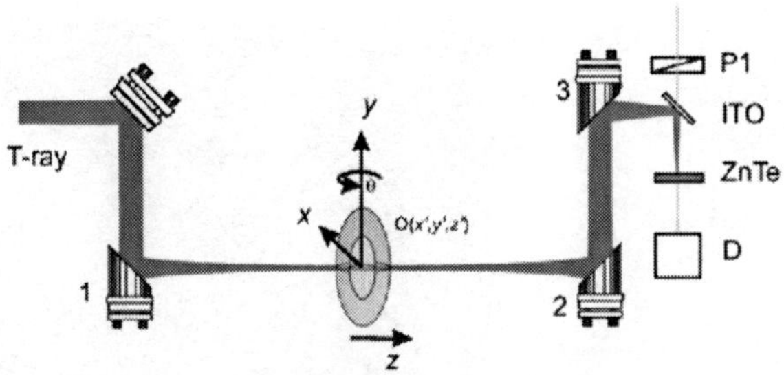

Fig. 6.2 Experimental setup of THz wave CT. THz wave is focused with an off-axis parabolic mirror onto the target. Temporal waveform of THz pulses transmitted through the target is recorded. The target is driven by a rotation stage and a linear translation stage. THz waveform at each rotation and lateral transition position is recorded

be well considered so that it provides sufficient depth of focus along with the high spatial resolution. In order to obtain a tomographic image, the target is set on a rotation stage which is able to rotate about the z axis and the target moves along l and z directions by a 2D translation stage. Thus, the target is scanned in three dimensions corresponding to variables of z, l and θ. Another off-axis parabolic mirror collects the transmitted THz waves and a THz wave detector records its temporal waveform.

Refractive index of the materials can be approximately treated as 1 for almost all common materials when performing X-ray CT. As a result, X-ray beam can be treated as a simple straight line without any beam bending in the CT process. However, different materials may have different refractive index in THz band. For example, refractive index of plastic is $\sim$1.5, of glass is $\sim$2, and of semiconductor materials could be as high as 3–4. As a result, refractive behavior of THz wave may not be negligible in CT process unless specific conditions are satisfied. Further, the wavelength of THz wave is usually comparable to spatial features of the target, such as textures, or microstructures, thus diffraction and scattering of the beams may also affect propagation of THz wave through the target. To reconstruct the THz wave tomographic image using Equation (3), the material's refractive index should have slow variation within a THz wavelength, otherwise, ray-tracing techniques need to be used in order to correctly retrieve 3D image of the target. In pulsed THz wave tomography, the entire THz waveform is recorded. Different physical quantities can be used to represent different features of the target. For example, the amplitude of transmission provides absorption coefficient, the phase delay represents refractive index of the target and the spectroscopy provides the composition information of the target.

Polystyrene foam material has very low absorption coefficient ($<0.001\ \text{cm}^{-1}$ @ 1 THz) and almost constant refractive index ($n = 1.02$) for THz waves, especially for THz wave with relatively low frequency. Thus, it qualifies to be a very good material to be used to investigate optical properties of THz waves. Figure 6.3 shows a photo of a target made from a piece of polystyrene foam that was used for THz

Fig. 6.3 Photo of a target used in THz wave CT imaging process. Cubic polystyrene foam is carved with "THZ" through its height

wave CT process. Three letters "T", "H" and "Z" were carved in the target. Since there is no change in z direction, the cross section at $z = z_0$ reflects the full 3D information of the target. In order to obtain the image, the sample was driven along the x-axis by a linear stage. For each x position, the sample was rotated within a range of $\theta = 0$–$180°$ with an interval of 1.8°. Imaging system recorded the THz waveform for each x and θ. Figure 6.4 shows 3 THz waveforms recorded in the experiment. Those waveforms were obtained for the lateral displacement $x = 0$, and rotation angles of 0°, 60° and 120°, respectively. Both amplitude and position of peak in THz waveform vary for different beam paths. Figure 6.5a shows time delay of THz pulses in the x–θ space. In the frequency domain, time delay of THz pulses corresponds to the phase delay. Figure 6.5b gives the phase distribution at 1 THz (the central frequency of THz pulses) in the x–θ space. Using time delay to replace absorption in Equation (3), the 3D image of the target can be reconstructed according to its refractive index distribution. Figure 6.6 shows the retrieved cross section image of the target of Fig. 6.3. The profile of target and the carved letters are reconstructed due to refractive index difference between air and polystyrene foam. Similarly, the phase distribution in the frequency domain can be used to retrieve 3D image of the target and the result is very similar to Fig. 6.6. Further, one can also use variation of THz amplitude to retrieve the absorption coefficient (or extinction coefficient) distribution of the target. The image based on amplitude attenuation of THz wave is less sharp due to scattering of the waves by the target.

When the sample is scanned in the vertical direction along with the lateral scanning and rotation, the cross section of the target can be obtained for each height that provides a true 3D image of the target. To save the scanning time, single pulse detection method discussed in Chapter 3 can be used to record THz waveforms. Figure 6.7 presents cross sections images of a ping-pong ball recorded by THz wave CT at different heights. When all these cross sectional images are combined the THz wave CT retrieves cavity structure of the ping-pong ball, which is presented in Fig 6.8. The cavity structure of the ping-pong ball can be clearly seen in the

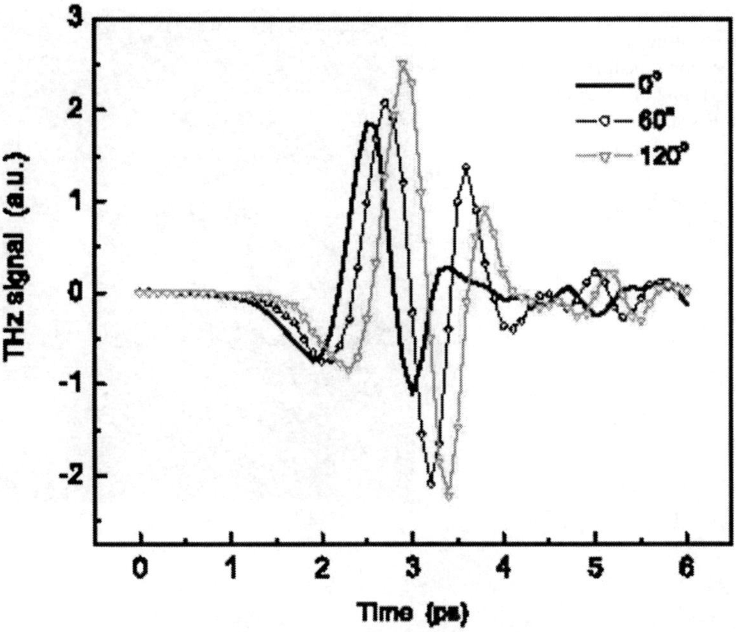

Fig. 6.4 Temporal waveforms of THz pulses transmitted through the target at 0°, 60° and 120°, respectively. THz pulse propagation through the target following different path does not only lead to different absorption, but also gives different phase delay. The phase delay reflects refractive index distribution of the target

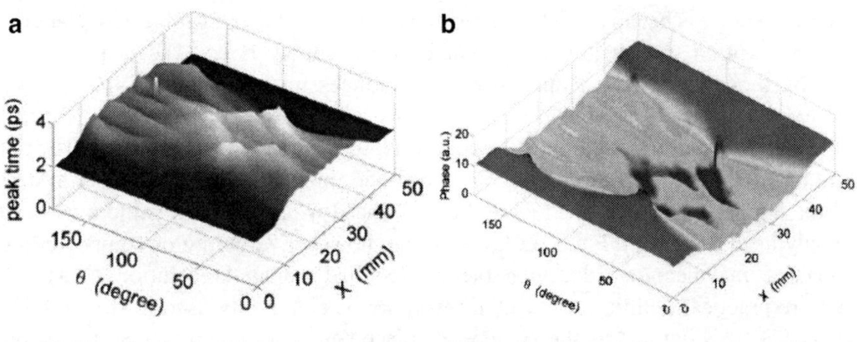

Fig. 6.5 The distribution of physical qualities in a THz wave CT measurement presented in x–θ space. (**a**) Distribution of THz peak timing, and (**b**). distribution of phase delay at 1 THz

THz wave CT; however, the thickness of the retrieved shell is much thicker than the real one. This is due to the fact that the physical model of THz CT is based on gradual change of refractive index; however, the real target has a sharp change at boundaries of the shell. Further, the shell of the ping-pong ball has relatively high refractive index and the Fresnel losses at its surface are not negligible. The Fresnel losses also affect the image quality. Figure 6.9 shows an optical photo of a piece

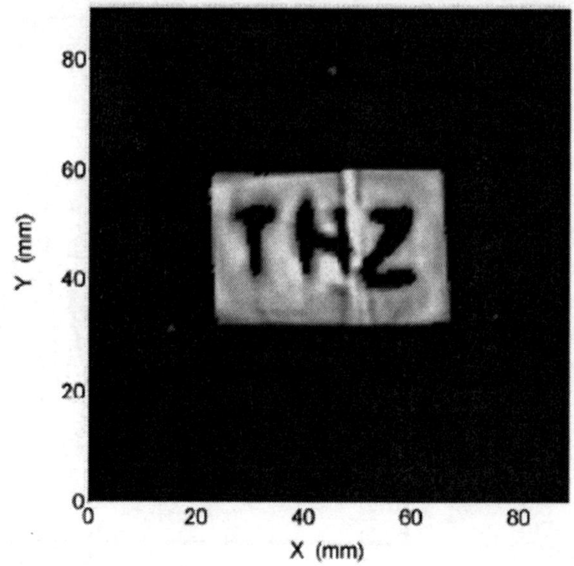

Fig. 6.6 3D image of the target reconstructed via inverse Radon transform of the peak timing distribution in the x–θ space

of dry and defat turkey bone and the corresponding THz CT image [3]. The optical photo only shows profile of the bone, while THz CT image shows the profile as well as the central cavity structure. However, there are rich microstructures in the bone, which are not retrieved in THz CT image. The inability of reconstruction to resolve microstructure is because the CT algorithm requires that the target has low scattering and absorption, which are not satisfied for the turkey bone. The heavy scattering by those complicated microstructures reduces image quality of the retrieved image.

Spatial resolution is the most important feature for an imaging system. The definition of spatial resolution may vary from one imaging system to anther system. If the imaging system is diffractive limited, the Rayleigh criteria is the most commonly used definition. For THz CT imaging, however, structure or composition of the target may become a dominant factor affecting the spatial resolution. As a result, a more practical definition of spatial resolution is commonly used in study of THz wave CT. It is defined as the smallest distance between two spots that can be distinguished in the retrieved image. Figure 6.10 shows photos of a cylindrical sample made using polystyrene foam. The sample contains eight holes with a diameter of 2.5 mm each and two slots with 2.5 mm width. The smallest distance between two holes lying on the line AA′ is 0.5 mm. Figure 16.11a gives the reconstructed cross section image of the sample using THz CT. The image indicates that THz CT provides a spatial resolution of at least 0.5 mm, because even the closest holes are distinguishable in the image. Ray approximation used in the reconstruction does not consider the diffractive behavior of THz wave limited by the imaging system aperture. In reality, the measured phase change is an involution between the sample

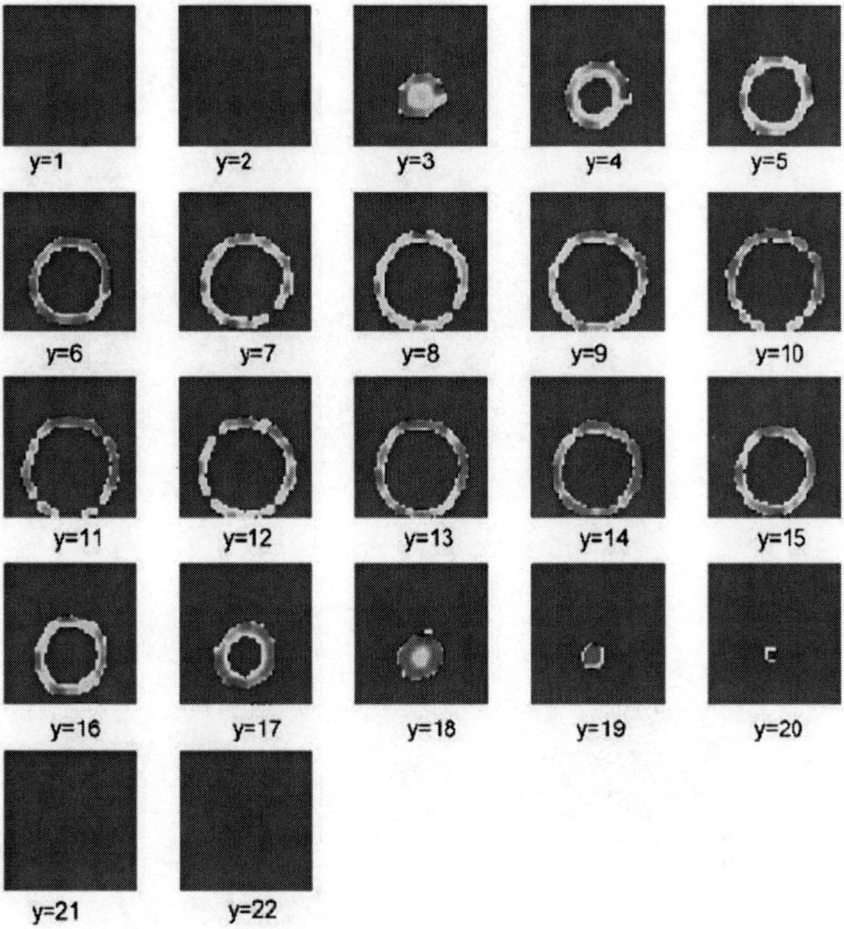

Fig. 6.7 THz wave CT images of a ping-pong ball

caused phase delay and the diffraction effect. Using Wiener filter, the diffraction induced phase change can be removed from the THz CT image to obtain the pure sample information. Figure 6.11b shows the enhancement in the sharpness of the structure of the target obtained by applying Weiner filter on the image in Fig. 6.11a.

T-Ray Diffraction Tomography

Diffraction phenomenon plays a strong role in THz wave imaging due to the relatively long wavelength of THz waves. Interaction between THz wave and the target can be described by the following Maxwell equation [4]:

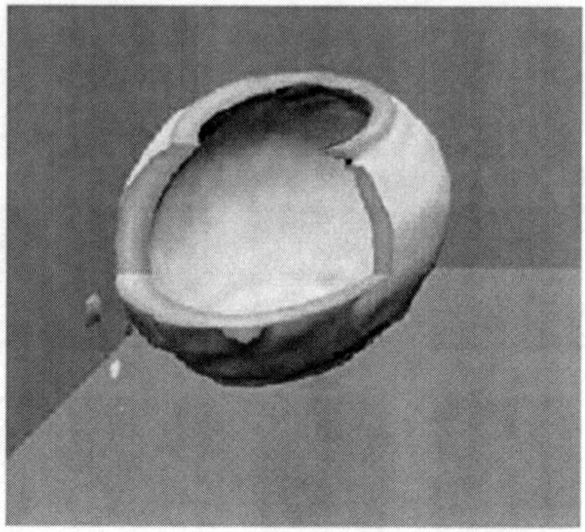

Fig. 6.8 3D image of the ping-pong ball cavity

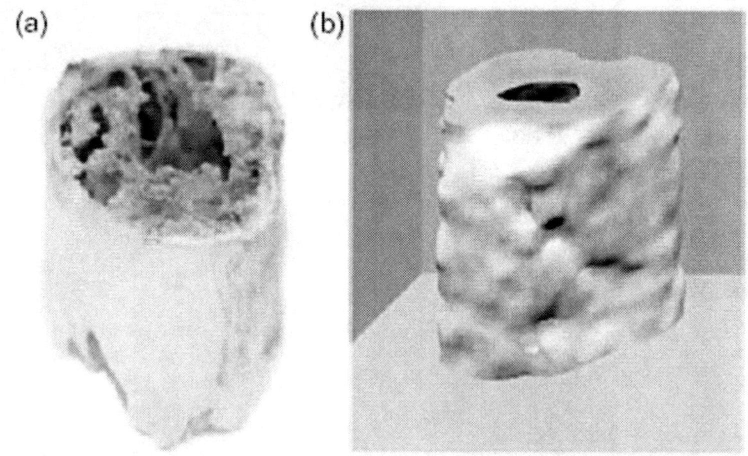

Fig. 6.9 Photo of a turkey bone (*left*) and its THz wave CT image (*right*)

$$\nabla^2 E + \frac{\mu(r)\varepsilon(r)}{c^2}\frac{\partial^2}{\partial t^2}E + [\nabla \ln \mu(r)] \times (\nabla \times E) + \nabla[E \cdot \nabla \ln \varepsilon(r)] = 0, \quad (4)$$

where r denotes a position in space, c is the speed of light, $\varepsilon(r)$ and $\mu(r)$ are the complex permittivity and permeability of the material, respectively. If the variation of both ε and μ is small within a wavelength, the above equation can be simplified to

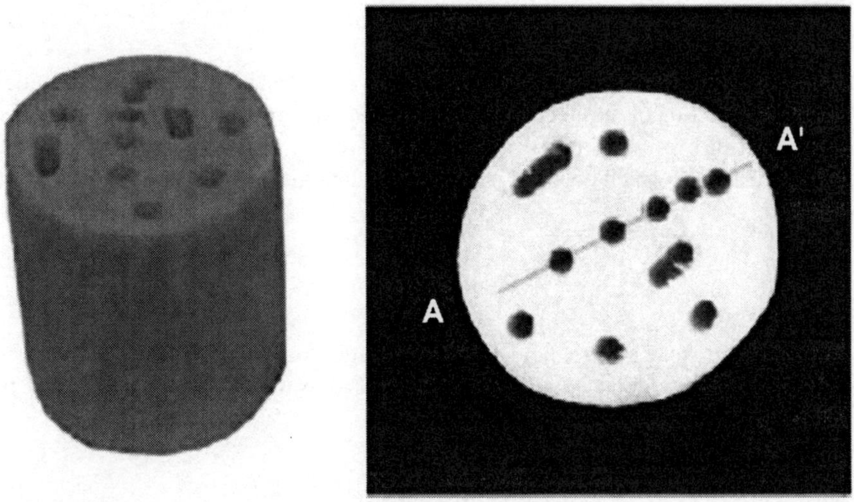

Fig. 6.10 Side view and top view of a target used in THz wave CT. The target is made of polystyrene foam

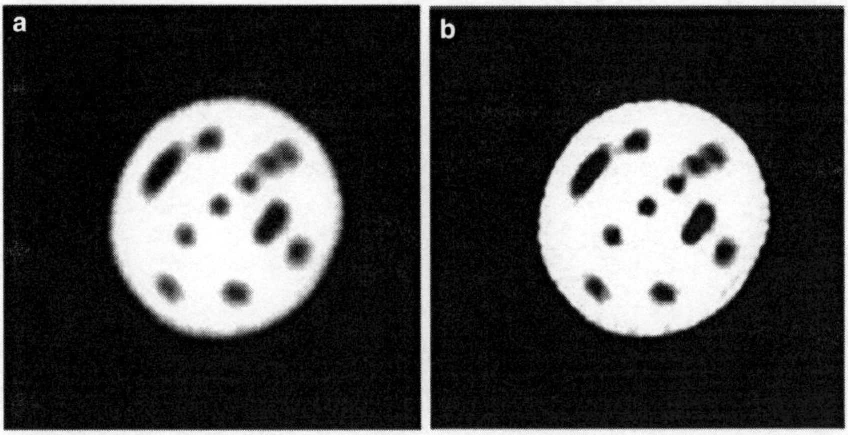

Fig. 6.11 (**a**) THz wave CT cross section image of the target shown in Fig. 6.10 (**b**) the cross section image with Wiener filter applied

$$\nabla^2 E + \frac{\tilde{n}(\omega,r)^2}{c^2}\frac{\partial^2}{\partial t^2}E = 0 \tag{5}$$

.

Here $\tilde{n} = \sqrt{\varepsilon\mu}$ is the complex refractive index of the target. If polarization effects of the EM wave are negligible, the vector equation can be further simplified to the Helmholtz scalar equation:

$$\nabla^2 u(r) + k_0^2 \tilde{n}(\omega,r)^2 u(r) = 0, \tag{6}$$

where $k_0 = 2\pi/\lambda$ is the wave number of EM wave in vacuum and $u(r)$ is the complex amplitude function of the electric field. In an imaging process, $u(r)$ can be described as the sum of two components: $u(r) = u_0(r) + u_s(r)$, where the former is electric field of EM wave which does not penetrate through the target or is a solution to the equation

$$\left(\nabla^2 + k_0^2\right) u_0\,(r) = 0, \tag{7}$$

while the later is the electric field of EM wave after transmission through the target.

If the target modulates only phase of the EM wave, the electric field can be written as

$$u(r) = \exp\left[\phi(r)\right] = \exp\left[\phi_0(r) + \phi_s(r)\right], \tag{8}$$

where ϕ is the complex phase of the electric field. Assuming that the scattering induced phase shift changes slowly, Equation (7) can be further simplified using the first order Rytov approximation to

$$\phi_s(r) = \frac{1}{u_0(r)} \int G(r - r')V(r')u_0(r')dr', \tag{9a}$$

where $G(r–r')$ is a Green function [5] and is described as

$$G(r - r') = \frac{\exp\left(ik_0 \left|r - r'\right|\right)}{4\pi \left|r - r'\right|} \tag{9b}$$

and $V(r)$ is defined as

$$V(r) = -k_0^2[\tilde{n}^2(\omega,r) - 1]. \tag{9c}$$

Compared to optical waves, diffraction of THz pulses shows its unique properties. The THz pulse contains broadband spectrum and its interference pattern is generated by sum of all frequency components. Figure 6.12 illustrates an experimental setup of Young's double slit interference measurement for THz wave. In this experiment, the sample was a 0.17 mm thick aluminum foil with double slits of width 1 mm each and the distance between those two slits being 6 mm. The distance from the double slits plane to the THz wave detector was 48 mm. Since THz pulse has a short pulse width, its interference pattern also varies in time. The number of peaks and valleys in the interference pattern increases with time delay. This

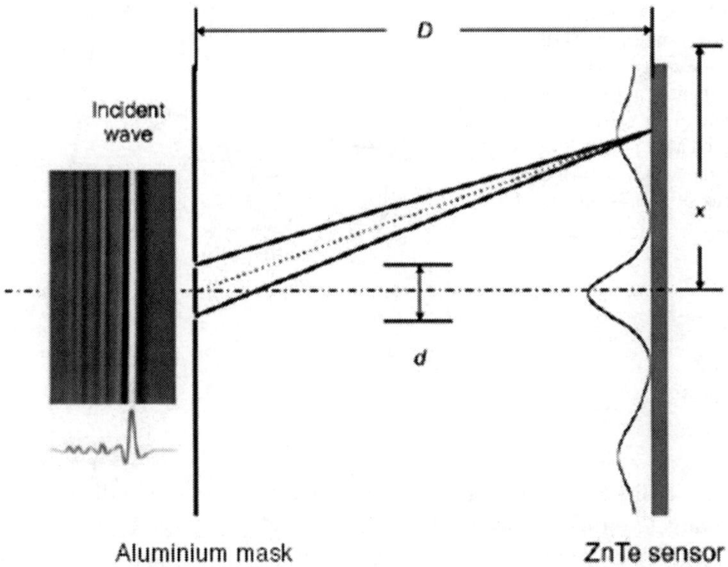

Fig. 6.12 Experimental setup of THz wave Young's double slit measurement. The double slits are carved on 0.17 mm thick aluminum foil with width of 1 mm each. The distance between those two slits is $d = 6$ mm the distance between the double slit and the detection crystal is $D = 48$ mm

phenomenon indicates that, after THz pulse pass through the double slits, lower frequency components propagate faster than the higher frequency components. Figure 6.13a shows the evolution of THz wave interference pattern along the direction perpendicular to the double slits. Figure 6.13b shows position of peak and valley in Fig. 6.13a, which presents the wavefront propagation of THz wave starting from each slit.

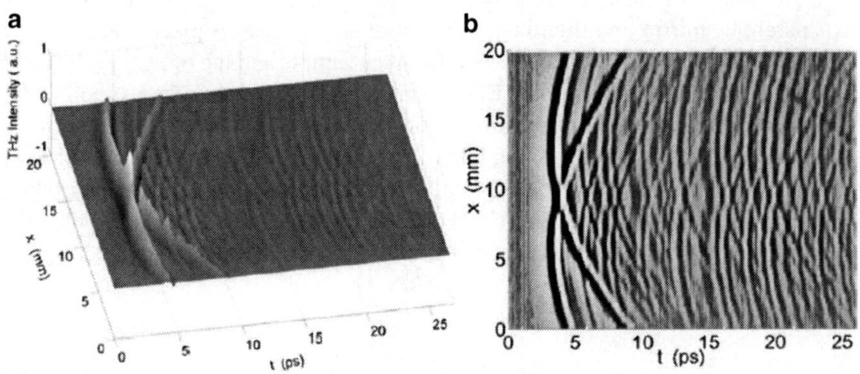

Fig. 6.13 The temporal evolution of THz field along x axis. (**a**) experimental result, and (**b**) the normalized electric field (according to peak of THz waveform)

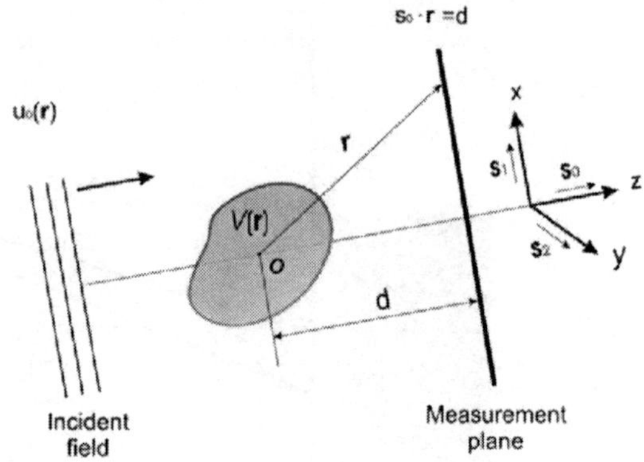

Fig. 6.14 The concept of diffraction tomography (DT). The carrier wave in DT measurement is a plane wave. Wavefront of the carrier wave is recorded after propagation through the target, and which is used to retrieve 3D image of the target

In THz diffraction tomography, a planar THz wave is used to illuminate a target and the target's structural information is extracted by measuring and processing the diffracted THz distribution by the target. Figure 6.14 shows the concept of diffraction tomography using a monochromatic wave. The light source in a diffraction tomography is a plane wave. The THz wave is detected in a detection plane perpendicular to its propagation direction after transmission through the target. The Fourier transform of the forward scattered electric field of the THz wave on the detection plane is proportional to 3D Fourier transform of the object function in a displaced hemisphere in the frequency domain. The displacement value equals to a negative incident wave number along the incident wave direction. Consequently, reverse Fourier transform of the measurement result leads to the object function. In THz CT, one needs to record THz field of an entire beam; however, in THz DT, THz wave distribution in the detection plane needs to be resolved. This can be achieved by using the 2D imaging technique discussed in Chapter 3. In THz DT set up, THz waves are generated from a point emitter via optical rectification process excited by high intensity fs laser pulses. The generated THz beam is collimated by a off-axis parabolic mirror and illuminates the entire target which is placed on a rotation stage. The forward scattered THz beam from the sample and the optical probe beam propogate collinearly through the EO crystal. This probe beam records the 2D distribution of the THz diffraction formed on the EO crystal. The 2D image of the target is obtained for each rotation angle by recording the forward scattered wave detected using a large aperture EO crystal. Figure 6.15 shows a target made of three plastic rectangular bars used in THz DT. If the target modulates only the phase of THz wave, the first order Rytov approximation can be used to reconstruct the 3D image:

$$u = u_0 \ln\left(\frac{u_s}{u_o} + 1\right) \tag{10}$$

.

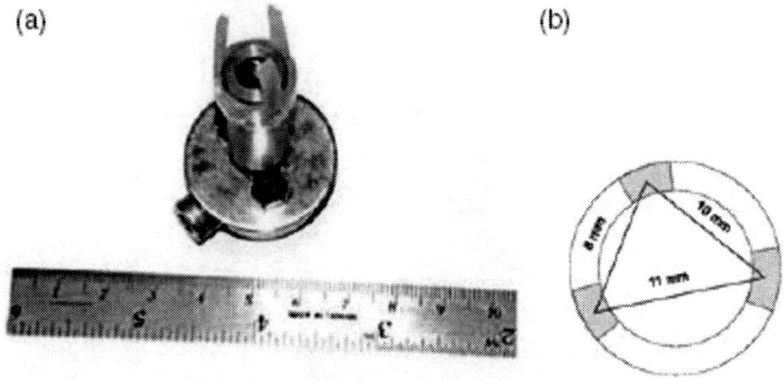

Fig. 6.15 Photo of the target in THz wave DT (*left*) and its top view illustration (*right*). The target consists of 3 plastic bars

Figure 6.16a gives a reconstructed refractive index distribution of the sample shown in Fig. 6.15 and the Fig. 6.16b shows the reconstructed image presented in 3D. The image is reconstructed based on 0.2 THz component.

The purpose of imaging is to obtain an image that closely resembles the target and provides the maximum structural information of the target. Therefore, the quality of image is evaluated by calculating similarity between the target and its reconstructed image. The imaging quality of THz wave tomography is defined as a normalized cross correlation function between a target and its image [6, 7]:

$$Q = \frac{\sum_{i,j=0}^{M,N} [O(i,j) - \overline{O}][I(i,j) - \overline{I}]}{\sum_{i,j=0}^{M,N} \left|O(i,j) - \overline{O}\right| \sum_{i,j=0}^{M,N} \left|I(i,j) - \overline{I}\right|}, \qquad (11)$$

where i,j are the pixel index, O is the object function of the target, $\overline{O}$ is mean of O, I is the reconstructed image function and $\overline{I}$ is its mean value. Figure 6.17 shows image quality of THz DT image presented in Fig. 6.16 as a function of the THz wave frequency. For low frequency wave (<0.2 THz), the imaging quality is relatively low due to the diffraction limited spatial resolution. At higher frequency range (> 0.45 THz), the imaging quality is limited by the lower Signal to noise ratio. Thus, the imaging quality of the THz wave diffraction tomography increases with frequency until 0.2 THz and has a plateau between 0.2 and 0.45 THz and further decreases with the frequency.

3Dimensional Imaging Using Fresnel Lens

A Fresnel lens is a Fresnel zone plate with phase or amplitude patterns formed by series of concentric circles. The Fresnel lens manipulates the optical wave via

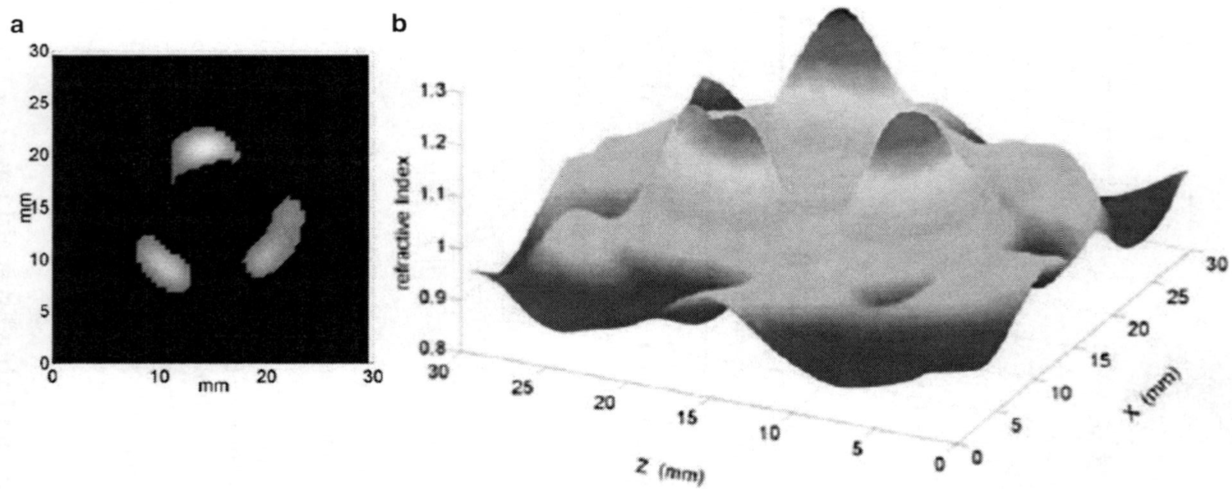

Fig. 6.16 The reconstructed THz wave DT image of the target shows in Fig. 6.15 (**a**) refractive index distribution of the target, and (**b**) 3D image of the refractive index distribution

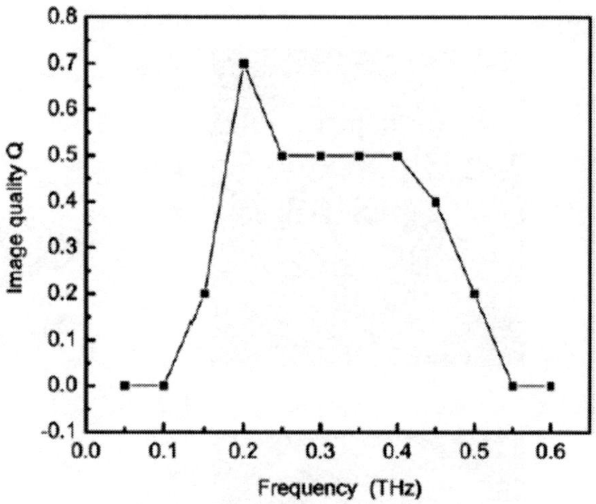

Fig. 6.17 The image quality of THz wave DT as a function of THz wave frequency

diffraction and interference. Fresnel lens can be considered as an example of diffraction optical components. Compared to traditional optical lens, Fresnel lens is more flexible in design and fabrication. Further, it can be much thinner than refractive lens with same aperture and focal length and thus can replace large aperture lens in order to reduce the weight. Binary Fresnel lens are fabricated by photolithography and etching on a transparent disk to make phase modulation features. *N* circles of photolithography and etching gives $L = 2^N$ levels of phase modulation depth. Figure 6.18 shows pictures of three THz wave binary Fresnel lens fabricated on an intrinsic silicon wafer. Levels of those lenses are 2, 4, and 8, respectively. Figure 6.18 illustrates the diffraction patterns of THz pulses focused by these three lenses.

Diffraction of THz wave by Fresnel lens along *z*-axis (normal of the lens) is [8]

$$u(z) = \sum_n A_n \iint exp\left[i2\pi \left(\frac{n}{r_p^2} + \frac{1}{2\lambda z} \right) (x^2 + y^2) \right] dxdy, \tag{12}$$

where $n = \pm 1, \pm 2, \ldots .., A_n = \sin c(n/L)$. r_p^2 is Fresnel zone of the lens with the dimensions of area and indicates a zone area where the phase delay is constant. For an incident plane wave, its focal spot is located at

$$z_n = -\frac{r_p^2}{2\lambda n}, \quad n = 1,2,\ldots\ldots . \tag{13}$$

Equation (13) indicates that Fresnel lens has a very high chromatic dispersion and its focal length is inversely proportional to the wavelength of the incident wave. Equation (13) also shows that a Fresnel lens has multiple orders of focal spots for a monochromatic plane wave. The diffraction efficiency for its focal point of each order is

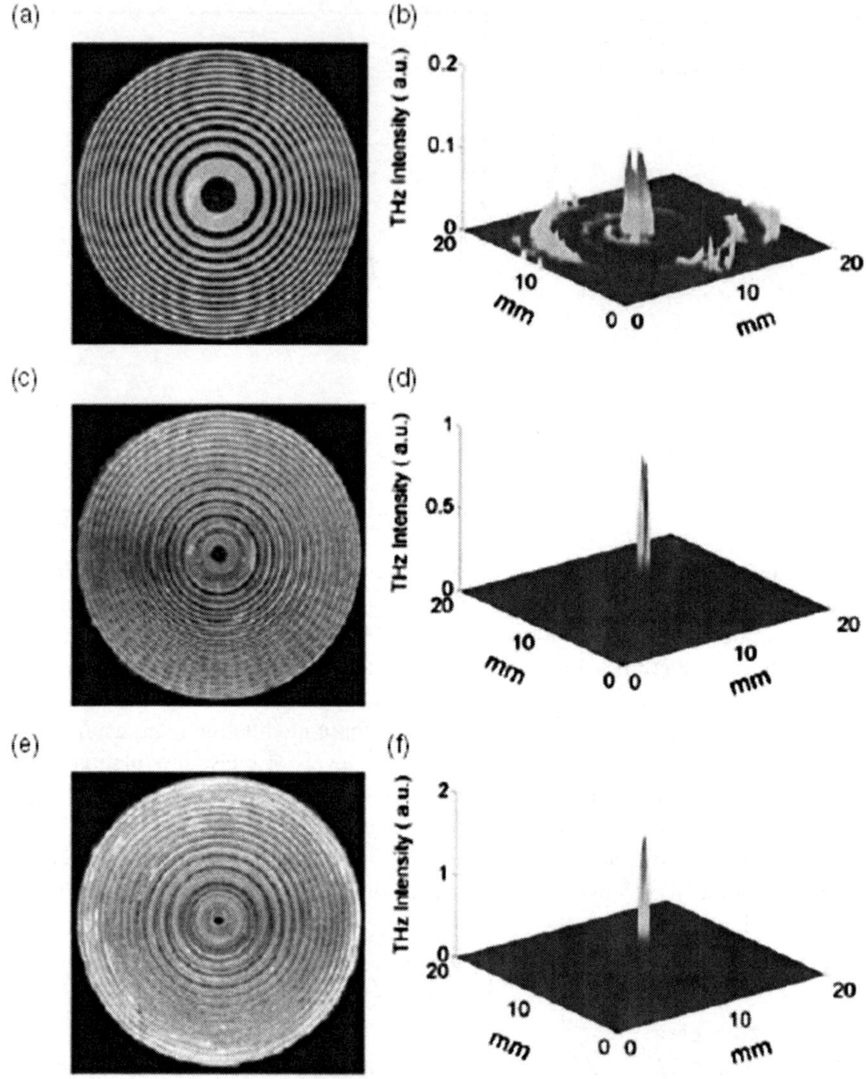

Fig. 6.18 THz wave Fresnel lenses made on silicon wafer with a level of 2 (**a**), 4 (**c**), and 8 (**e**), respectively. The corresponding THz wave intensity distribution at the focal plane (**b**, **d**, and **f**)

$$\eta(n) \equiv |A_n|^2 = \sin c^2(n/L). \tag{14}$$

Equation (14) indicates that the lower order of focal spot gives the higher diffraction efficiency. Table 6.1 summarizes the diffraction efficiency for the first order focal spot of those three silicon based Fresnel lenses. An aluminum zone plate is used as comparison.

Table 6.1 Diffraction efficiency of the first order focal spot for Fresnel lenses with different binary levels

	L			
Diffraction efficiency η	2	4	8	Al zone plate
Calculated result (%)	41	81	95	41
Experimental result (%)	11	75	90	38

Fresnel lens can focus on a plane wave and image the target like a refractive lens and follows the following lens formula:

$$\frac{1}{d_0} + \frac{1}{d_1} = \frac{1}{F_\lambda}. \tag{15}$$

Where d_0 is the distance between the lens and the object and d_1 is the distance between the lens and the image and F is the focal length. When Fresnel lens is used to process image, its first order focal length is usually used for the highest diffraction efficiency. Combining Equations (13) and (15), one has the following formula giving the object distance.

$$d_0 = \frac{r_p^2 d_1 \nu}{2d_1 c - r_p^2 \nu}, \tag{16}$$

where ν is frequency of the THz wave. Equation (15) indicates that, in an imaging system, if the imaging plane is fixed, images made by carrier waves with different wavelengths reflect profile of target at different longitudinal location. In pulsed THz wave imaging process, the entire THz waveform is recorded. A Fourier transform of the measured THz waveform provides the frequency dependent response of the Fresnel lens. Figure 6.19 shows that the focal length of a Fresnel lens increases with THz wave frequency. Thus, the linearly frequency-dependent focal length of a Fresnel lens allows tomographic imaging of a target using multiple frequencies. Objects at various positions along the beam propagation path are uniquely imaged on the same imaging plane using a Fresnel lens at different frequencies of the imaging beam. This procedure allows the reconstruction of an object's tomographic contrast image by assembling the frequency-dependent images, providing a new tomographic imaging modality. This becomes extremely useful when both the target and the imaging geometry cannot be moved during the experiment. Figure 6.20 shows amplitude of THz pulses after being focused by a Fresnel lens [9]. It shows an interference pattern due to interference between different frequency components. The dispersion induced aberration leads to a low imaging quality.

The Fresnel lens induced THz wave tomographic imaging setup is similar to the 2D THz wave imaging system discussed in Chapter 3, except that the 30 mm diameter silicon based Fresnel lens replaces the imaging lens. This binary lens has a focal

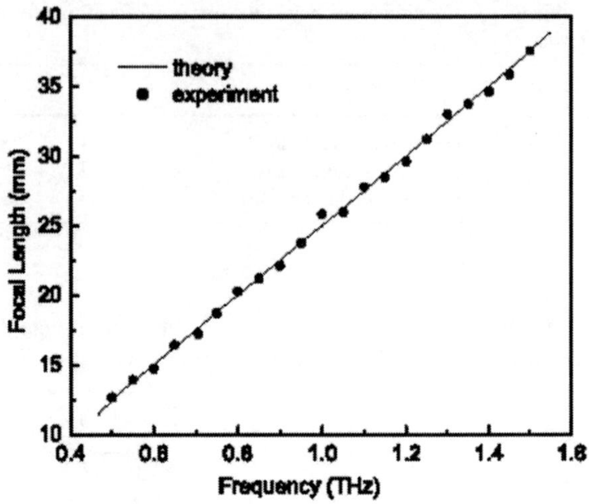

Fig. 6.19 The first order focal length of a Fresnel lens as a function of THz frequency

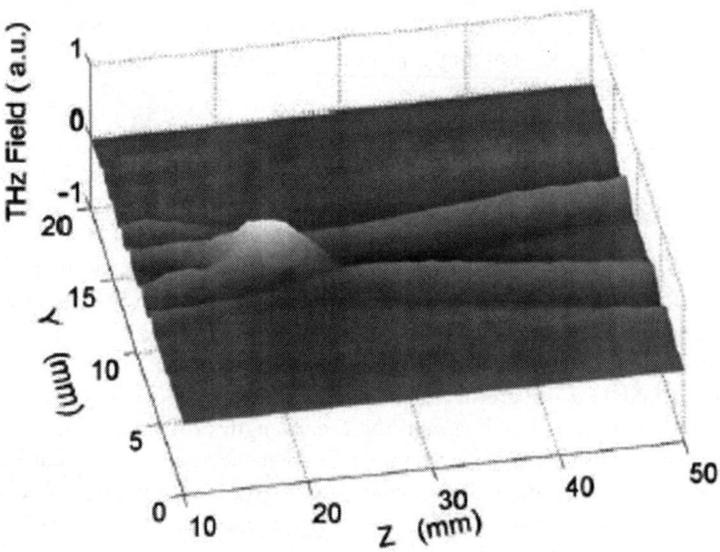

Fig. 6.20 The axial evolution of THz amplitude focused by a Fresnel lens

length of 2.6 cm at 1 THz. By scanning the time delay between the THz and optical probe beam, a temporal waveform of the THz wave at each pixel on the image plane was recorded using a CCD camera. Fourier transformation of the temporal waveforms provides the THz field amplitude (or intensity) distribution on the image plane at each frequency. The measured two-dimensional THz field distribution at each frequency provides images of the THz field transmission of a target at each corresponding position along the z-axis. Figure 6.21a, b, and c show illustration of

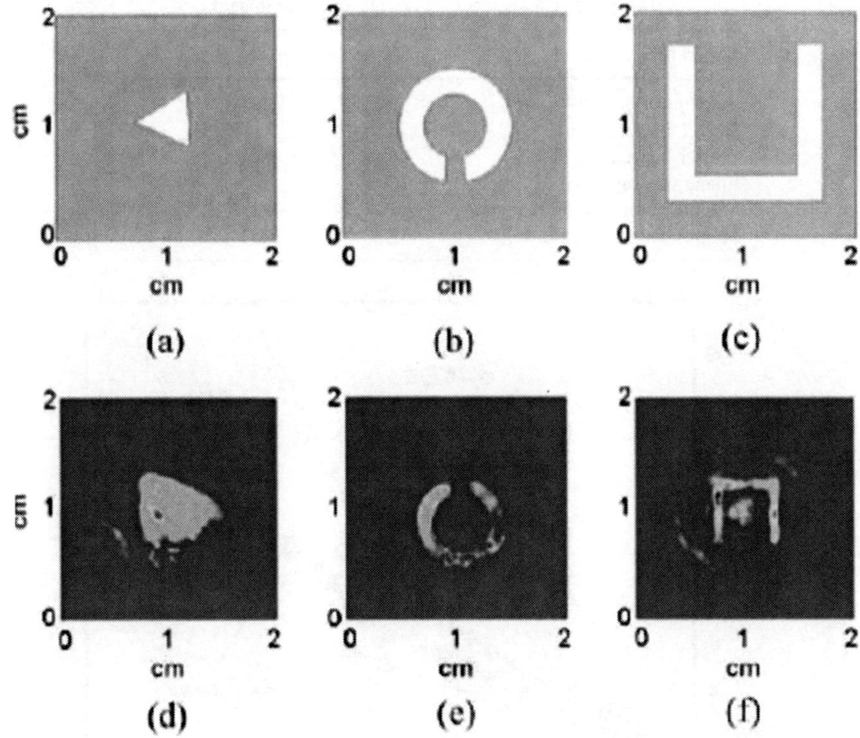

Fig. 6.21 Illustrations of three targets used in Fresnel lens based THz wave tomographic imaging (a), (b), and (c). Their images made with THz wave at different frequencies (d), (e), and (f)

three objects used in the pulsed THz wave tomographic imaging system. The three targets (OU, OC and OT) were made from 2 mm thick polyethylene sheet with the dimensions of 60 by 40 mm with three different shapes carved on these three samples. These targets were placed along the THz beam path with their distances to the lens, corresponding to d_0 in Equation (15), being 3, 7 and 14 cm, respectively. Images of patterns on the sensor plane at distance $d_1 = 6$ cm are measured at frequencies of 0.74, 1.24, and 1.57 THz, respectively. The corresponding images are shown in Fig. 6.21d, e, and f. At each frequency, a Fresnel lens images a different plane section of a target object while images from other depths remain blurred. Each point in the different object planes along the z-axis is mapped onto a corresponding point on the image plane (sensor plane) with the magnification factor $- d_1/d_0$ at their corresponding frequencies. Table 6.2 summarizes the properties of those three objects and their images.

Figure 6.22 shows target distance d_0 as function of TLHz wave frequency for three different image distances of d_1 equal to 4.6, 5.6 and 7.4 cm respectively and shows a strong agreement between the experimental (points) and the measured (curves) values. Lateral spatial resolution of THz wave image created by a

Table 6.2 Characters of 3 targets and their images

Object (Fig. 6.22)	▲	∩	∪
Size (mm)	5	10	14
Object distance (cm)	3	7	14
Carrier frequency (THz)	0.75	1.24	1.57
Image size (mm)	10	8	6
Magnification (experimental)	–2	–0.8	–4.2
Magnification (calculation)	–2	–0.85	–4.3

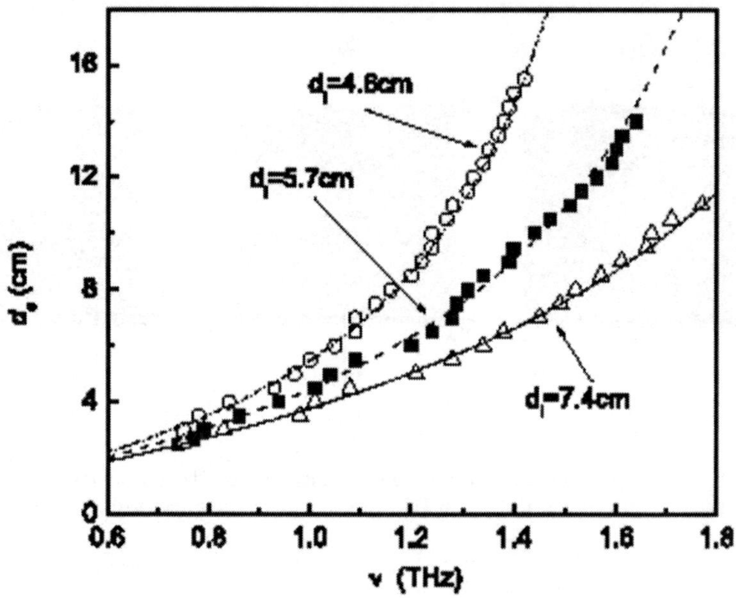

Fig. 6.22 The object distance in Fresnel lens based THz wave tomographic imaging system as functions of THz frequency for three different image distances

Fresnel lens is limited by diffraction of the imaging system. The longitudinal spatial resolution is determined by the spectral resolution of the THz wave detector. The relationship between longitudinal range of the object plane and the bandwidth of THz wave can be derived from Equation (16) as

$$\delta d_0 = \left(\frac{d_0^2}{z_\nu} \right) \frac{\delta \nu}{\nu}. \tag{17}$$

Equation (17) shows that the longitudinal resolution increases with higher THz wave frequency. The longitudinal resolution also quickly decays with the object distance. Figure 6.23 shows longitudinal spatial resolution in Fresnel lens induced tomographic image as a function of THz frequency. The depth of field of the imaging

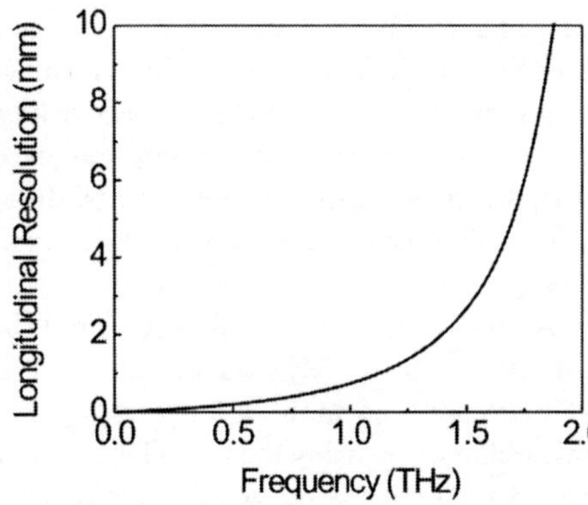

Fig. 6.23 The longitudinal resolution in Fresnel lens based THz wave tomographic imaging system as a function of THz wave frequency (corresponding to the object distance)

system, provided by Equation (21) of Chapter 3 also affects the longitudinal resolution of the system The distance between the two distinguishable planes should be larger than the depth of field for a better resolution.. If the spatial resolution on the imaging plane equals to the Airy disk size and is much smaller than aperture of the lens, depth of field of the imager is defined as

$$L = 2.44\lambda \frac{l^2}{D^2}. \tag{18}$$

Equation (18) indicates that the depth of field is also proportional to square of the object distance and inversely proportional to the carrier wave frequency.

Similar to THz DT, image quality of Fresnel lens induced THz tomographic imaging is also related to frequency of THz wave. The image quality is best between

Fig. 6.24 Photo (a) and cw THz wave CT image of an air freshener

0.5 and 1.25 THz and reduces at higher and lower frequency ends. At the lower frequency end, the lateral resolution is limited by diffraction of the carrier while the longitudinal resolution is limited according to Equation (17). It is also worth noticing that, since Fresnel lens creates image following the imaging formula in Equation (15), the magnification factors will be different for targets with different object distances. This needs to be taken into account in order to correctly retrieve the 3D image.

Above discussions are focused on THz wave tomography techniques utilizing pulsed THz wave. THz wave tomographic images can also be obtained using cw THz. Figure 6.24 shows a tomographic image (displayed in 2D) of an air freshner case obtained by using the CW Gas Laser at 1.63 THz with a power of 180 mW. The case is made of plastic, which is transparent to THz wave. The THz wave tomography image shows both the external pattern of the shell profile and the internal structure of the case.

References

1. S. Wang, and X.-C. Zhang, "Pulsed terahertz tomography," *J. Phys. D: Appl. Phys.* **37**, R1 (2004).
2. S. R. Deans, *The Radon Transform and Some of its Applications*. New York, John Wiley & Sons (1983).
3. B. Ferguson, S. Wang, D. Gray, D. Abbot, and X. C. Zhang, "T-ray computed tomography," *Opt. Lett.* **27**, 1312 (2002).
4. M. Born, and E. Wolf, *Principles of Optics*. Cambridge, Cambridge University Press (1997).
5. P. M. Morse, and H. Feshbach, *Methods of Theoretical Physics*. New York, McGraw-Hill (1953).
6. J. C. Russ, *The Image Processing Handbook*. Michigan, CRC Press (1992).
7. D. N. Bhat, and S. Nayar, "Ordinal measures for image correspondence," *IEEE Trans. Pattern Anal. Mach. Intell.* **20**, 415 (1998).
8. G. B. Jin, Y. B. Yang, and M. X. Wu, *Binary Optics*. Beijing, National Defense Industry Press (1998).
9. S. Wang, T. Yuan, E. D. Walsby, R. J. Blaikie, S. M. Durbin, D. R. S. Cumming, J. Xu, and X. C. Zhang, "Characterization of T-ray binary lenses," *Opt. Lett.* **27**, 1183 (2002).

Chapter 7
THz Wave Near-Field Imaging

THz waves offer innovative imaging and sensing capabilities for applications in material characterization, microelectronics, medical diagnosis, environmental control, and chemical and biological identification. However, the spatial resolution of conventional THz imaging technique is limited by diffraction of THz waves to be in the same order as THz wavelength (1 THz = 300 μm). This diffraction limit is an obstacle for using THz technology in probing the electronic and optical properties of semiconductor and bimolecular nanostructures. Several approaches have been used to obtain a sub-wavelength spatial resolution based on near-field techniques. One way to overcome diffraction is to use a sub-wavelength size aperture to limit the detection or generation area. This technique is known as apertured THz wave near-field microscopy. The aperture could be a static aperture made on a metallic screen or a dynamic one excited by an optical beam. Localized THz wave emitter or sensor based on real or virtual instant photocurrent excited by a highly focused optical beam can also provide spatial resolution much finer than THz wavelength. Another way, called apertureless THz near-field microscopy, use a sharp tip as local field enhancer which scatters the evanescent light in the near-field region of the target to make it detectable in the far field, and provide a spatial resolution well below the diffraction limit. Last but not least, THz wave emission microscope based on STM technique can achieve a nanometer resolution. A pulsed laser is used to generate photo-carriers on the semiconductor surface and a biased scanning-tunneling-microscope (STM) needle is used to modulate the localized electric field in the Schottky barrier under the tip. The transient photo-carriers driven by the modulated field emit THz waves, which can be detected at the modulated frequency in the far field. THz wave near-filed microscopy described above represents a milestone toward THz wave spectroscopic imaging of materials and devices at nanometer, sub-nanometer, and even atomic scales.

Spatial Resolution in Near-Field Imaging

The object distance in a traditional optical imaging system is much longer than the optical wavelength and called far field imaging, as shown schematically in

DOI 10.1007/978-1-4419-0978-7_7, © Springer Science+Business Media, LLC 2010

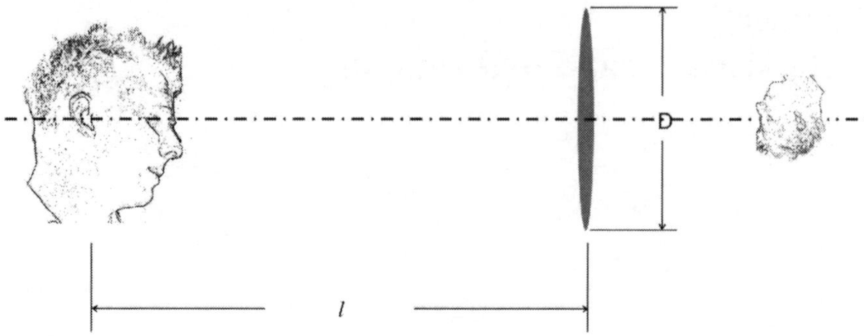

Fig. 7.1 Far field imaging. Its spatial resolution is limited by the optical wavelength and aperture of the imaging lens

Fig. 7.1. Spatial resolution of far field imaging is limited by diffraction of the optical wave. In most imaging systems it is the numerical aperture of the employed lens that determines the image resolution. The spatial resolution of a diffraction limited system is

$$\Delta = 1.22\lambda\frac{l}{D}, \tag{1}$$

where λ is the optical wavelength, l is the object distance and D is the diameter of the imaging lens. Equation (1) indicates that the spatial resolution of a far field imaging system cannot be much higher than the optical wavelength. The spatial resolution of THz wave far field imaging is limited to a sub-millimeter range by the THz wave wavelength and is inadequate for some applications. For instance, the dimension of many biological cells is in micron or sub micron scale; size of microstructures in semiconductor devices could be even smaller. In order to investigate spectral features of biomolecules inside cells or carrier dynamics in semiconductor devices using THz wave spectroscopic imaging, the desired spatial resolution is much shorter than THz wavelength.

The spatial resolution can be also understood in the frequency domain. Wave number component of the carrier wave along a certain direction must be larger than the spatial frequency of the target along the same direction in order to resolve the target. This condition is described as $k_{//} \geq 2\pi/a$, where $k_{//}$ denotes the carrier wave number component parallel to surface of the target, and a is size of the target. Wave number of the carrier wave is given by $k = 2\pi/\lambda$. As a result, the wave number component along the direction perpendicular to the target plane is described as $k_\perp^2 = k^2 - k_{//}^2 = 2\pi\sqrt{a^2 - \lambda^2}/(a\lambda)$. If $a < \lambda$ is satisfied, which indicates a sub-wavelength spatial resolution, the perpendicular wave number component is imaginary. Propagation of the carrier wave along the perpendicular direction is

$$E = E_0 e^{ik_\perp z} = E_0 e^{-|k_\perp|z}. \tag{2}$$

Equation (2) indicates that, when a carrier wave encounters a sub-wavelength object along one direction, the electric field exponentially decays along the perpendicular direction. Most energy is confined within a region close to the structure instead of propagation into the far field. To obtain sub-wavelength spatial resolution, one needs to utilize those non propagation waves and detect them within the near-field distance. In near-field imaging, one usually uses a sub-wavelength scatter, either an aperture or a tip, to couple the non propagation wave. In this case the spatial resolution is not limited by Equation (1) but by dimension of the sub-wavelength scatter.

Scattering of a plane wave by a spherical particle has been rigorously discussed by Mie and others. Scattering of an EM wave by an irregular object with a dimension similar to or larger than the wavelength can also be approximated solved by Mie's scattering solution. When the scatter size is much smaller than the carrier wavelength, the scattered electric field can be expressed by Rayleigh scattering:

$$E,H \propto \frac{a^3}{\lambda^2}\frac{e^{ikr}}{r}. \tag{3}$$

Here a is radius of the spherical particle, r is the distance between a field point to center of the sphere. Equation (3) shows that in far field, amplitude of the scattering EM field is proportional to cubic of the scatter dimension, and is inversely proportional to square of the wavelength. Since dimension of the scatter is much smaller than the carrier wavelength, the scattering cross section is very low.

A typical problem for EM wave interaction with target of sub-wavelength size is EM wave propagation through a small aperture. When the aperture is made on an ideal metal film, which is infinity thin, transmission of EM wave through the small aperture is [1]

$$\begin{aligned} t &= 1 + J_1(2\,ka)/ka - (1/ka)\int_0^{2\,ka} J_0(t)dt \\ &\cong 1, \;\; ka >> 1 \\ &\cong (ka)^2/6 \;\; ka << 1 \end{aligned} , \tag{4}$$

where $J_1(x)$ and $J_0(x)$ are the first and zeroth order Bessel function, respectively. Considering the area of aperture is proportional to a^2, the overall amplitude transmission also has a format of a^3/λ^2, where a^2/λ^2 is contributed by the transmission in Equation (4) and the additional a is contributed by the aperture area. The low scattering coefficient of a sub-wavelength scatter can also be understood from diffraction. The angular distribution of diffraction broadens when diameter of the aperture decreases. The diffraction angle is larger than π when radius of the aperture becomes much smaller than the optical wavelength. Since only diffraction components with diffraction angle smaller than π can be forward propagating, aperture smaller than wavelength prevents larger portion of diffraction wave from detectable in the far field. Figure 7.2 shows schematic of angular distribution of EM wave diffracted by an aperture with sub-wavelength radius. The diffraction angle is approximately inversely proportional to radius of the aperture. This leads to a

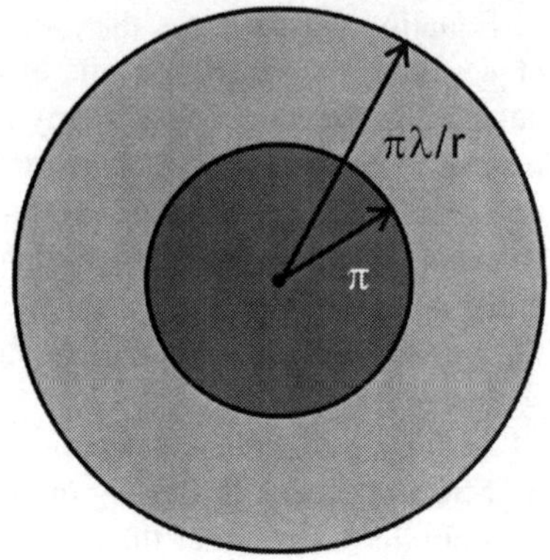

Fig. 7.2 Angular distribution of diffraction components from a sub-wavelength aperture. The *outer circle* indicates angular distribution of the entire diffraction components and inner circle indicates the propagation components

transmission coefficient through the aperture which is proportional to a^2/λ^2. Further considering the aperture area, the THz field amplitude transmission is proportional to a^3/λ^2.

When EM wave is diffracted (scattered) by a sub-wavelength target, the electric field can be categorized into two parts: the far field propagation wave and the evanescent wave, which does not propagate. Electric field of the far field propagation wave is proportional to cubic of the target size. Since its wave number in the lateral direction is smaller than spatial frequency of the target, it does not carry the spatial information. On the other hand, the evanescent wave, as described in Equation (2) carries the spatial resolving information; however its electric field exponentially decays with the distance from the target. As a result, the evanescent wave is not detectable in the far field. When $a \ll \lambda$ is satisfied, the decay constant of the evanescent wave can be extracted from Equation (2) as $d = 2\pi/a$. The range within such feature distance is considered the near-field. Figure 7.3 shows two configurations for near-field optical microscopy methods, which can be considered as near-field excitation mode and near-field probing mode. In the former configuration, the incident optical beam is limited by a sub wavelength aperture, which could be a metal screen with a pin hole, a taped fiber or a waveguide tip. The optical wave transmits through such a sub-wavelength aperture and generates evanescent wave. The target locates within the near-field range from the aperture and scatters the evanescent wave to the far field for detection. In the later configuration, the incident beam is directly illuminated on the target. The scattering light by sub-wavelength structures on the target forms evanescent wave around it. One can either place the detector at the near-field region for direct detection or pick up the evanescent wave by an aperture within the near-field region for far field detection.

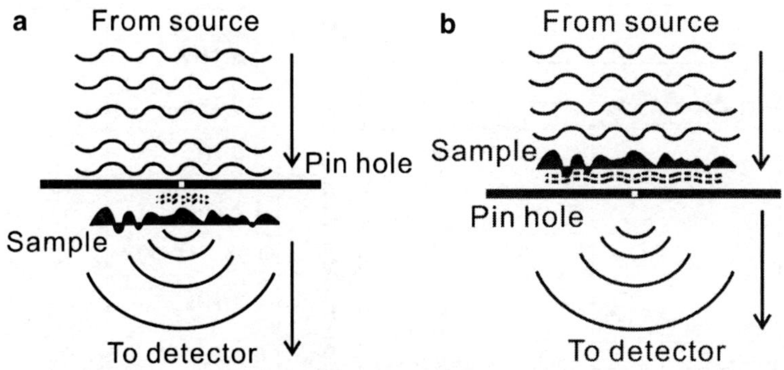

Fig. 7.3 Two basic near-field imaging configurations. (**a**) An aperture is placed just in front of target. (**b**) An aperture is placed just after the target

Apertured THz Near-Field Imaging

THz near-field imaging can be realized by limiting the incident THz wave using a sub-wavelength pin hole [2–3, 5–6] on a metallic screen as presented in Fig. 7.4. Here the THz wave detector is a GaAs wafer based photoconductive dipole antenna. A layer of metal film is coated on back side of the wafer, and a sub-wavelength pin hole is fabricated on the metal film opposite to the dipole antenna. To enhance THz wave coupling through the sub wavelength pin hole, the GaAs substrate extrudes into the pin hole. The target is placed very close to the pin hole. When the target is illuminated by THz wave, its microstructures scatter THz wave and form evanescent waves, which contains spatial information. When the pin hole is placed within near-field region of the target, it couples the evanescent waves through the pin hole and the electric field is detected by the dipole antenna locating at opposite side of the

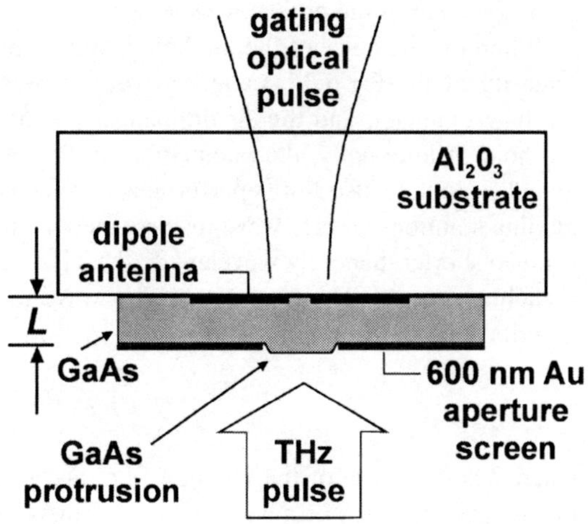

Fig. 7.4 Experimental setup of THz wave near-field imaging by a metallic sub-wavelength aperture (Courtesy of Dr. Mitrofanov)

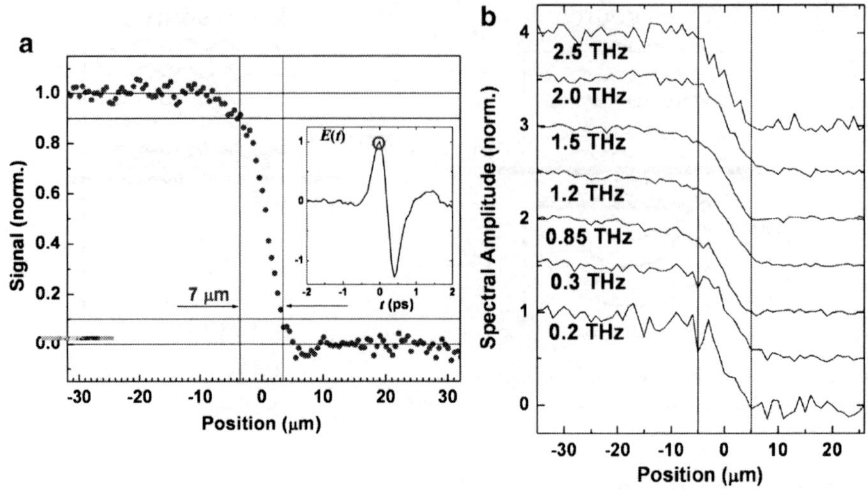

Fig. 7.5 Spatial resolution of THz wave near-field imaging setup in Fig. 7.7. *Left*, THz field as a function of scanning distance when a THz wave detector is placed after a 5 μm diameter pin hole scanning across a metal semiconductor boundary. Inset gives the measured THz waveform. *Right*, spatial resolution of THz wave near-field imaging with different frequency components when 10 μm diameter pin hole is used (Courtesy of Dr. Mitrofanov)

GaAs wafer. Scanning the pin hole across the target achieves imaging of the target with a spatial resolution limited by size of the pin hole. Figure 7.5a shows the THz field amplitude when scanning such a near-field imager cross a metal/semiconductor boundary, where the diameter of pin hole is 5 μm. A lateral spatial resolution of 7 μm was obtained, which is much shorter than THz wavelength. Figure 7.5b shows the THz field transition across a metal/semiconductor boundary for THz waves with different frequencies, when the pin hole diameter is 10 μm. The result indicates that spatial resolution of the THz wave near-field imager is limited by the pin hole and is independent of the optical wavelength.

When the diameter of the pin hole is much smaller than the optical wavelength, intensity of the far field scattering wave is proportion to the sixth power of the pin hole diameter, and thus is dramatically reduced with the shrinking size of the pin hole. Additionally, the attenuation of THz wave when propagating inside the metallic pin hole also limits performance of the near-field imager. Since there is no guiding solution for THz wave inside a metallic tunnel when diameter of the tunnel is much shorter than THz wavelength, the THz wave pass through the tunnel via the tunneling process. The transmitted THz wave intensity in the tunneling process is described as

$$I_t = I_0 e^{-\pi l/d}, \tag{5}$$

where l is thickness of the metallic film, and d is diameter of the pin hole. Since the metallic film is not made by an ideal metal, the THz wave can penetrate into the metal for a certain depth. Intensity of THz wave inside the metal also shows an exponential decay format:

$$I_m = I_0 e^{-l/a}. \tag{6}$$

Here a is penetration depth of the THz wave in this metal. Combining Equations (5) and (6) one has the smallest possible pin hole which can be made on that metal film.

$$d = a\pi. \tag{7}$$

When a pin hole even smaller than described by Equation (7) is made on a metal film, attenuation of the pin hole is comparable or even larger than attenuation by the metal itself and thus THz wave propagation cannot be limited by the aperture. The typical penetration depth for THz wave through metal is in the order of 100s nm. This makes it hard to achieve spatial resolution better than 100s nm by using a sub-wavelength aperture alone.

Use of small pin hole not only affects the contrast between the open area and the metal area, but also limits the overall dynamic range in the measurement. One way to improve dynamic range is to reduce thickness of the metal film. However, one cannot unlimitedly reduce thickness of the metallic film to lower than the penetration depth; otherwise THz wave will transmit through metal film. A practical thickness of the metal film is in the same order as its penetration depth. As a result, the aperture dimension is also limited by the dynamic range of the imaging system. This limitation is presented as

$$d \geq \frac{\pi l}{\ln(D/D')}, \tag{8}$$

where D denotes measurement dynamic range of the imaging system without pin hole and D' is the required dynamic range in order to present an acceptable image.

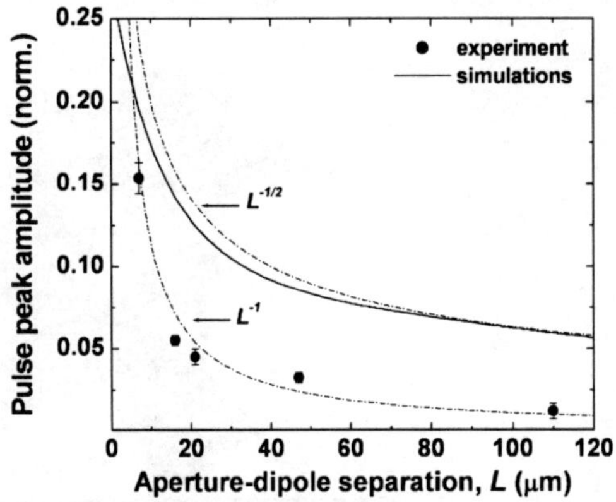

Fig. 7.6 Detected THz field as a function of the distance between the sub-wavelength aperture and the antenna (Courtesy of Dr. Mitrofanov)

The THz field of waves reaching to the detector is also affected by the distance between the antenna and the pin hole. Figure 7.6 shows the THz field as a function of distance between the metallic film and the THz wave detector. Figure 7.6 shows that THz field is reversely proportional to distance between the pin hole and the detector. Use of thin GaAs crystal shortens the distance and increases detection sensitivity of the imager. According to Equation (8) it allows to use smaller pin hole for higher spatial resolution.

THz Near-Field Imaging with a Dynamic Aperture

Besides the real aperture made on a metallic screen, the size of THz field can also be limited by a sub-wavelength aperture which is generated by an excitation optical beam [7, 8]. Such an aperture is controlled (generated, removed, or sized) by the excitation optical beam, and thus is named a dynamic aperture. In THz wave near-field microscopy, such a dynamic aperture can be excited by focusing a laser beam on a semi-insulating semiconductor wafer. Figure 7.7 shows a schematic concept of a THz wave near-field imaging using dynamic aperture. This system is very similar to a traditional raster scanning type THz wave imaging system. While the sample stander is made by a piece of semi-insulating GaAs wafer. The target is a thin chip which is attached on one side of the GaAs wafer. Besides a THz wave, an excitation laser beam, called the sampling beam, is focused on the GaAs wafer. Without the sampling beam, the GaAs wafer is transparent for THz wave. When the sampling beam is illuminated on the GaAs wafer, it generates free carriers in the wafer and induces a local conductive zone. Transmission of THz beam from this localized conductive zone is reduced due to exist of photo carriers. As a result, the sampling beam generates a sub THz wavelength size aperture on the GaAs wafer. Transmission of the aperture is lower than its surrounding area, thus the aperture is a negative aperture. The sampling beam is modulated by an optical chopper,

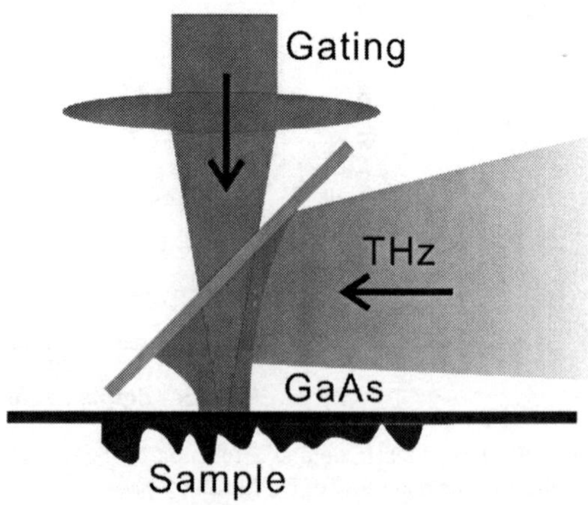

Fig. 7.7 Experimental setup of THz wave near-field imaging with a dynamic aperture

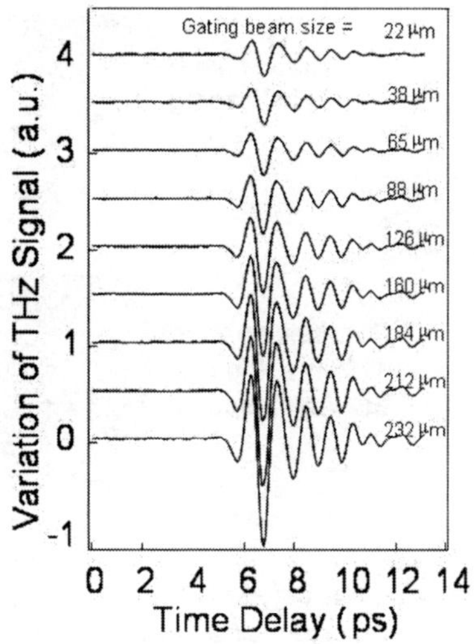

Fig. 7.8 Waveform of THz pulses "passing through" dynamic apertures of different diameters

whose frequency is used as the reference of a lock-in amplifier and only THz field modulated by the aperture is recorded in the lock-in amplifier. Figure 7.8 shows THz temporal waveforms after "passing through" dynamic apertures with different diameters. The transmission of THz wave decreases with decreasing dynamic aperture, while the temporal waveform remains the same. Fourier transform of those waveforms shows that the transmission spectrum is independent of the aperture size. THz waveform is clearly observed even if the size of dynamic aperture is smaller than 1/10th of THz wavelength (the central wavelength was 333 μm in the experiment).

When a target is closely attached onto a thin GaAs wafer, it locates within the near-field region of the dynamic aperture. As a result, the dynamic aperture can result in sub-wavelength spatial resolution by utilizing the evanescent waves. Figure 7.9 gives THz wave images of two targets. Comparing the images of the first sample obtained by near-field imaging and far field imaging, one can clearly see that the dynamic aperture technique dramatically increases the spatial resolution of the image. If the target is placed at the back side of the GaAs wafer, the spatial resolution will be reduced due to larger distance between the target and the sub-wavelength aperture. In principle the spatial resolution of dynamic aperture induced near-field image is only limited by focal spot size of the sampling beam, which could be sub micron. However, smaller aperture will reduce detection sensitivity. Additionally, one needs to focus the sampling beam and at the same time avoid blocking the incident THz wave. This presents challenge to using a focal lens with very high numerical aperture. As a result, the best spatial resolution obtained in experiment is 14 μm.

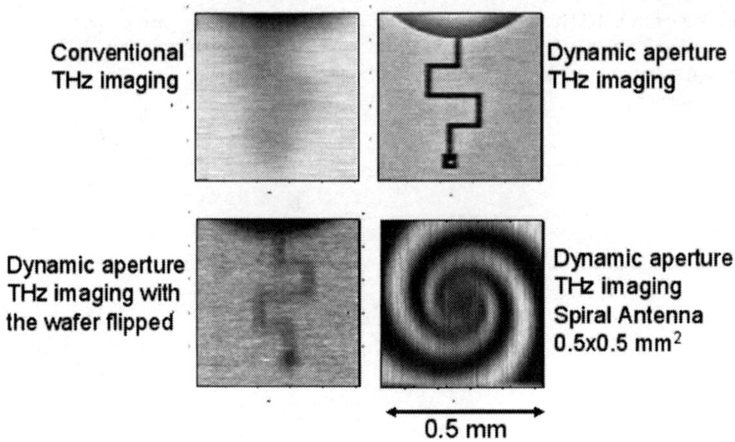

Fig. 7.9 THz wave images of two targets. *Up left*, far field THz wave image for target #1; *up right*, near-field image of target #1 when it locates at front surface (facing to the sampling beam) of the GaAs wafer; *bottom left*, near-field image of target #1 when it locates at back surface of the GaAs wafer; *bottom right*, THz wave near-field image of the target #2

THz Near-Field Imaging with Small Emitter or Detector

THz wave near-field microscopy induced by a sub-wavelength aperture provides spatial resolution much smaller than THz wavelength, while dynamic range of the imaging system suffers low throughput. Pulsed THz radiation is generated by exciting semiconductor or nonlinear materials using ultrafast laser beams at near infrared frequency band, which can be focused to sub-micron size. Apply such a sub-wavelength emitter close to the target maps the target with a spatial resolution limited by the emitter. Or if the target itself generates THz wave with optical excitation, localized emitters directly maps profile of the target. One example is THz wave generated through optical rectification process, where size of THz wave emitter is limited by spot size of the excitation optical beam in the electro-optical crystal [9, 10]. THz field generated via optical rectification process is proportional to power density of the excitation optical beam. Due to multi photon absorption and photo-carrier screening, THz wave generation saturates when a very high excitation density is applied. Further increase of excitation power may damage the EO crystal. The saturation of THz wave generation limits the excitation power density in the optical rectification emitters. On the other hand, diffraction of THz source also affects THz field received by the detector in THz wave near-field imaging system. The affect is more severe when the emitter size is much shorter than THz wavelength. As presented in Equation (4), diffraction of the THz source limits forward propagation of generated THz radiation. When optical rectification occurs within a region much smaller than THz wavelength, THz wave intensity in the far field obeys the following equation:

$$P_{\mathrm{THz}} \propto I_0^2 r^2. \tag{9}$$

Equation (9) indicates that the far field THz power is proportional to area of the emitter when the excitation laser intensity is fixed. The far field THz power as a function of the emitter size in the optical rectification process is presented in Fig. 7.10. The relationship between the THz emitter size and the far field THz power has three regimes. When the emitter size is much larger than THz wavelength, the far field THz power is inversely proportional to the emitter area due to higher excitation intensity for smaller emitter. When the size of emitter is much smaller than THz wavelength, the far field THz power proportional to the emitter area following Equation (9). Between those two extremes, THz wave generation has the highest conversion coefficient.

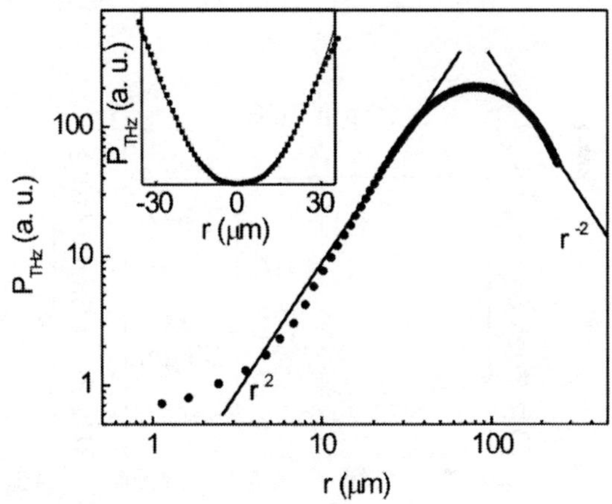

Fig. 7.10 THz wave far field emission power as a function of laser focal spot size

According to Equation (9), the far field THz power is proportional to the emitter area, which is less affected by reducing the emitter size comparing to THz wave transmission through a sub-wavelength aperture, where the far field THz power is proportional to cubic of the aperture area. In addition, THz wave emitted from a sub-wavelength emitter suffers the tunneling lose through a sub-wavelength hole with finite thickness. Figure 7.11 shows the concept of THz near-field imaging via optical rectification. An excitation laser beam is focused into a thin EO crystal to generate THz wave. The target is attached on the backside of the EO crystal, which is within the near-field region of THz emitter. THz wave transmitted through the target is collected by an off-axis parabolic mirror and then focused on to a THz wave detector. THz waveform is recorded when scanning excitation spot across that target, and thus obtains THz wave image of the target. Spatial resolution of the THz wave image is limited by size of the THz wave emitter, which could be sub THz wavelength. Figure 7.12 shows THz peak field when scanning across the boundary of a metallic film. Spatial resolution in those images is about 20–30 μm. To get higher spatial resolution, one need to focus the excitation laser beam to smaller focal spot size and use thinner THz emitter. A thin emitter is necessary to make sure that

Fig. 7.11 Schematic of THz wave near-field imaging using localized THz emitter via optical rectification process

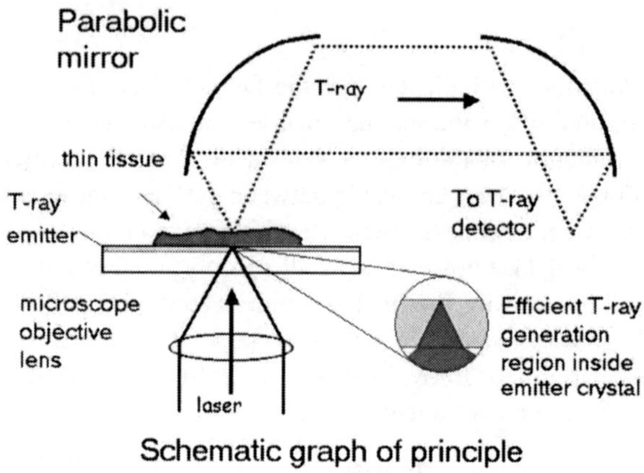

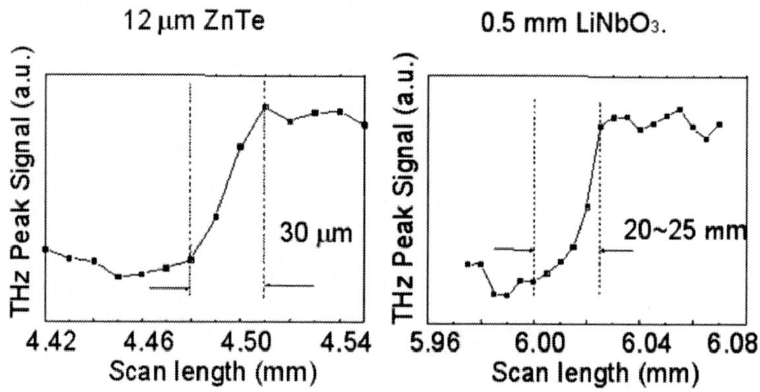

Fig. 7.12 Spatial resolution of THz wave near-field imaging in Fig. 7.12 using ZnTe crystal (*left curve*) and $LiNbO_3$ crystal (*right curve*) as the emitter

the beam size is small throughout the emitter the target is within near-field range of the emitter. To generate a thin THz source, one can either use a thin EO crystal, such as ZnTe crystal, or on the other hand, one can use an EO crystal short phase matching length to form a thin THz source in a thick EO crystal, such as $LiNbO_3$ crystal. Usually the thickness of THz source should not be larger than its diameter. An alternative way to solve this problem is to use optical rectification process with resonant enhancement such as using GaAs or InAs crystal. The resonant absorption limits the thickness of the THz source due to the short absorption depth; it also enhances the EO coefficient due the resonant enhancement. If the target itself, such as semiconductor devices, can generate THz wave with optical excitation, one can generate localized emitter on the target and image the target using the localized emitter. The spatial resolution is determined by size of the localized emitter, i.e. the focal spot size of the excitation laser beam. Figure 7.13 shows THz emission microscopy images of two biased IC chips [11]. Since a damaged chip (b) has different electric

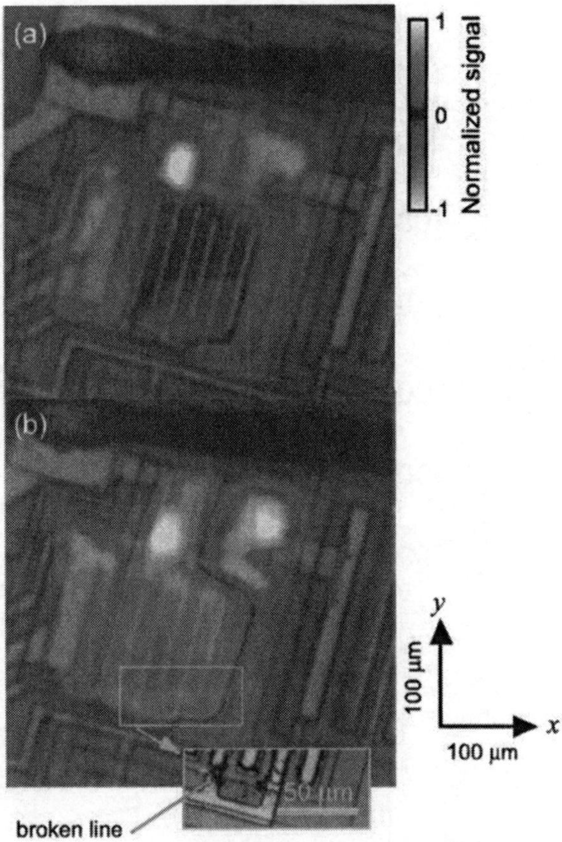

Fig. 7.13 THz wave emission microscopy images of two IC circuits. (**a**) a normal chip and (**b**) a damaged chip with a broken wire (Courtesy of Dr. Kawase)

field distribution comparing to a normal chip (a), the localized THz wave emitter generates THz pulses with different waveform. Therefore it allows THz emission microscopy to distinguish damaged chip from normal ones.

Sub-wavelength spatial resolution can also be obtained by a localized THz wave detector [12, 13]. When detecting THz wave through EO sampling process, THz field is recorded by modulating polarization of the probe beam in an EO crystal. Similar to a localized THz wave emitter, one can focus the probe beam to form a localized THz wave detector to sub THz wavelength size. By placing the target within near-field distance to that localized detector one can record THz wave image with spatial resolution determined by size of the THz wave detector. Figure 7.14 shows THz wave image of a pin hole on a metallic film taken with a localized THz wave detector [13]. Chapter 3 introduced 2D imaging technology, which records 2D distribution of THz wave using an extended probe beam simultaneously. A similar technique can be used to record a 2D image for detector size limited near-field imaging. Figure 7.15 presents a 2D near-field imaging system. The THz wave detector in such a system is a thin EO crystal. The target is attached onto the detector, THz

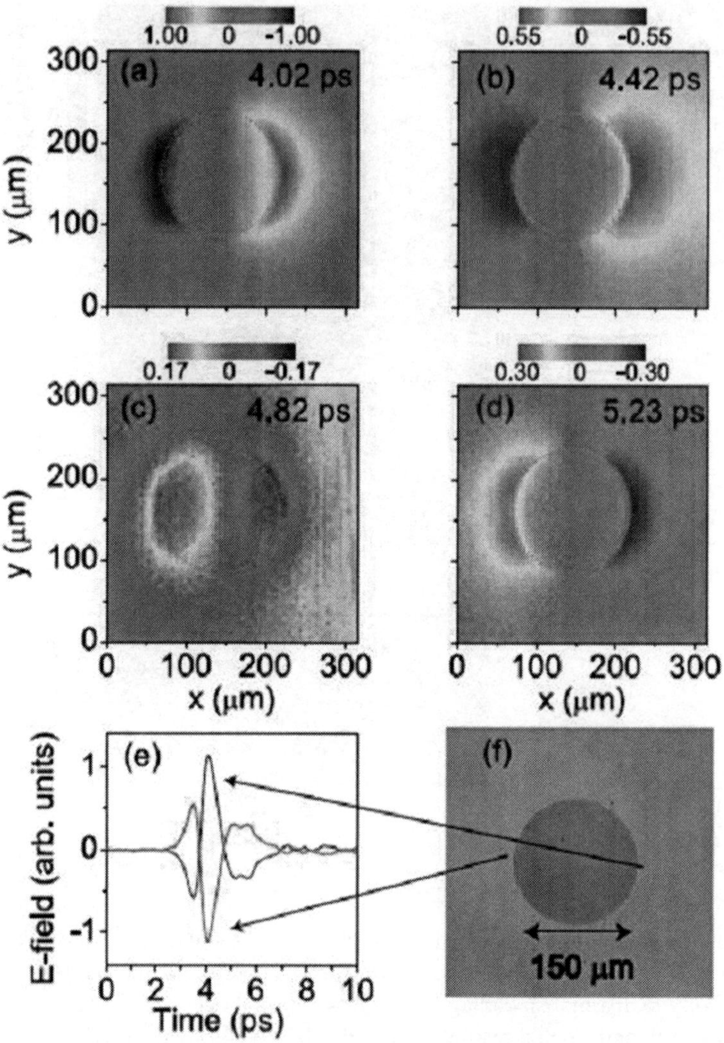

Fig. 7.14 THz wave near-field images of a pin hole on a metallic film with THz field at different time delays (**a**–**d**). THz waveforms at different spot of the pin hole are illustrated in (**e**) and (**f**). (Courtesy of Dr. Planken)

wave transmitted through or scattered by the target propagates in the EO crystal. Collimated probe beam with linear polarization is illuminated into the EO crystal from the opposite direction and it is reflected by the front surface of the crystal. Reflected probe beam is modulated by the THz wave including those evanescent waves with spatial resolution information. The modulated probe beam is recorded using a CCD camera after passing through the polarization analyzer. Since the probe beam is near infrared wave which has much shorter wavelength than THz wave,

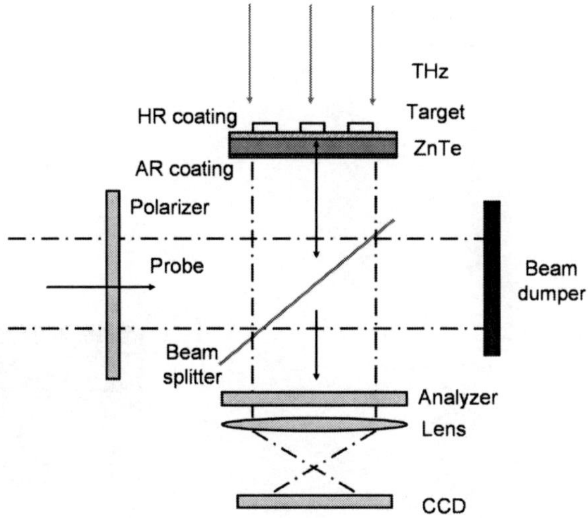

Fig. 7.15 Concept of 2D THz wave near-field imaging using EO sampling

thus the spatial resolution in the CCD camera could be much finer than THz wavelength. Using this method one can achieve both sub-wavelength resolution and high imaging speed.

THz Near-Field Imaging by Tip Scattering

Due to throughput limitation, it is difficult to realize near-field imaging by a sub-wavelength aperture with a spatial resolution shorter than a hundredth of the optical wavelength. In order to avoid the throughput limitation by the sub-wavelength aperture, one can use a sub-wavelength tip to locally influence the interaction between the EM wave and the target, and thus obtain a spatial resolution in the imaging process better than the optical wavelength. Figure 7.16 illustrates the concept of a metallic tip coupling with EM wave when it is close to the target surface. The tip can be approximately considered as a metallic sphere with the same diameter when the tip is very close to the target surface. Interaction among the metallic sphere, the target and the EM wave can be treated as the metallic sphere and its image within the target surface interaction with the EM wave. The local EM field is enhanced by the dipole formed by the sphere and its image. If steady field approximation is used, the propagation component can be ignored. When the incident EM field is perpendicular to the surface of the target, the effective polarization is [14]

$$\alpha_{\perp}{}^{\mathrm{eff}} = \frac{\alpha(1+\beta)}{1-\frac{\alpha\beta}{16\pi r^3}}, \tag{10}$$

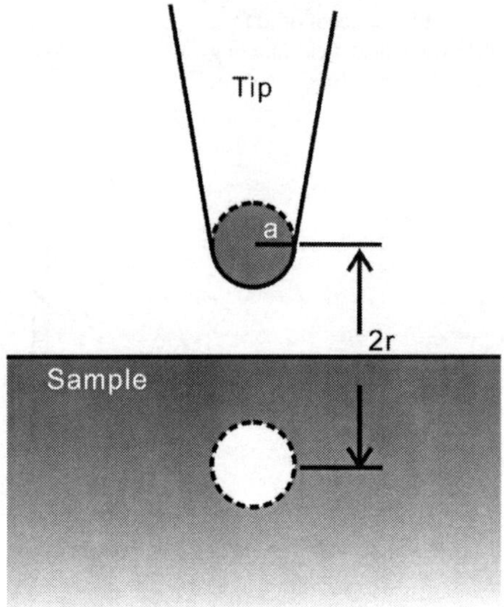

Fig. 7.16 Metallic tip coupling with the target

where $\alpha = 4\pi a^3(\varepsilon_p - 1)/(\varepsilon_p + 2)$, $\beta = (\varepsilon_p - 1)/(\varepsilon_p + 1)$ and r is the distance between tip and target. Here ε_p is complex dielectric constant of the tip material. When the electric field is parallel to the sample surface, one has [6]

$$\alpha_{//}^{\text{eff}} = \frac{\alpha(1-\beta)}{1 - \frac{\alpha\beta}{32\pi r^3}}. \tag{11}$$

The tip/sample absorption and scattering of the incident field can be extracted by Mie's scattering theory. When the carrier wavelength is much larger than diameter of the tip, the scattering and absorption cross section are [14]

$$\begin{aligned} C_{\text{sca}} &= \frac{k^4}{6\pi}\left|\alpha^{\text{eff}}\right|^2 \\ C_{\text{abs}} &= k\text{Im}(\alpha^{\text{eff}}) \end{aligned}. \tag{12}$$

When $\alpha\beta/16\pi r^3 << 1$ is satisfied, Equation (12) is approximated to be $C_{\text{sca}} \propto a^6/\lambda^4$, which consists of the scatting cross section of a sub-wavelength sphere described in Equation (3), and $C_{\text{abs}} \propto a^3/\lambda$. According to Equations (10) and (11), the effective polarization is stronger when electric field of the incident wave is perpendicular to the sample surface. Since no aperture is associated, the tip induced near-field imaging avoids suffering from the transmission loss through a sub-wavelength aperture.

The apertureless near-field technique can be adapted to THz waves. Figure 7.17 shows an experimental setup of THz wave near-field imaging system utilizing a metallic tip to scatter THz wave close to the target [15]. A copper tip with a 5-μm

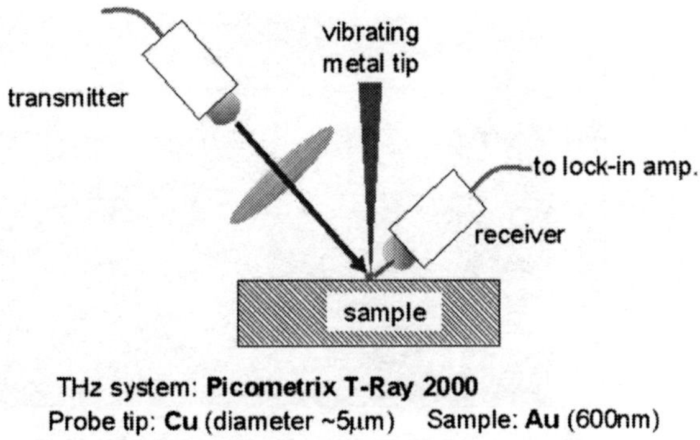

Fig. 7.17 Setup of apertureless THz wave near-field imaging with a metallic tip (Courtesy of Dr. Mittleman)

diameter is hanging above the target, and its apex is very close to surface of the target. THz wave is focused onto the target beneath the tip. The THz wave scattered by the tip is collected and fed into a THz wave detector. The electric field perpendicular to the target surface is described as [15]

$$E_{\perp} \propto \frac{\mu_0 c^2}{4\pi} \frac{p}{r^3}, \tag{13}$$

under near-field condition. Here p is the polarization formed by coupling between the tip and the target, and r is radius of the tip. Equation (13) indicates that the near-field electric field is proportional to polarization of the system. While on the other hand the far field THz radiation is proportional to second-order temporal derivative of the polarization. Figure 7.18 compares near-field THz waveform and integral of

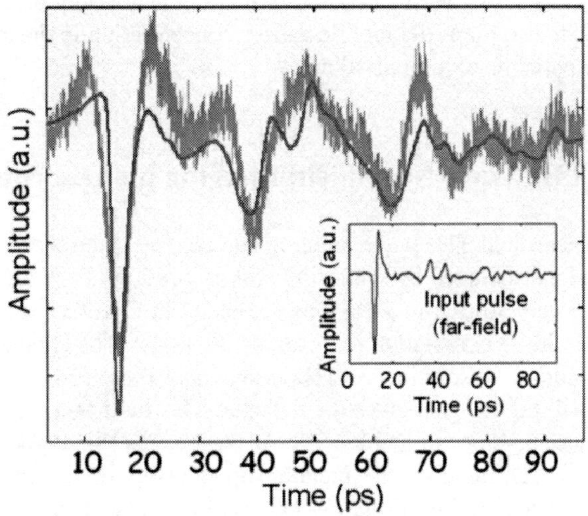

Fig. 7.18 Comparing of THz wave temporal waveforms for scattering wave and integral of the far field THz waveform. Inset gives THz waveform detected in far field (Courtesy of Dr. Mittleman)

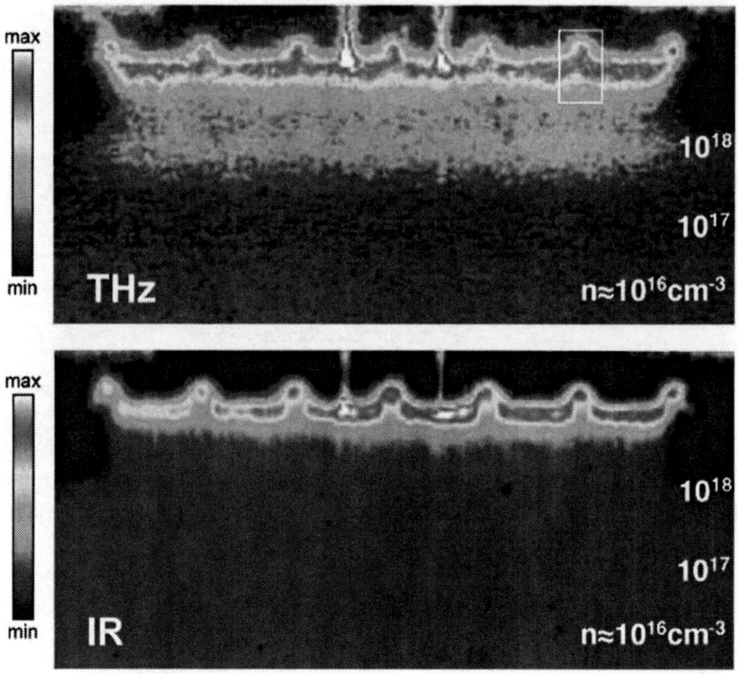

Fig. 7.19 THz wave near-field image (*upper image*) and IR image (*lower image*) of a multi-transistor structure (Courtesy of Dr. Keilmann)

THz waveform measured in the far field. They have similar waveforms. Figure 7.19 shows a THz wave near-field image of a multiple-transistor device structure using a CW THz laser with 118 μm wavelength [17]. Comparing to the near-field imaging using an IR wave, THz wave imaging is not only presents the profile of transistors but also maps the mobile carrier concentration in the target. Experiment indicates a spatial resolution of 40 nm.

THz Wave Near-Field Imaging by Absorption in Metallic Tip

Near-field THz wave imaging can also be taken by recording the reduction of THz signal induced by scattering and absorption of the metallic tip. The imaging setup is very similar to what was presented in Fig. 7.17, except that the detector detects entire THz radiation reflected by the target. The coupling between tip and the target induces absorption of THz waves and thus leads to modulation of the THz signal. Figure 7.20 gives the detected THz field as a function of the distance between the metallic tip and surface of the target [18]. It shows that modulation depth of the THz field by the metallic tip is about 10^{-3}, which is much higher than what is calculated by Equation (12). On the other hand the modulation depth does not

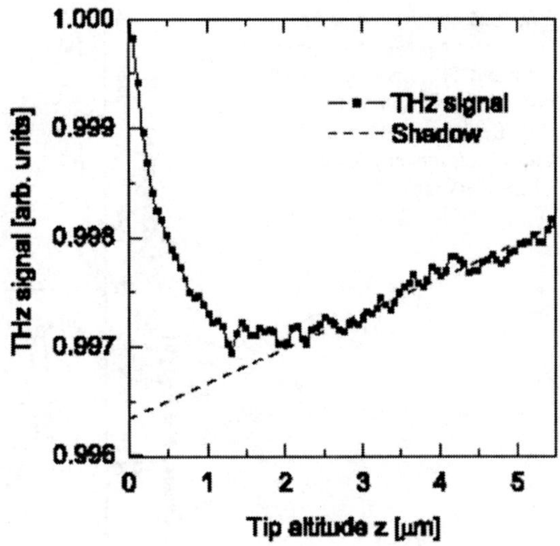

Fig. 7.20 Detected THz field as a function of the distance between tip and sample. *Dashed curve* indicates the signal reduction caused by shielding of the tip (Courtesy of Dr. Kersting)

monotonically increase when decreasing of the distance between the tip and the target. The modulation increases with the distance within the initial 2 μm. This unexpected phenomenon can be explained by the resonant absorption of THz waves by tip/target circuit. The tip/target circuit can be considered as a circuit consisting of resistors, inducers and capacitors. The current induced by the incident EM wave in the circuit is [18]

$$I = \frac{E(\frac{D}{2})}{2\sqrt{R^2 + \left(\omega L - \frac{1}{\omega C}\right)^2}}, \tag{14}$$

where D is diameter of the THz focal spot, ω is circular frequency of the incident wave, R, C, and L are resistance, capacitance and inductance of the circuit, respectively. The power loss induced by the current flow is [9]

$$\eta = \frac{Z_0 R}{4\pi\left(R^2 + \left(\omega L - \frac{1}{\omega C}\right)^2\right)}. \tag{15}$$

Here Z_0 denotes the free space impedance. When the loss is small, the electric field reduction approximately equals to $\eta/2$. The power loss reaches to the maximum when $\omega = 1/\sqrt{LC}$ in Equation (15). In this condition, the modulation depth of detected signal reaches its maximal. The resistance and inductance can be considered as constants in that circuit while the capacitance is determined by the gap between the tip and target. When the tip is approximately considered as a sphere, the capacitance between the tip and the target can be extracted by the following series [18]

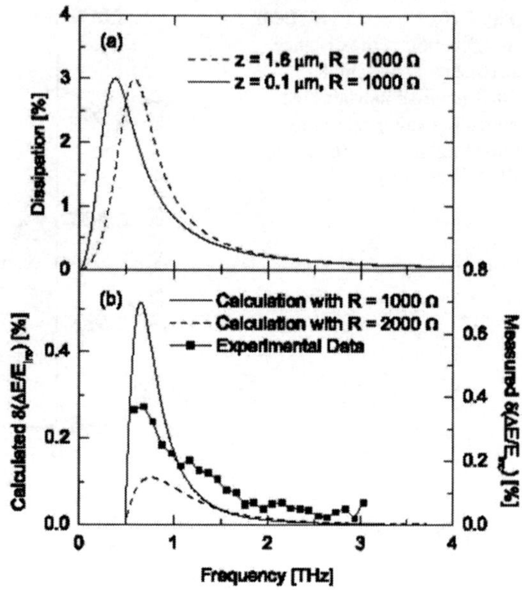

Fig. 7.21 (**a**) Relationship between the power extinction ratio and THz frequency, and (**b**) the relationship between THz field modulation depth and THz frequency (Courtesy of Dr. Kersting)

$$C = 4\pi\varepsilon_0 R\left(1 + r + \frac{r^2}{1 - r^2} + \cdots\right), \quad (16)$$

where R is radius of the sphere, $r = R/2z$, and z is the distance between center of the sphere and surface of the target. Figure 7.21a presents the power extinction coefficient as a function of frequency with different distances between the tip and surface of the target. Figure 7.21b shows the modulation of the THz field as a function of frequency.

The resonant absorption from the metallic tip increases modulation depth to THz field. When a near-field imaging system is limited by its dynamic range, increasing of modulation depth means better spatial resolution. Figure 7.22a is a near-field image of a metal grating obtained using this technique. The sample is fabricated by coating a layer of golden film on a piece of silicon substrate. Period of the grating is 20 μm and thickness of the golden film is 1.2 μm. The platinum tip hangs above the target, and the distance between the platinum tip and the golden film is fixed at 10 nm. Figure 7.22b shows the amplitude of THz signal when the 100 nm diameter tip scans across the gold/silicon boundary. The experiment gives a 150 nm spatial resolution, 2000th of the central THz wavelength.

Tip Enhanced THz Emission Near-Field Imaging

If the incident beam in the setup shown in Fig. 7.17 is fs laser pulse instead of THz wave, and a biased voltage is applied between the semiconductor and the

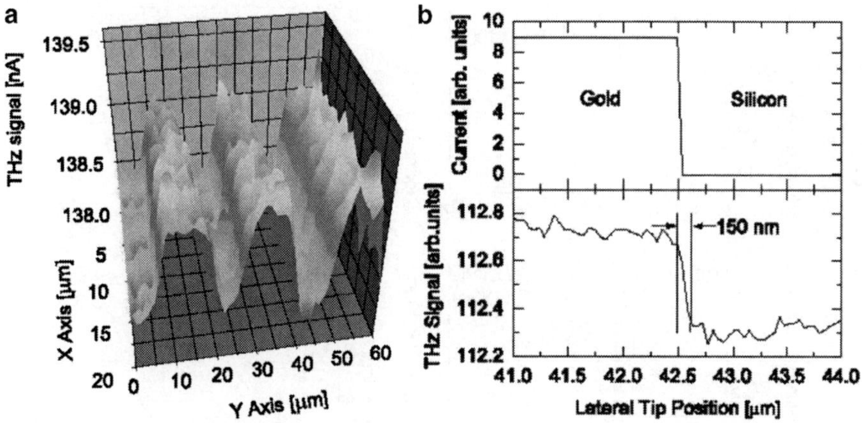

Fig. 7.22 THz wave near-field image obtained by tip absorption of THz field. (**a**) 2D image of a metallic grating on silicon substrate; (**b**) THz field (*lower curve*) and tunneling current (*upper curve*) as tip scanning across a metal/semiconductor boundary (Courtesy of Dr. Kersting)

metallic tip, a THz emitter similar to a photoconductive dipole antenna is formed [18, 20, 21]. The pulsed laser generates photo-carriers at the semiconductor surface and the biased scanning-tunneling-microscope (STM) needle modulates the localized electric field in the Schottky barrier under the tip. The transient photo-carriers driven by the modulated field emit THz waves, which can be detected at the modulated frequency in the far field. To distinguish THz wave generated from the tip/semiconductor coupling from what is generated from the wafer surface itself, the former is named the tip signal and the later is named the wafer signal. If AC voltage with certain modulation frequency is applied across the Schottky barrier and the frequency of AC voltage is used as the reference frequency of a lock-in amplifier, only the tip signal is detected although wafer signal is stronger. Due to the differences in permittivity, doping density, Schottky potential and carrier mobility of various semiconductor materials, the tip signals generate different THz field transients. This unique property is used to distinguish different components in semiconductor hetero- or quantum structures. Figure 7.23 compares THz waveforms for both tip signal and wafer signal when different materials are used. Waveforms of THz pulses differ considerably when different semiconductor materials are used to generate THz wave with biased tip. Similar to THz wave generation from larger emitters, tip generates stronger THz field coupling with semiconductor which has higher carrier mobility, such as p-type InAs. A common character for the tip signal is that it usually contains more low frequency component than wafer signal generated from the same semiconductor. However when the semiconductor material has higher doping density and the doping induced free carriers has high mobility, such as n-type InAs, the tip signal is very low. When the semiconductor has high doping density and high carrier mobility, contact between the metallic tip and the semiconductor is more close to an Ohm contact rather than a Schottky contact, and thus no bias voltage is applied. This is confirmed by the current-voltage (I-V) curve of

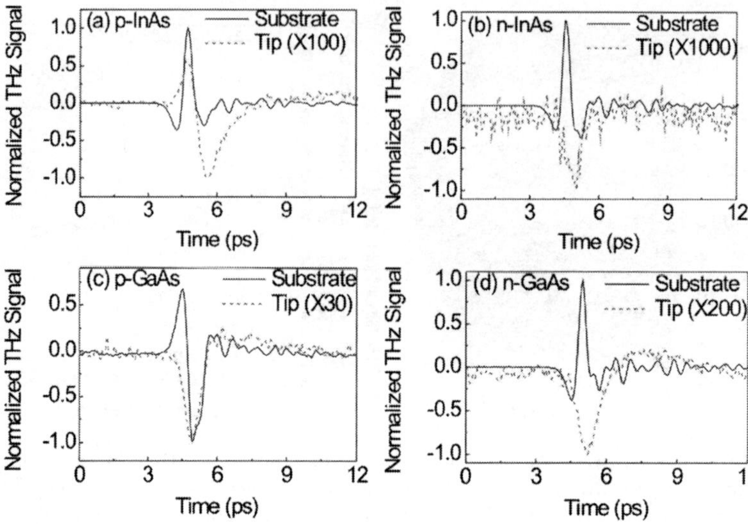

Fig. 7.23 THz waveforms of tip and wafer signals generated from different semiconductors. (**a**) p-type InAs, (**b**) n-type InAs, (**c**) p-type GaAs, and (**d**) n-type GaAs

the tip/semiconductor system. Applying a DC bias in addition to the AC voltage between the tip and semiconductor can bias the Schottky junction working at different conditions. Figure 7.24a shows tip induced THz field generated from a p-type InAs wafer as a function of the DC bias. Figure 7.24b shows the I-V curve for such a system. This curve clearly shows that the tip-semiconductor interface has a Schottky contact.

The tip-semiconductor contact is presented in Fig. 7.25. Due to their different work functions, a Schottky barrier is generated under the contact point between the metallic tip and the semiconductor wafer, and is approximately considered as a hemisphere region centered at the contact spot. Properties of the Schottky barrier

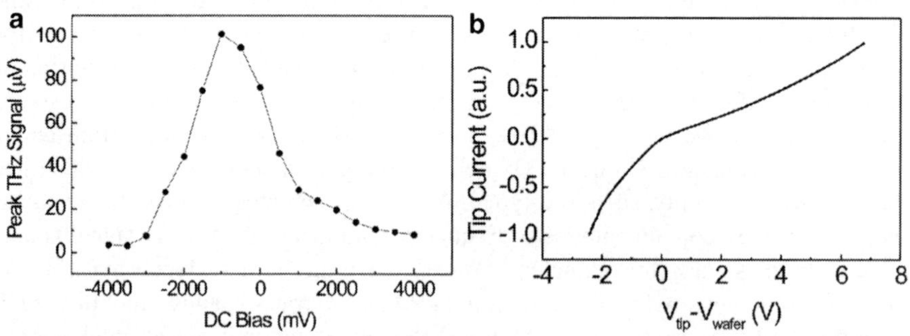

Fig. 7.24 (**a**) THz field as a function of biased voltage when a piece of p-type InAs wafer and a tungsten tip is used, where the AC voltage is fixed at 1 $V_{p\text{-}p}$. (**b**) I-V curve between semiconductor and the metallic tip

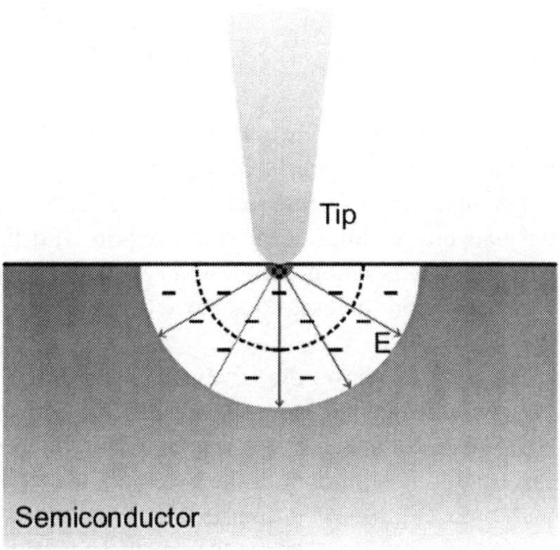

Fig. 7.25 Localized Schottky junction formed at the tip/semiconductor contact point

can be calculated using the following approximations. (1) the depletion region is a hemisphere with a radius of R centered at the contact spot; (2) all free carriers are eliminated from the depletion region, thus the net charge density is the doping density N_D in the depletion region; (3) there is another hemisphere underneath the contact spot with a radius of r_0, which contains same amount of net charge with the depletion region but with the opposite sign; (4) the electric field has homogenous angular distribution in the depletion region. Here r_0, the radius of the contact spot, is named the effective radius of the metallic tip. Since the target has a flat surface, r_0 could be much smaller than radius of the tip. If radius of the tip is 10 nm, its effective radius could be much smaller than 1 nm. Electric field in the depletion region is

$$E = \frac{N_D q(R^3 - r^3)}{3\varepsilon_S r^2}, \tag{17}$$

where q denotes charge of an electron, r the distance between the contact spot to a field point, and ϵ_S the permittivity of the semiconductor. The potential in the depletion region is extracted by integrating its electric field along the radius. Therefore one can calculate the radius of the depletion region using biased voltage applied on the Schottky barrier:

$$R = \sqrt{[3]\frac{3\varepsilon_S r_0 U}{N_D q}}. \tag{18}$$

The radius is proportional to cubic root of the barrier potential. When a metallic tip contacts with a piece of InAs crystal with 1 nm effective radius, R is approximately 30 nm. The potential in the depletion region is

$$U = -\frac{N_D q}{3\varepsilon_S}\left(\frac{R^3}{r_0} - \frac{R^3}{r} - \frac{r^2}{2} + \frac{r_0^2}{2}\right). \quad (19)$$

Figure 7.26a and b show the electric field and potential as a function of the radius r in the depletion region. Although radius of the depletion region is usually much larger than the effective radius, the majority change of electric field and potential happens within a very small region with the similar scale of r_0. This limits THz wave generation within a very smaller region and allows tip enhanced THz wave emission imaging to achieve a spatial resolution even smaller than radius of the tip.

Figure 7.27 shows the schematic setup of the tip enhanced THz emission imaging. A femtosecond laser illuminates the semiconductor surface at the Brewster angle with a diameter of a few hundreds of microns. A STM needle with a tip diameter of 40 nm is brought to the laser spot via a piezoelectric stage. The tip is biased with both DC and AC voltages with amplitudes V_{DC} and V_{AC} up to several volts. When the distance is larger than 1 nm, a capacitor is formed between the tip and the target, which is conductive for AC current. When the tip moves closer to the target and the distance is shorter than 1 nm, tunneling current is observable. Finally when the tip contacts the target, it passes contacting current. THz wave could be

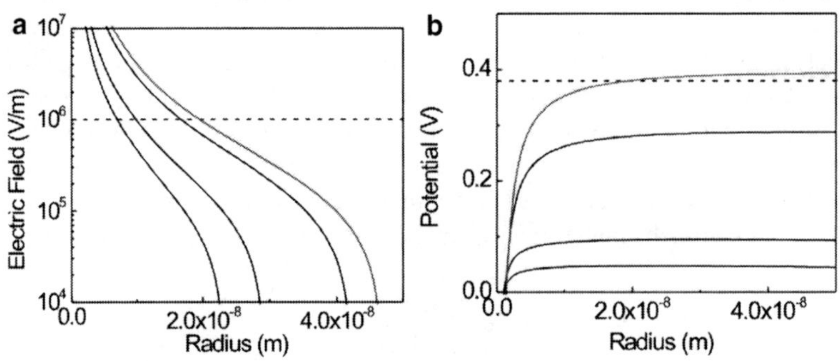

Fig. 7.26 Electrical field and potential in the Schottky junction as a function of radius

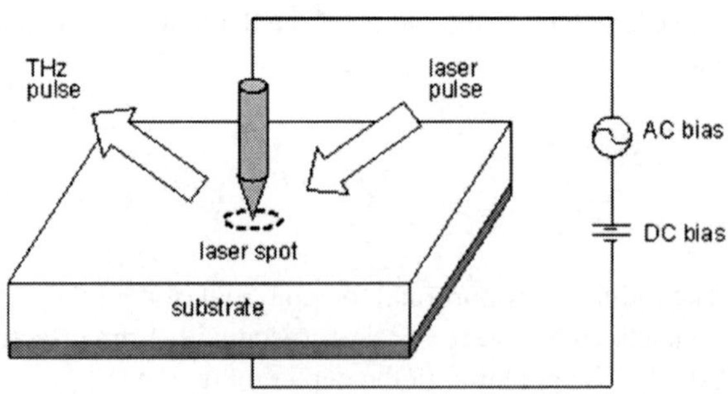

Fig. 7.27 Schematic of tip enhanced THz wave emission microscopy

emitted under all three conditions. By scanning tip across the semiconductor surface, one can map nanostructures on the semiconductor via emission of THz field. Figure 7.28 shows THz wave image across InAs and golden film boundary taken by tip enhanced THz near-field imaging. THz signal gives clear transition within a 1 nm scanning distance. Table 7.1 compares different kinds of THz wave near-field microscopes.

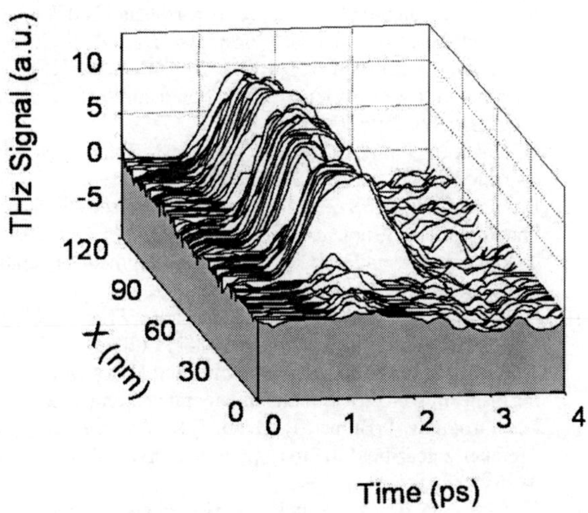

Fig. 7.28 Time-resolved THz pulses when scanning the metallic tip across a metal/semiconductor boundary using tip enhanced THz wave microscopy

Table 7.1 Comparing of THz wave near-field microscopes

THz wave near-field microscopy	Method to obtain spatial resolution	Spatial resolution	Note
Metallic aperture	Transmission through sub-wavelength aperture	7 μm	
Dynamic aperture	Transmission through sub-wavelength aperture	14 μm	
Localized THz wave emitter or sensor	Emitter or sensor size	20 μm	2D imaging capability
Tip scattering	Scattering by sub-wavelength tip	40 nm	
Tip absorption	Absorbing by sub-wavelength tip	150 nm	Detect the electronic property coupling between tip and sample
Tip enhanced THz emission microscopy	Tip induced local emitter	1 nm	Sample needs to be a THz wave emitter

References

1. H. Levine, and J. Schwinger, "On the theory of diffraction by an aperture in an infinite plane screen," *Phys. Rev.* **74**, 958 (1948).
2. S. Hunsche, M. Koch, I. Brener, and M. C. Nuss, "THz near-field imaging," *Opt. Commun.* **150**, 22 (1998).
3. O. Mitrofanov, R. Harel, M. Lee, L. N. Pfeiffer, K. West, J. D. Wynn, and J. Federici, "Study of single-cycle pulse propagation inside a terahertz near-field probe," *Appl. Phys. Lett.* **78**, 252 (2001).
4. O. Mitrofanov, M. Lee, J. W. P. Hsu, I. Brener, R. Harel, J. F. Federici, J. D. Wynn, L. N. Pfeiffer, and K. W. West, *IEEE J. Select. Topics Quant. Electro.* **7**, 600–607 (2001).
5. M. Berta, S. Danylyuk, F. Kadlec, P. Kuzel, and N. Klein, "THz near-field spectroscopy based on metal-dielectric antennae," *IRMMW-THz*, 373 (2006).
6. K. Ishihara, K. Ohashi, T. Ikari, H. Minamide, and H. Yokoyama, "Terahertz-wave near-field imaging with subwavelength resolution using surface-wave-assisted bow-tie aperture," *Appl. Phys. Lett.* **89**, 201120 (2006).
7. Q. Chen, Z. Jiang, G. X. Xu, and X.-C. Zhang, "Near field THz imaging with dynamic aperture," *Opt. Lett.* **15**, 1122 (2000).
8. B. Gompf, N. Gebert, H. Heer, and M. Dressel, "Polarization contrast terahertz-near-field-imaging of anisotropic conductors," *Appl. Phys. Lett.* **90**, 082104 (2007).
9. T. Yuan, J. Xu, and X.-C. Zhang, "Development of terahertz wave microscopes," *Infrared Phys. Technol.* **45**, 417 (2004).
10. R. Lecaque, S. Gresillon, and C. Boccara, "THz emission microscopy with sub-wavelength broadband source," *Opt. Express* **16**, 4731 (2008).
11. T. Kiwa, M. Tonouchi, M. Yamashita, and K. Kawase, "Laser terahertz-emission microscope for inspecting electrical faults in integrated circuits," *Opt. Lett.* **28**, 2058 (2003).
12. O. Mitrofanov, I. Brener, R. Harel, J. D. Wynn, L. N. Pfeiffer, K. W. West, and J. Federici, "Terahertz near-field microscopy based on a collection mode detector," *Appl. Phys. Lett.* **77**, 3496 (2000).
13. N. C. J. van der Valk, and P. C. M. Planken, "Electro-optic detection of subwavelength terahertz spot sizes in the near field of a metal tip," *Appl. Phys. Lett.* **81**, 1558 (2002).
14. B. Knoll, and F. Keilmann, "Enhanced dielectric contrast in scattering-type scanning near-field optical microscopy," *Opt. Comm.* **182**, 321 (2000).
15. K. Wang, D. Mittleman, N. Valk, and P. Planken, "Antenna effects in terahertz apertureless near-field optical microscopy," *Appl. Phys. Lett.* **85**, 2715 (2004).
16. A. J. L. Adam, J. M. Brok, M. A. Seo, K. J. Ahn, D. S. Kim, J. H. Kang, Q. H. Park, M. Nagel, and P. C. M. Planken, "Advanced terahertz electric near-field measurements at sub-wavelength diameter metallic apertures," *Opt. Express.* **16**, 7407 (2008).
17. A. J. Huber, F. Keilmann, J. Wittborn, J. Aizpurua, and R. Hillenbrand, "Terahertz near-field nanoscopy of mobile carriers in single semiconductor namodevices," *Nano Lett.* **8**, 3766 (2008).
18. H.-T. Chen, S. Kraatz, R. Kersting, and G. C. Cho, "Identification of a resonant imaging process in apertureless near-field microscopy," *Phys. Rev. Lett.* **93**, 267401 (2004).
19. R. Kersting, H. Chen, N. Karpowicz, and G. Cho, "Terahertz microscopy with single submicrometre resolution," *J. Opt. A: Pure. Opt.* **7**, s184 (2005).
20. Y. Tao, H. Park, J. Xu, H. Han, and X.-C. Zhang, "THz wave near-field emission microscope," *Springer Ser. Chem. Phys.* **79**, 759 (2004).
21. Y. Tao, H. Park, J. Xu, H. Han, and X.-C. Zhang, "Field-induced THz wave emission with nanometer resolution," *Proc. SPIE*, **5649**, 1 (2005).

Chapter 8
THz Technology in Nondestructive Evaluation

The Nondestructive Evaluation and Testing, in short NDE, discipline includes technologies and methods with the goal to examine objects and materials (samples) without impairing their future use. For example, ultrasounds and X-rays have been used in NDT applications for a long time such as material inspection, medical diagnostics, manufacturing, and quality control. On the other hand, in destructive testing, the sample is damaged during testing process. The destructive testing could be an extreme testing, where the selected samples are tested up to a failure point, then, the behavior of similar samples is statistically extrapolated. Or it can be a nonextreme test, where the sample is dissembled for a better investigation. Examples of destructive testing are found in mechanical elasticity and stress, heat insulation, and corrosion resistance measurements. NDE involves mechanical, optical, or chemical analysis, by use of ultrasonic waves, thermal waves, and electromagnetic waves. The results of applying NDE have a very broad impact on many fields, such as helping the aeronautics industry to ensure the integrity and reliability, and supporting cancer research by finding tumors. The implementation of NDE techniques must include, at least, the following components:

- A source that generates the signal.
- A detector or device to pick up the signal.
- A method to combine both emission and detection signals.
- A device to record and process the signal.
- A method to interpret and analyze the signal.

The application of THz technology in NDE utilizes the transparent property of THz wave through most of dielectric materials as shown in Fig. 8.1. Both time-domain (TD) and Continuous-Wave (CW) technologies can be used in NDE applications. In general, TD technology is used in the first place to explore the spectral response (refractive index and attenuation) of the sample as a function of frequency. The spectral response information resulting from time-domain measurement is used to select the CW frequency that is more appropriate for the particular purposes of the inspection. Many sources and detectors exist in both CW and pulsed modes that can be used in NDE. The performance of each of those systems can be

X.-C. Zhang, J. Xu, *Introduction to THz Wave Photonics*,
DOI 10.1007/978-1-4419-0978-7_8,

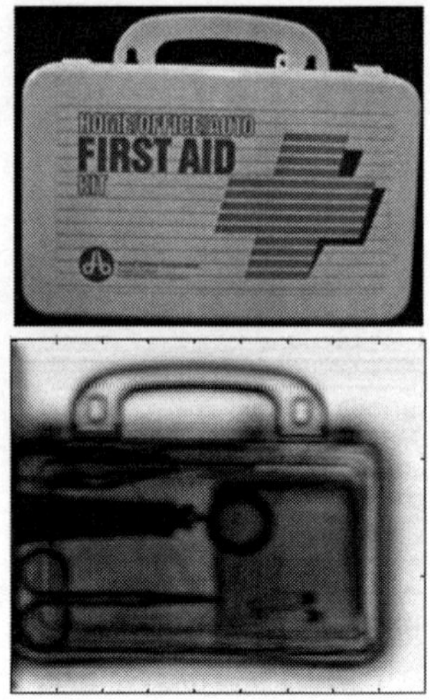

Fig. 8.1 The THz image of "First Aid Kit" using 200 GHz CW THz imaging system

compared in terms of Signal-to-Noise-Ratio (SNR), Dynamic Range (DR), Noise Equivalent Power (NEP), and the responsively. SNR is defined as the ratio between the signal and the noise measured within the system bandwidth. DR is defined as the ratio between the lowest and the highest detectable signal. The lowest signal is usually related to the noise floor, and the highest signal is determined by the maximum power that the source or detector can handle without damage or saturation. The NEP is defined as the input power that produces an SNR equal to 1 at the output of the detector with a particular modulation frequency, wavelength, and bandwidth. The unit is usually expressed in power or power per square root bandwidth. This last definition is usually used for broadband detectors and involves the bandwidth of the detector. The responsively is the ratio of the electrical output to the excitation signal and it is usually expressed in units of voltage or current divided by input power. The selection of emitter, detector, and approach (CW or time-domain) depends strongly on the characteristics of the samples and the purposes of the inspection.

Carrying on NDE with THz Waves

Any study intended to explore the application of a particular technology is fundamentally based on following three aspects: (i) the technology, (ii) the application, and (iii) the experimental setup (Fig. 8.2). The relationship among these aspects could be application-driven (market-driven) or technology-driven (research-driven). In a technology-driven approach, the technology is pushing to find a suitable application or to demonstrate the feasibility of an application (proof of concept). On the other hand, an application-driven approach is seeking the most suitable technology to solve a specific challenge. In any case, the experimental setup is strongly conditioned by the driver approach, whether it is application-driven or technology-driven. Keeping optimum balance among the three aspects is challengeable and

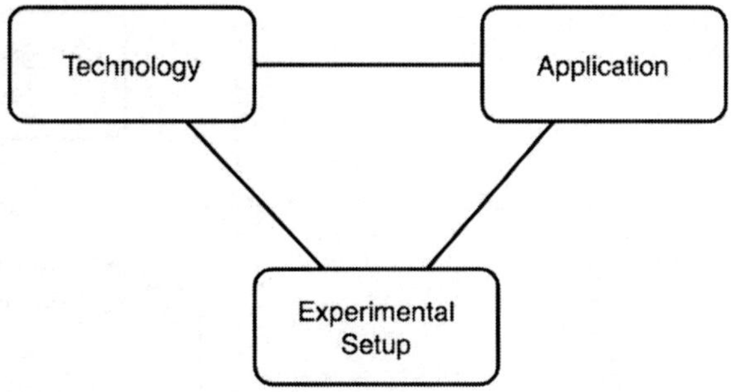

Fig. 8.2 The relationship of three aspects in application development

requires a good understanding of the dominant approach and the main objectives of the application. For example, in an application-driven approach, the priority is the development of a prototype capable to operate in similar conditions as the final application. Therefore, the important parameters are speed, reliability, durability, false and positive alarm rates, and good performance in blind tests. On the other hand, the priority in a technology-driven approach is the confirmation of proof of concept under laboratory conditions, including a demonstrator and the validation with control samples. In both approaches we can define a common workflow to design the experiments and the experimental setup (Fig. 8.3).

The workflow intends to answer these questions: What do we want to see? How do we want to see it? Can we see it? How can we see it? How does the technology perform compared to other alternatives? The first question aims to focus on the scope of the inspection. It is seldom the case that the same technology and setup can be applied to detect different types of defects, thus, we must specify the features of the defects as precisely as possible. The second question aims to define the ideal experimental conditions. For example, is transmission geometry preferred or reflection? How fast do we want the system to operate? Whether or not those ideal conditions are met is to be answered by the third question. In case defects can be detected under desired conditions, then, we can move forward. However, that does not happen very often and many times conditions must be adjusted in order to obtain a reasonable result, which lead us to the fourth question. In case ideal conditions do not work, how do we have to change them so defects can be detected? Finally, in case the defects can be detected, we need to benchmark our technology against other possible alternatives, if any exists. For instance, the technology may not be able to detect all sorts of defects but it may be the only technology to perform certain detection, which would make it very interesting. On the other hand, the technology may be able to detect all defects but at a higher cost than other alternatives, which would have less competitiveness. In detail, the goal of the first step, the preliminary testing, is to determine whether or not the sample is transparent to THz waves. The result will determine if transmission geometry is possible. The procedure for this step is to put the sample in a TDS or CW system and measure the signal being transmitted. The result is not conclusive in the sense that the measurement may be performed in conditions far from ideal but it will give an idea how difficult it is to run measurements in transmission. The measurement of the refractive index and attenuation is useful in transmissive samples. In nontransmissive samples, it is still

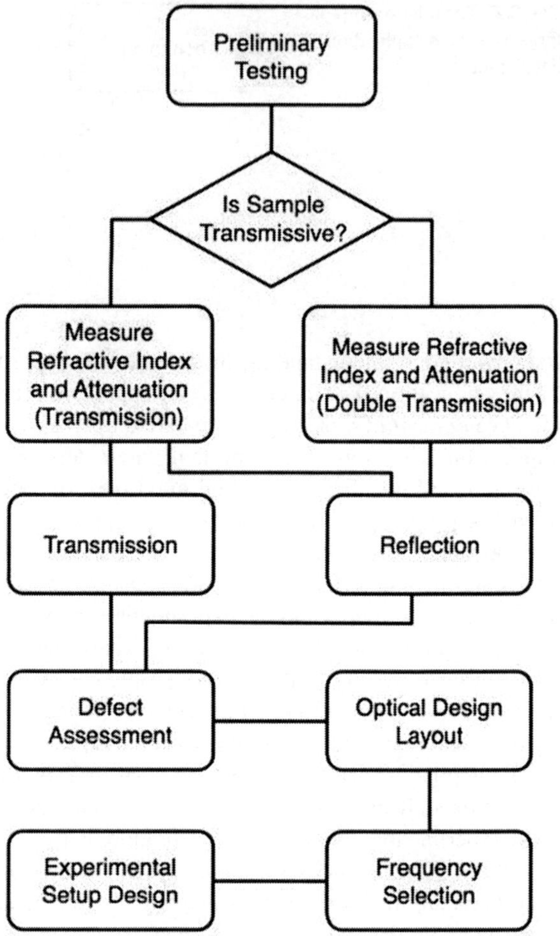

Fig. 8.3 Design of experiment decision tree

possible and convenient to measure the refractive index and attenuation coefficient using reflection geometry. The estimation of the optical properties is important to assess the dynamic range and setup characteristics. It can help to evaluate the performance versus resolution in imaging application. The best way to measure the optical properties is with a TDS system.

The next step is to determine the type of defect that we want to detect. Is the defect a morphological feature? Or is it a chemical feature? Does it show different optical properties? Depending on this assessment, specific technological alternatives are more appropriate than others. For instance, morphological and material discontinuities are often easily detected with CW system, whereas defects due to variation in chemical composition are often efficiently detected with TD system. The size of the defect is also very important because it has an effect on the frequency and the optical design of the inspection system. Typically, the spatial resolution of the system must be comparable or smaller than size of the defect needs to be detected.

The spatial resolution in a diffraction limited imager is, as described in Chapter 3, $\Delta z = 1.22\frac{\lambda \cdot d}{\Phi}$. With all these data, we can plot the sample's attenuation and the resolution of the system as functions of the carrier wave frequency. In the same figure, we can also plot the dynamic range of the inspection system (Fig. 8.4). Such plot will tell us the frequency range in which a good measurement is possible. In Fig. 8.4, the shaded area indicates this range. For instance, the useful range is that in which the dynamic range of the system is above the sample attenuation. Within that range we can estimate the resolution of the system and compare it with the size of the defect. Often, the best frequency is where the difference between dynamic range and sample attenuation is highest. The selection criterion is similar. Usually, CW system works better for imaging purposes while TD system is best suited for chemical analysis and depth information in which time of flight is important. If broadband spectrum is not required for the inspection, then CW imaging system is the most applicable choice. On the other hand, if spectroscopic or broadband spectral information is needed, then, TD system can be the only choice.

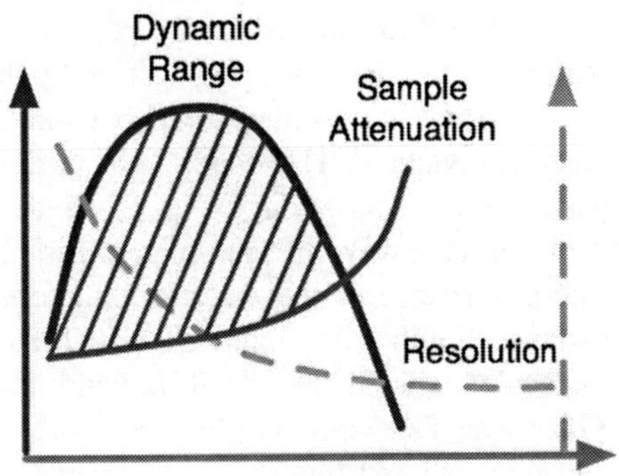

Fig. 8.4 Properties of THz wave inspection system

Attenuation of THz Wave – The Physics Behind Reorganization

In order for THz wave to "see" an object, the object has to influence THz wave in propagation. Interaction between THz wave and material can be precisely described using Maxwell equations. In most common cases however, it can be simplified to solve a problem of a monochromatic plane wave penetration through a (locally) homogenous material. The electric field is

$$E = E_0 \tilde{t}_1 \tilde{t}_2 e^{-ik\tilde{n}l} \left\{ 1 + \tilde{r}_1 \tilde{r}_2 e^{-2ik\tilde{n}l} + (\tilde{r}_1 \tilde{r}_2 e^{-2ik\tilde{n}l})^2 + \cdots \right\} = \frac{\tilde{t}_1 \tilde{t}_2 e^{-ik\tilde{n}l}}{1 - \tilde{r}_1 \tilde{r}_2 e^{-2ik\tilde{n}l}} E_0, \quad (1)$$

where E_0 is the incident electric field, $\tilde{n} = \sqrt{\tilde{\varepsilon}}$ is the complex refractive index of the material, $\tilde{t}_1$,$\tilde{t}_2$ are transmission coefficient of EM wave through both surfaces of the target, $\tilde{r}_1, \tilde{r}_2$ are reflection coefficient of EM wave at both surfaces of the target, $k = 2\pi/\lambda_{\text{air}}$ is the wave vector in the air, respectively. The transmission and reflection coefficient is governed by Fresnel principle,

$$\begin{aligned} r_{//} &= \frac{-n_2\cos\theta_i + n_1\cos\theta_t}{n_2\cos\theta_i + n_1\cos\theta_t}, \quad t_{//} = \frac{2n_1\cos\theta_i}{n_2\cos\theta_i + n_1\cos\theta_t} \\ r_{\perp} &= \frac{n_1\cos\theta_i - n_2\cos\theta_t}{n_1\cos\theta_i + n_2\cos\theta_t}, \quad t_{\perp} = \frac{2n_1\cos\theta_i}{n_1\cos\theta_i + n_2\cos\theta_t} \end{aligned} \tag{2}$$

Here $r_{//}$ and $r_{\perp}$ are reflection coefficient of EM wave with p and s polarization; $t_{//}$ and $t_{\perp}$ are transmission coefficient of EM wave with p and s polarization; θ_i and θ_t are incident angle and transmission angle; n_1 and n_2 are refractive index of media at each side of the boundary, respectively. When broadband wave is used, its propagation can be described as the sum of monochromatic waves. Equation (1) indicates that material modulates propagation of THz wave through three different formats, namely: the reflection, the absorption, and the scattering.

For materials with high refractive index, the THz wave is strongly reflected from its surface and is hardly penetrating into the material. A typical material which blocks THz wave propagation by the surface reflection is metal. Since metal has very high permittivity in THz band, it highly reflects THz wave. As a result, metal is opaque in THz waves. Absorption presents energy transmission from THz wave to the material during its propagation through the material. The penetration depth of THz wave in such material is limited due to the continuous energy loss. The absorbance of the material is determined by its energy state structures. For instance, liquid water highly absorbs THz wave, because rotation transition of water molecules is located in the THz band. Even if the material has low absorption of THz wave, THz wave may also be highly attenuated due to scattering if the material contents rich and fine structures whose sizes are comparable to THz wavelength, the scattering is more severe when variation of refractive index is large across those fine structures. The scattering is equivalent to increase propagation length inside the material, thus results in high extinction of THz wave, although the material has relatively low absorption. In fact, most of opaque materials block light due to the scattering. The typical THz wavelength is $\sim 300\ \mu\text{m}$, which is much longer than lots of common fine structures, such as dust (see Fig. 1.6 in Chapter 1). Therefore, THz wave is less influenced by scattering when propagation through most common target than visible wave, or near/mid infrared waves. Additionally, the energy of THz photon is lower than most of the chemical bonds. Low absorption and low scattering make THz wave transparent in most dielectric materials. This is the key why THz wave is promising in NDE applications.

Transmittance of material in the THz band can be characterized using THz wave time-domain spectroscopy. THz pulses are able to penetrate through lots of daily materials. Figure 8.5 presents waveform of THz pulses after penetration through different materials, where in Fig. 8.5(a) stones, (b) woods, (c) other constructive materials, and (d) packaging materials. THz waveform is still detectable after passing through about 1 cm of those materials. The extinction spectrum of those materials can be calculated through Fourier transform of time-domain THz waveforms. The extinction spectra of selected materials are shown in Fig. 8.6a and b.

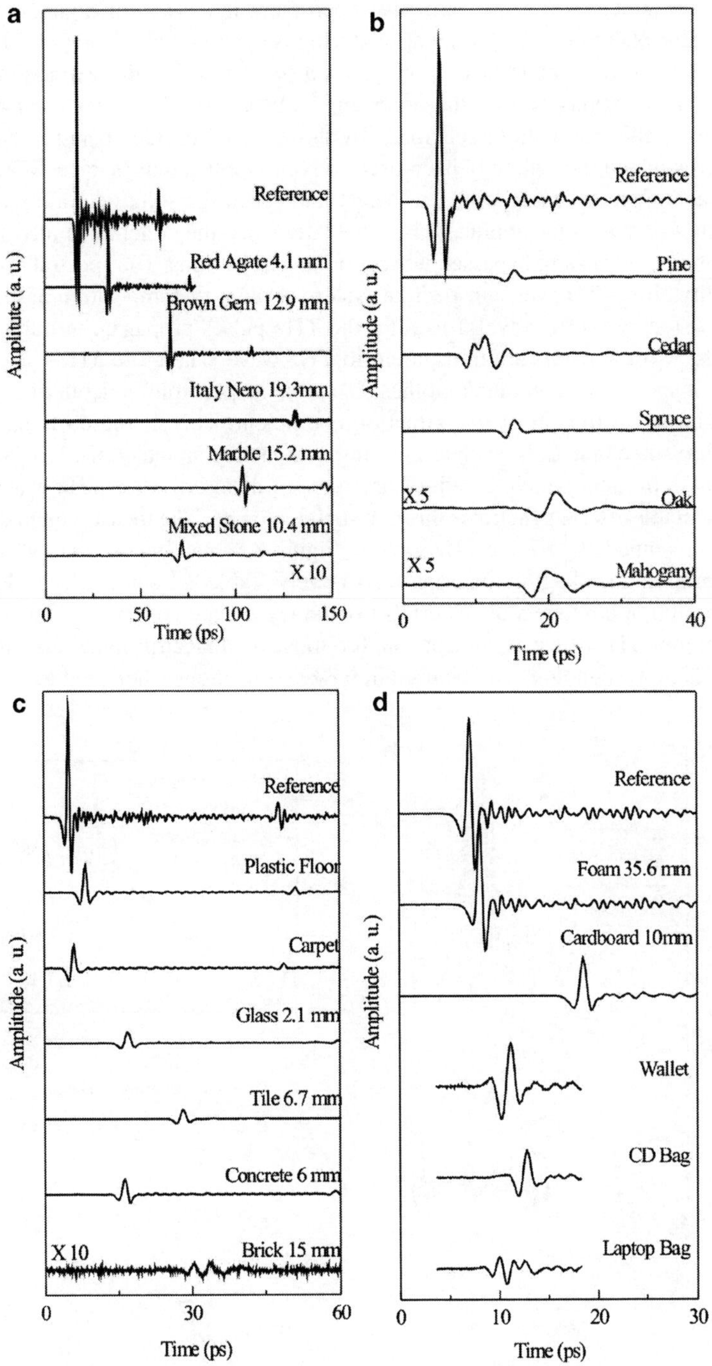

Fig. 8.5 Waveforms of THz pulses after penetration through various of constructive and packaging materials. (**a**) stones, (**b**) wood, (**c**) other constructive materials, and (**d**) packaging materials

Figure 8.6 indicates that the extinctive coefficient increases for higher frequency waves. This phenomena implies that scattering is the dominant factor for THz wave attenuation in those materials. Figure 8.6 also shows that the extinction coefficient of those materials is in the order cm^{-1}. As a result, THz wave can be used for NDE application with targets made by those materials. Compared to other carrier waves which have already been used in NDE applications, such as X-ray, THz wave has unique advantages such as safety and spectral resolution. Safety is essentially important to some applications, especially when the target to be inspected is a human being, for example passenger screening applications. The spectral resolvable capability allows THz wave inspection system to identify composition of the target. Figure 8.7 shows THz waveforms after the THz pulses propagate through various apparels. Most apparels are transparent to THz wave. Therefore THz wave can be used to inspect the target under clothes. Usually, the penetration depth of a material is defined by reciprocal of its extinction coefficient. This definition is hard to be applied to some materials, such as clothing and packaging materials. Here we use a practical definition, which is defined as layers of material, when THz waveform is still detectable after it penetrates through such materials. The measurement dynamic range is assumed 10^5 for the THz system. Figure 8.8a and b give penetration depth of clothing and packaging material, respectively. Table 8.1 summarizes THz wave transmission of homogenous materials (a) and layer materials (b).

Although THz wave is transparent for most of dielectric materials, and THz wave imaging technology can be used in NDE applications when the target consists

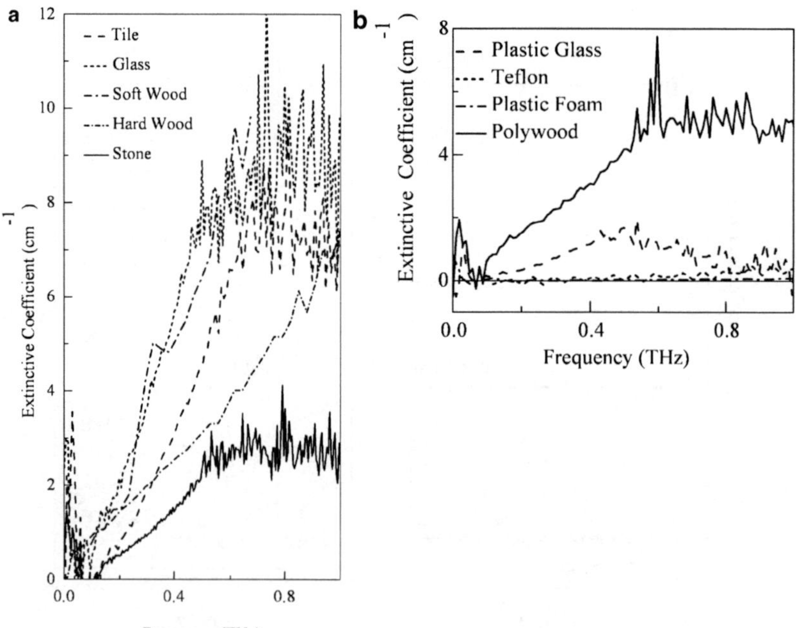

Fig. 8.6 Transmittance spectra of different (**a**) constructive and (**b**) packaging materials measured using THz TDS

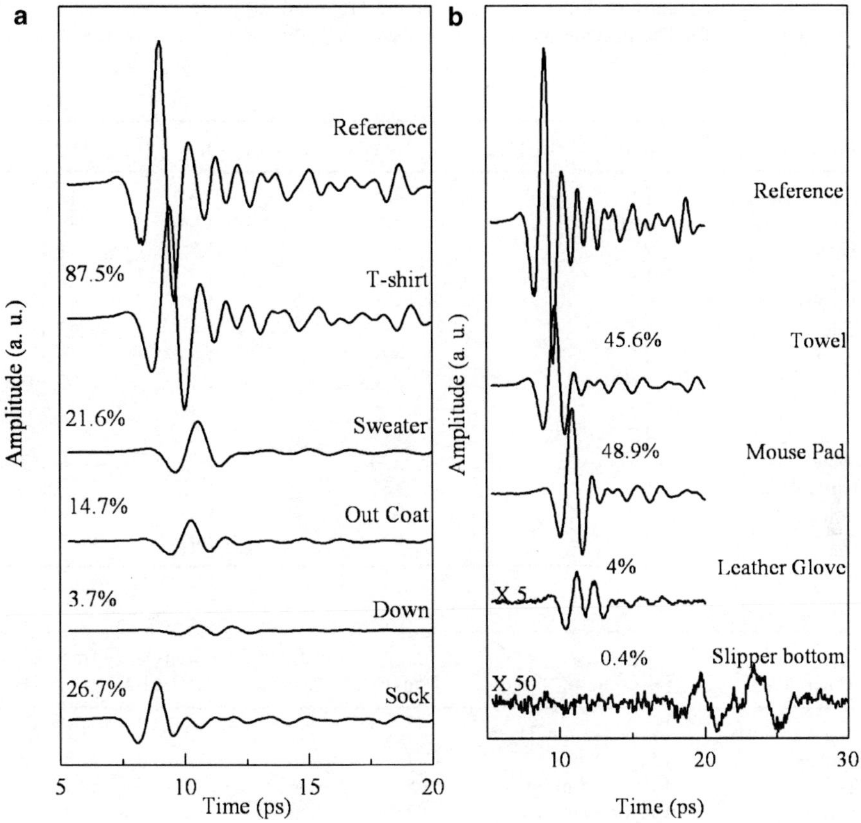

Fig. 8.7 Waveforms of THz pulses after penetration through (**a**) apparel and (**b**) other daily materials

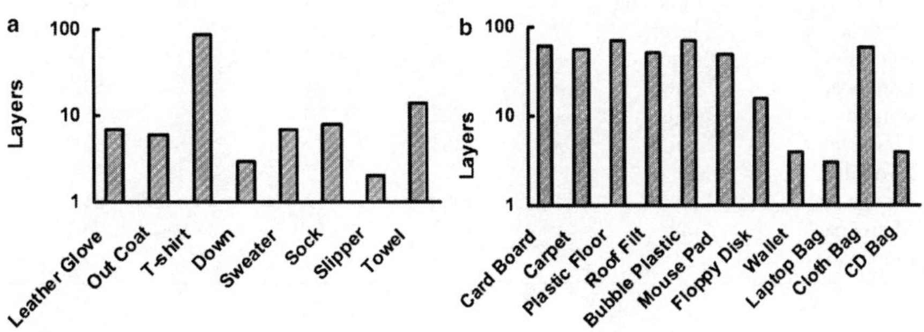

Fig. 8.8 The penetration depth of THz wave through (**a**) apparel and (**b**) other layer materials. The penetration depth is estimated according to a measurement dynamic range of 10^5

Table 8.1 Transmittance of different materials in the THz band. (a) homogenous materials, and (b) layer materials. The penetration depth is estimated according to the measurement dynamic range of 10^5

(a)

Material	(cm^{-1}) @ 0.5 THz	(cm^{-1}) @ 1 THz	N	D (cm)
Plastic glass	1.6	3.8	1.3	11.3
Polystyrene foam	<0.02	<0.04	1.004	288
Glass	7	~8	1.8	1.3
Quartz	1.9	3.5	1.5	3.6
Pottery	6.3	–	1.5	3.5
Tile	4.5	~7	1.6	3.5
Stone	2.5–5.8	4.6–8.5	1.6	6.1
Wood	2.6–6.2	> 3.5	1.1	2.9 – 5
Brick	4.5	~8	1.4	2.6
Concrete	4.7	9	1.2	3
Water	150	220	2.2	0.057
Chloroform	17	16	1.52	1.8
Toluene	1	3	1.51	4.2
Oil	3.7	9	1.42	2

(b)

Material	Description	Trans %	Trans layers	Trans % @ 1 THz	Trans % @ 0.5 THz
Leather Glove	2 layers, 1 mm/layer	4	7	2.8	3.2
Outer Coat	1 layer, 5 mm	14.7	6	3.8	22.8
T-shirt	1 layer 0.5 mm	87.5	86	70.6	87.4
Down	1 layer 20 m	3.7	3	1	5.7
Sweater	1 layer, 2 mm	21.6	7	1.9	40.8
Sock	2 layers, 2 mm/layer	26.7	8	13.2	35.5
Slipper	Bottom, 15 mm	0.4	2	0.1	0.04
Tower	Cotton, 1.5 mm	45.6	14	19.8	65.9
Wallet	Half part, empty	9	4	0.6	18.3
Laptop bag	Whole bag	3	3	0.8	4.1
Cloth bag	2 layers	68.7	60	79.2	64.6
CD bag	Empty	5.6	4	0.7	10
Card board	2.1 mm	62.5	24	9.2	53.7
Carpet	3 mm	56.7	20	0.3	35.5
Plastic floor	2.5 mm	70.3	32	11.6	52.6
Roof filt	0.6	52	17	1.4	35
Bubble plastic	4 layers	68.8	32	49.8	81.9
Mouse pad	3 mm	48.9	16	18.9	74.7
Floppy disk	–	15.4	6	20	9.6

of those materials, it is worth to note that THz wave transmission has its limitations. First of all, THz wave cannot penetrate through metal. Since the plasmon oscillation frequency of metal is higher than THz frequency, it shows high reflection and attenuation at THz waves. Secondly, THz wave is highly absorbed by liquid with polar molecules, due to interaction with their vibration and rotation transitions. For instance, THz wave only penetrates for a few hundreds microns into liquid water. Additionally, THz wave is highly absorbed by phonon bands of crystals if their energy level locates in THz band. Only THz wave whose photon energy is far away from those phonon bands can be used in NDE of such crystals. To employ THz wave imaging technology for NDE applications, one needs to consider the effect of those materials.

Space Shuttle Foam Inspection

The detachment of a piece of foam from the external fuel tank during the lift off of the Space Shuttle Columbia caused the tragedy on February 1, 2003. NASA engaged in a research to seek possible technologies to inspect the foam panels in order to avoid further detachments that could cause another tragedy. The follow-up investigation shows that the detachment was caused by the presence of defects (voids and delaminations) within the layers of the foam that reduces its structural performance. The Sprayed-On Foam Insulation (SOFI) is an excellent material for THz imaging because it has a low absorption coefficient and a low index of refraction at frequencies below 1 THz. The index of refraction and extinction coefficient of SOFI material was presented in Fig. 8.9, which were measured by THz wave time-domain spectroscopy. The extinction ratio allows estimating the maximum thickness of a panel as a function of the dynamic range available in a given experimental setup. For example, the typical dynamic range of a CW THz system in direct detection at 200 GHz is 30 dB, which is equal to a maximum thickness of 12″ one way (transmission), or 6″ roundtrip (reflection), approximately.

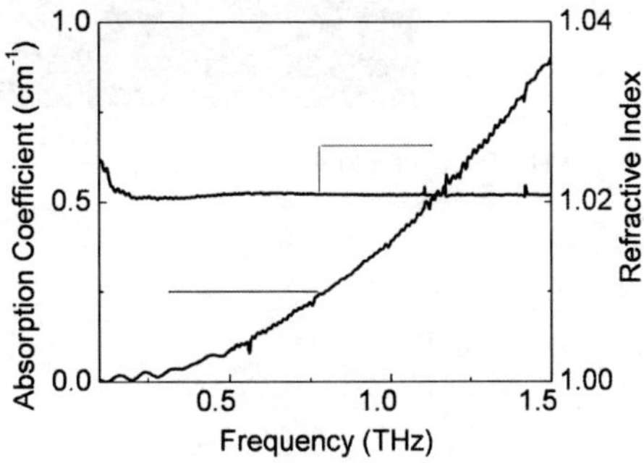

Fig. 8.9 Extinction coefficient and refractive index spectra of polyurethane foam in THz band

The SOFI panel is made by spraying polyurethane foam, layer by layer, onto an aluminum substrate. Therefore, the inspection of the panel is only possible in reflection geometry because aluminum does not transmit any THz wave. The assessment of the type of defects reveals that we seek morphological features such as voids and delaminations, which are translated in discontinuities in the density and the THz properties of the foam. These discontinuities will cause either a difference in attenuation of the THz radiation or a scattering at the discontinuity surface that will be translated in a difference in the power collected from a location with a defect compared to a defect-free location. The minimum size of the defects being sought is 0.25″, which will determine the design of the optics. Figure 8.10 shows an example of testing panels, where the defects are artificially made by introducing a circular polyurethane slice on the substrate or solid foam before spraying another layer on the top. Some of the defects are also made by injecting air into the foam while it is curing. The size of the panels is typically 2×2 feet and their thickness ranges from 1″ to 9″. When pulsed THz imaging system is used for the inspection of the foam, the existence of defects results in distortion of THz waveforms. One example of THz waveform distortion is shown in Fig. 8.11. Image of defect can be extracted by following peak amplitude of THz waveforms or it can be retrieved by variation of time delay. While most effectively, the distortion of THz waves can be calculated using cross correlation between THz waveforms as presented in the following equation.

Fig. 8.10 Photo of a SOFI testing sample

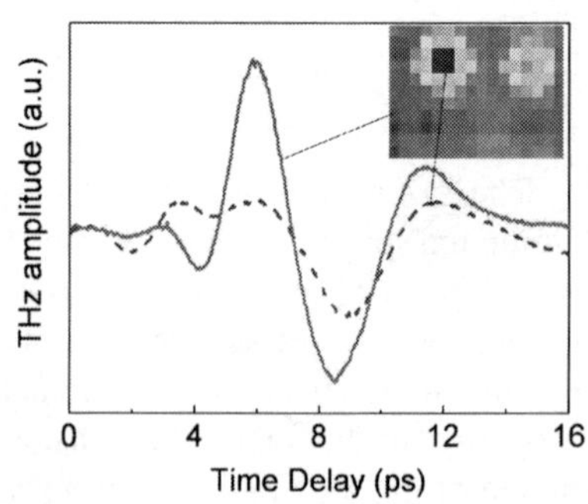

Fig. 8.11 The modulation of THz waveforms by presenting of defect in SOFI sample. Inset shows defects imaged according to the modulation

$$r_d = \frac{\sum_i \left[(X_i - \bar{X}) (Y_{i-d} - \bar{Y}) \right]}{\sqrt{\sum_i (X_i - \bar{X})^2} \sqrt{\sum_i (Y_{i-d} - \bar{Y})^2}}, \tag{3}$$

where X and Y are signal and reference waveforms, which are two array of numbers. By using cross correlation, each pixel in the image is presented by the entire THz waveform rather than just a single value of peak amplitude or time delay. As a result, it dramatically increases the imaging dynamic range. Using time-of-flight imaging technique, pulsed THz wave image is able to tell depth of the defect.

On the other hand, a CW THz wave imaging system could be simpler in construction, more compact, more flexible in operation and easier to analyze the result. In the evaluation of the optical design, a high frequency will provide a better resolution than a low frequency. However, the attenuation (extinction coefficient) grows exponentially as the frequency increases, thus the thickness of the panel has to decrease because the setup has a constant dynamic range. After several studies at different frequencies (200, 400, and 600 GHz), the best trade-off between resolution and panel thickness was found to be 200 GHz. For example, when a 30 dB measurement dynamic range is considered, the maximum thickness of the panel at 200 GHz could be 6″, while the maximum thickness is only 2.5″ for 400 GHz, therefore, 400 GHz will not be capable to inspect such thick panels. These conditions determine the design of the optical system and the rest of the experimental setup. Applying the Rayleigh formula with 200 GHz (1.5 mm), 6″ of thickness, and 0.25″ resolution target, the result is a minimum aperture of 45 mm. Figure 8.12 shows experimental setup of a CW THz wave imager, which uses a Gunn diode as the source and a Schottky diode as the detector. The experimental setup additionally comprises two focusing lenses, and a beam splitter and everything is designed to work in reflection geometry, which could be collinear or with a deflection angle, or pitch-catch. The second reflector is used to relieve the standing wave interference problem. In

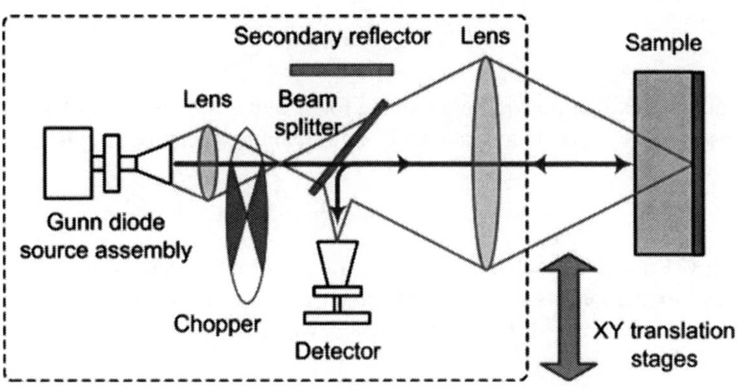

Fig. 8.12 Setup of CW THz wave imaging system

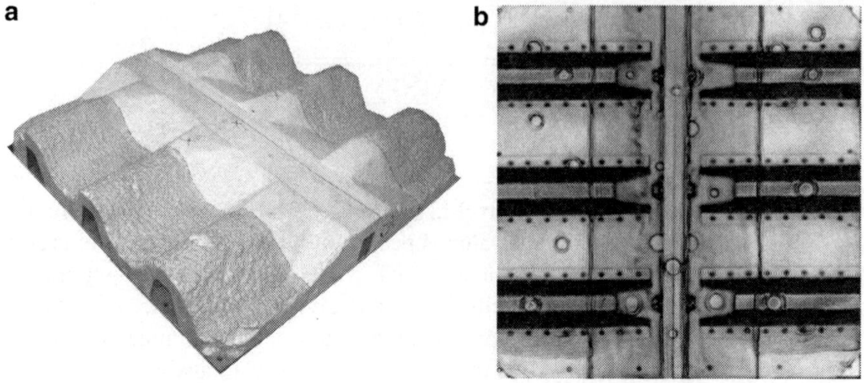

Fig. 8.13 (**a**) Optical photo of a 2′ × 2′ panel sample and (**b**) its THz wave image

the THz images, the defects appear as dark boundaries with light interiors, corresponding to the scattering and interference at the edge of the defect and enhanced transmission due to the lack of material in the interior. Figure 8.13 shows a THz wave image of defects in a testing panel and a photo of the testing panel as comparison. It has been observed that most of the defects appear in the vicinities of structural features such as stringers, stiffeners, and rivets. The sample in Fig. 8.13 shows it has six stringers and the foam is sprayed following the resulting geometry with an average thickness of 2″. The THz images show the position of big (> 0.5″) and medium (0.25″–0.5″) defects very clearly. Natural defects such as rollovers are also detected in the vicinities of the rivets. The system is not very sensitive to the surface condition and it is very tolerant to the depth of the defect. The system complies with the most desired characteristic criteria given by NASA.

If the standing wave interference modulation is ignored, the effect of defects can be considered as changing extinction of THz waves. The contrast of image is

$$\begin{aligned} C &= \tfrac{1-T}{1+T} \\ T &= e^{-2\alpha(\omega)d}, \end{aligned} \tag{4}$$

here α denotes extinction coefficient of the foam, and d is thickness of the defect. In most cases, $-\alpha d<<1$ is always satisfied, and Equation (4) can be simplified to

$$C \approx \alpha(\omega)d. \tag{5}$$

Equation (5) indicates that the image contrast is proportional to thickness of the defect and extinction coefficient of the foam material. However, there is another important factor of the THz wave imaging, which is the measurement dynamic range, which is described as:

$$D = D_0(\omega)e^{-2\alpha(\omega)H} \tag{6}$$

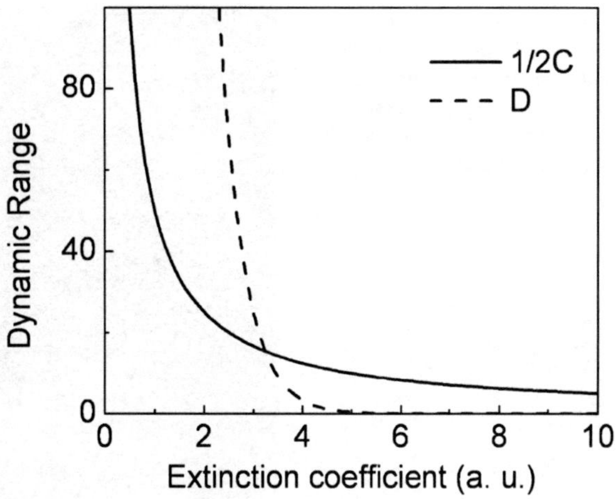

Fig. 8.14 The dynamic range and reciprocal of image contrast as functions the extinction ratio

where D_0 is the system dynamic range without a sample, and H is the thickness of the foam sample. Only when dynamic range of the image (D) is greater than the reciprocal of the image contrast (1/2C), the defect can be recognized. Figure 8.14 gives D and $1/2C$ as a function of α. Only when α locates at right side of the crossing point, THz wave image can identify the defect.

A real defect usually does not have a clear boundary and regular profile as those manmade ones shown in Fig. 8.13. A criteria needs to be setup in order to decide if an area has a defect or not. The criteria can be defined using statistic distribution.

$$z = \frac{|\sigma - \overline{\sigma}|}{\Theta_\sigma}. \tag{7}$$

Here σ is standard deviation of THz signal within a testing area, $\overline{\sigma}$ indicates the mean of standard deviation of the reference samples, and Θ_σ gives standard deviation of σ in reference samples. $z < 1$ means the testing area is similar to the normal (reference) area and is less possible to be a defect. While, $z > 1$ means the testing area is irregular comparing to the reference area and is likely contenting a defect. Figure 8.15 shows a THz wave image of a testing sample with premade and natural defects. The areas with boundary of solid squares are used as reference areas, and circled by dashed squares are testing areas. The reference samples result in a mean standard deviation of THz signal $\overline{\sigma} = 0.0337$, and $\Theta_\sigma = 0.0038$. Table 8.2 summarizes analysis results of those testing areas.

Armor Plate Inspection

In this example, THz wave imaging technology is used to evaluate bullet impact of a Kevlar© composite bulletproof plate used by the Belgian Military troops deployed

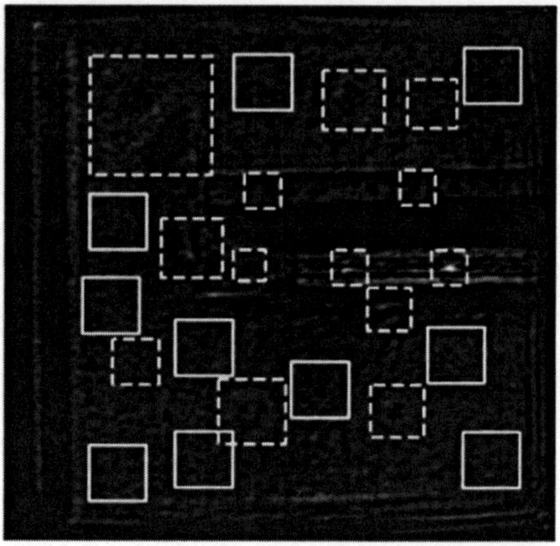

Fig. 8.15 THz wave image of a SOFI panel. Areas with *solid square boundary* were used as reference, and areas with *dashed square boundary* were defect areas

Table 8.2 The reliability analysis of defect identification in Fig. 8.14

ID	Type	Δ_B	z	Uncertainty (%)
1	1″ mid-plane delamination (F)	0.0634	7.74	0.1
2	0.5″ mid-plane delamination (E)	0.0552	5.60	0.1
3	1″ substrate delamination (C)	0.0464	3.32	0.3
4	0.5″ substrate delamination (B)	0.0456	3.11	0.6
5	1″ substrate delamination (C)	0.0545	5.41	0.1
6	0.25″ substrate delamination (A)	0.0603	6.95	0.1
7	0.25″ substrate delamination (A)	0.1199	22.49	0.1
8	0.5″ substrate delamination (B)	0.1697	35.49	0.1
9	0.25″ substrate delamination (A)	0.0593	6.68	0.1
10	0.25″ substrate delamination (A)	0.0559	5.78	0.1
11	0.25″ mid-plane delamination (D)	0.0396	1.53	39.7
12	Omitted, natural defect	0.0395	1.52	39.7
N1	Natural defect	0.0540	5.30	0.1

in different scenarios around the world. These bulletproof plates endure a wide range of stress situations, from direct impacts of bullets and projectiles to mechanical stress caused by sudden body movements during combat operations. There is an interest to obtain a safe, durable/reliable, affordable, and user-friendly technology to inspect the mechanical integrity of these plates. The plate has received the impact of a bullet. The impact was made at the ballistic laboratory located at Royal Military Academy in Brussels. This laboratory permits controlling the ballistic parameter of the projectiles and impact conditions. In this particular case, an 8 g 9 mm bullet was fired at the plate at 419 m/s, measured at 2.5 m before the impact location. Developing THz wave imaging in bulletproof plate inspection can be used as a typical example of a technology-driven application. Although there is a clear long-term interest to bring this technology to a real application, the goal at the first stage is to assess whether or not THz technology is capable to provide an inspection result comparable to those available technologies such as X-ray images or infrared image can provide. In consequence, the priority is to provide a proof of concept result, which is to see the defects rather than to perform reliable and fast inspections. For this purpose, the target is firstly inspected by X-ray imaging system and infrared imaging system. The defect information obtained from the X-ray and infrared images indicate that the geometrical features have a size in the order of millimeter. Therefore, the optical setup must be designed so the spot size is around a few mm or less. The spatial resolution is feasible using a 0.2 THz carrier wave (wavelength 1.5 mm) and imaging with a large NA lens.

The bulletproof plate has very high extinction at THz waves, so that neither a pulsed THz system nor a CW THz system using direct detection method will provide enough dynamic range to image such a bullet proof plate with transmission geometry. Utilizing heterodyne detection method, a CW THz system (presented in Fig. 8.16) with 0.2 THz wavelength provides a dynamic range of 60 dB, whereas such a CW system in direct detections only has a typical dynamic range of 30 dB. Transmitted THz wave can be recorded with heterodyne detector. Therefore, THz wave image of such a bulletproof plate can be taken. The result of the THz images, both from the amplitude (Fig. 8.17a) and phase (Fig. 8.17b) channels, in transmission geometry does not only show the impact spot and but also the features in the surrounding area, i.e., 6 radial cracks and concentric stress lines. Similar image can be taken with X-ray imager as presented in Fig. 8.17c. X-ray image provides higher spatial resolution and gives crystal clear image of the impact spot and stress lines; however the image system is not portable and represents a health hazard for humans.

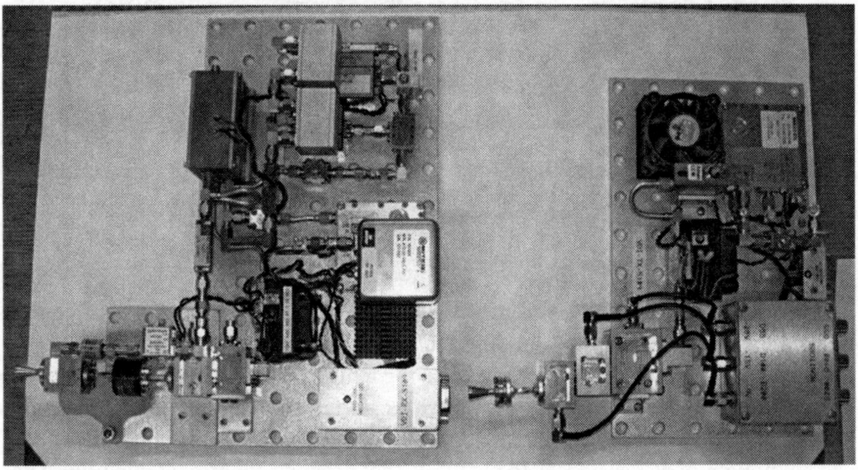

Fig. 8.16 A 200 GHz CW heterodyne detection system. The receiver is on the left and the emitter is on the right

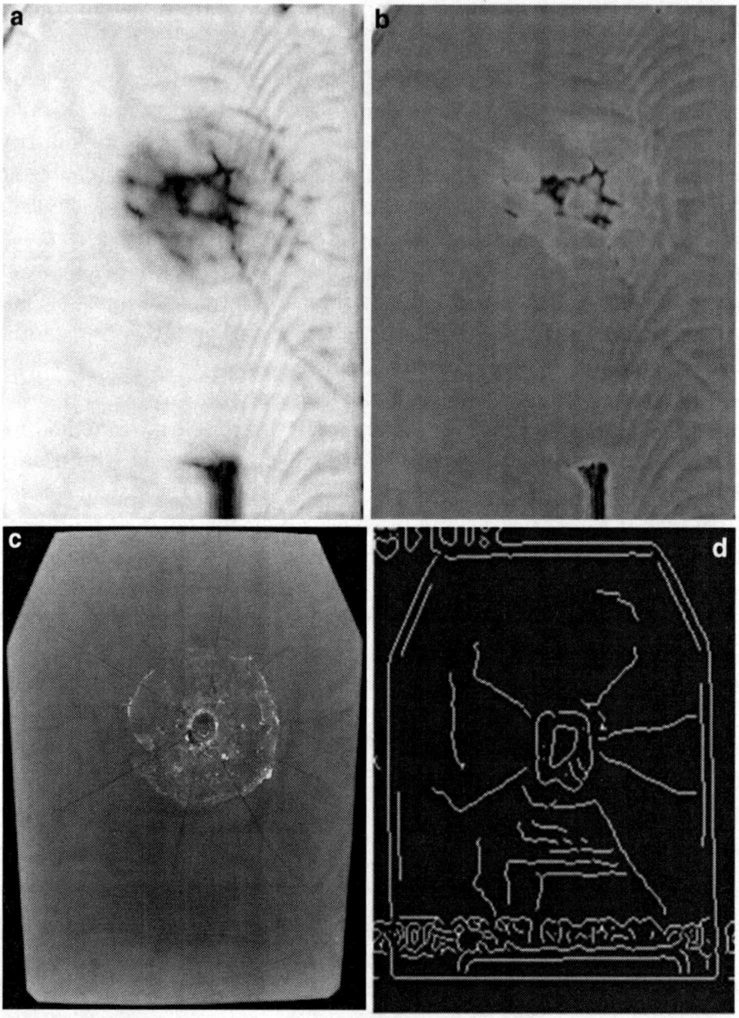

Fig. 8.17 THz images of the bulletproof plate from the (**a**) amplitude and (**b**) phase channel. (**c**) and (**d**) are comparison images of the sample plate imaged by X-ray and thermal-graphic approach

Therefore it cannot be deployed near the conflict area. The plate can also be evaluated using a pulsed thermal-graphic technique, where the surface of the plate is exposed to a heat pulse of a few second, using a high power source such as lamps or hot air blowers. Thermal waves travel from the surface into the bulk of the plate after the thermal front comes into contact with surface. Subsurface discontinuities (flaw) can be thought of as resistances to heat flow that produce abnormal temperature patterns at the surface, which can be recorded with an infrared camera. Figure 8.17d shows thermal-graphic image of the bulletproof plate taken with 8–12 μm IR wave.

Imaging process techniques such as time, spatial filter and digital edge detection filter need to be applied in order to achieve a better visualization of the near surface defects created immediately after the impact. As comparison, those features are more readily seen, with no image processing at all, in THz wave images. The only processing is that the amplitude information is displayed in log scale. Log scale usually enhances the information present at low values. An interesting feature is the dark area around the impact in the amplitude information, which could be related to a local higher density of the material due to the compression generated by the pressure waves after the impact. The feature also appears in the phase channel, indicating that the change in density also influences the optical path. The waved pattern is caused by the interference between the beam reflected off the front and back surface of the plate. Among the three techniques, the best results are obtained with the X-ray system, followed by THz and IR imaging system. THz images offer an easier to interpret data than IR images and do not require intensive image processing and analysis, thus, reducing the risk of false positives and the training of the operator. On the other hand, the IR system is faster in acquiring the data than the THz system. However, this may change in the future as rapid scan systems and THz array detectors are being developed than can reduce the acquisition time from several minutes to few seconds.

Rust Under Paint

Evaluation of rust under paint is very attractive in metal protection applications, such as body of automobiles. THz wave is transparent in painting material, thus could be used to evaluate surface under paint. THz wave imaging in rust under paint evaluation is also a technology-driven example and, therefore, the goal is to investigate the potential of THz technology to detect the degree of rust of painted metal surfaces. The sample used in evaluation is a rectangular piece of steel plate with different degree of rust with different paints on top. One surface of the plate is previously painted with a primer. The other surface is not primed before painting it. The different degree of rust depends on the time the surface has been submerged into a sea-salt bath. Five degrees of rust are simulated: no rust, 2-days rust, 4-days rust, 7-days rust, and 11-days rust.

Because the substrate is a metal, the sample is not transmissive to THz wave and the only possible geometry is reflection imaging setup. From a defect assessment perspective, THz wave imaging is seeking chemical differences that will change the reflectivity of the sample. Therefore, spatial resolution is of a lesser importance than in the previous examples. Figure 8.18a is photos of sample for both surfaces. The paint of one surface of the sample (top figure) is applied onto a primer base, M is the surface without paint, PG is the painted area with gold paint (metallic) over primer, PB is the area painted with nonmetallic blue paint over a primer layer. The paint of the other surface of the sample is applied directly onto the surface without primer, M is the surface without paint, G is the area painted with gold paint

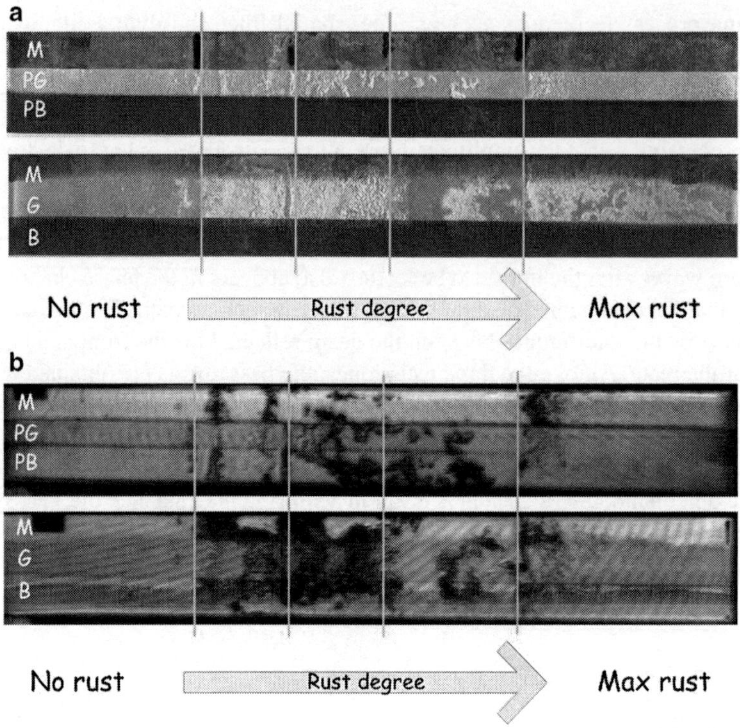

Fig. 8.18 (**a**) optical images and (**b**) THz wave images of both surfaces of the painted sample. *Upper image* corresponds to the paint applied onto a primer base, and *lower image* corresponds to the paint applied directly onto the surface without primer

(metallic), and B is the area painted with blue paint (nonmetallic). The sample is imaged using THz wave reflection imaging setup with 1.63 THz wave from a gas laser. THz wave images of both surfaces of the testing sample are shown in Fig. 8.18b. Due to roughness of rust, the area with rust has lower reflection to THz waves, thus it looks darker in THz wave image. THz wave image of rust pattern matches well with the visual inspection and most importantly, it is able to see the rust pattern under paint, which is hardly analyzed by visible light. However, it is difficult to use THz wave image to get a clear relation to the degree of the rust.

Carbon Fiber Composites Inspection

Composite materials are becoming very prominent in many industries, especially in transportation, aeronautics, and aerospace. Composite material with carbon fiber forming a network and filling with resin has been widely used because of its high strength and light weight. However this kind of material is not resistant to heat damage, and can be damaged when heated up to 200°C. To ensure their performance, technologies need to be developed to differentiate safe and unsafe materials. Evaluation of carbon fiber composite material using THz wave imaging focuses on composites that have suffered damages caused by intensive heat. These materials represent a challenge for THz, because they exhibit a high THz reflectivity, thus, limiting the capabilities to perform inspection in transmission geometry. This is another technology-driven example and, as in the previous examples, the goal here is to assess the potential of THz to see the different kinds of defects caused by heat treatments. For instance, intensive heat, such as caused by a flame or torch, can deteriorate the resin and/or change the orientation and integrity of the yarns. At a later stage, the study would become more application-driven upon successful results.

Despite the high reflectivity of the sample measured in a preliminary test, the polarization of the radiation is important to determine the penetration depth because the yarn structure is highly anisotropic. As discussed in Chapter 4, the conducting fibers will reflect the incoming THz radiation depending on its polarization. If polarization of THz wave is parallel to the fiber, the reflection is optimum because the electrons can move along the fiber easily. In case the polarization of THz wave is perpendicular to the fiber, the electrons cannot follow the excitation well and the radiation can travel further into the sample. Figure 8.19 shows THz wave images of 3 burned carbon fiber composite samples and photos of those samples are used as comparisons. THz wave images were made with 0.6 THz wave generated from a Gunn diode and the reflected THz wave was detected by a Golay cell. Three carbon fiber composite samples were preburned by propane flame with different degrees of heat-induced damages. It can be seen that the appearance of the defects depends on the polarization of the radiation with respect to the direction of the fibers. For instance, structure is more visible when the polarization is perpendicular to the fiber direction due to increased penetration of the radiation into the sample. It can also be seen that the contrast in the THz images is much higher than that of the optical image, although the resolution is lower.

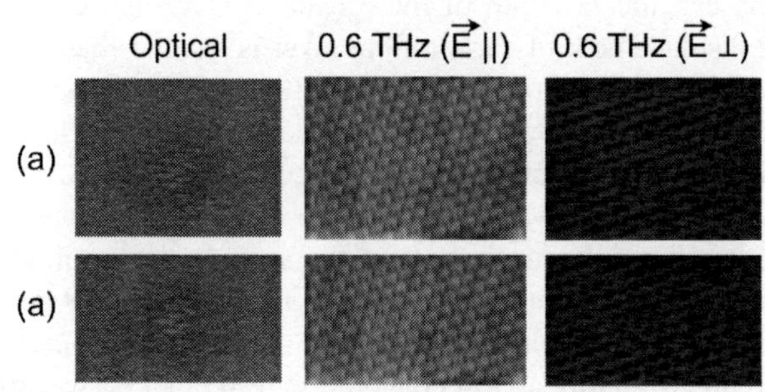

Fig. 8.19 0.6 THz wave images of three carbon fiber composite samples with different polarization orientation. Optical images of those samples are used as comparison. (**a**) sample with surface burning (**b**) sample with large area and deep burning, and (**c**) sample with small area but deep burning

The information provided by the THz wave images gives a more complete outline of the damaged area than what is apparent in a visible light image. The CW THz wave images reveal features that are important for application: damaged areas show a strong change in reflectivity, and the reflectivity is highly dependent on the polarization of the incoming radiation. The latter is quite similar to the function of a wire grid polarizer, which reflects waves whose electric field is parallel to the grid and allows those whose field is perpendicular to pass. Thus, the image with parallel electric field shows mainly information about the topmost surface, while the perpendicular field image penetrates several layers further. An important consideration for an NDE is the rate of false calls. This is mitigated in this imaging modality by the fact that the images provided are easily recognizable to human vision. This is illustrated with Fig. 8.19a, which shows a minor damage that does not have an effect on the strength of the material but it is still apparent in a visual inspection of the material. It can be seen that the scorch mark that is apparent in the optical image does not appear in the THz image since the burning did not affect the underlying structure. That the severity of the damage in terms of effect on the physical strength of the material is correlated with the apparent effect in the THz images was confirmed in a separate work.

The principle of using THz wave imaging to evaluate heat damage in carbon fiber composite can be studied using THz wave time-domain spectroscopy. Figure 8.20a compares waveforms of THz pulses reflected by burned and unburned areas on the composite material. The result shows that, the reflection of THz wave is reduced after the sample is burned, and the waveform is also broadened. Measurements with a TD system yield additional information about the material. Since the measurement is time-resolved, it is possible to extract the presence and depth of multiple reflections from the surface. Figure 8.20b shows spectra of those reflected THz pulses. Reflection spectrum of burned area shows reduction of amplitude, red shift and most importantly, it has oscillation structure in the spectrum. This oscillation indicates there is multireflection by different layers of the composite material when THz pulses interact with it. This multi-reflection indicates that there is delamination after the material is burned. As a result, utilizing the spectral oscillation, THz wave inspection can locate depth of the reflection layer and evaluate severity of burning. Since displacement of carbon fiber layer is smaller than THz wavelength, the multireflection does not directly lead to splitting of THz pulse. To identify THz pulses reflected from each layer, one can use deconvolution technique, to retrieve THz waveform from the reflected signal. By deconvolving the waveform reflected by a suspect location with a waveform from an undamaged location, it is possible to precisely locate the reflection events at the surfaces in the time domain. This information provides a description of important types of material deformation such as delamination, wherein the multiple layers of the sample become separated from one another. Figure 8.21 compares deconvolution result of THz waveforms reflected from burned (a) and unburned (b) areas. There are multiple peaks in the burned area reflection signal. Those two peaks locating at 200 and 80 μm from the main peak indicate displacement of multiple layer structures. No multiple peaks appear in the unburned area reflection signal, which means there is no delamination in such area.

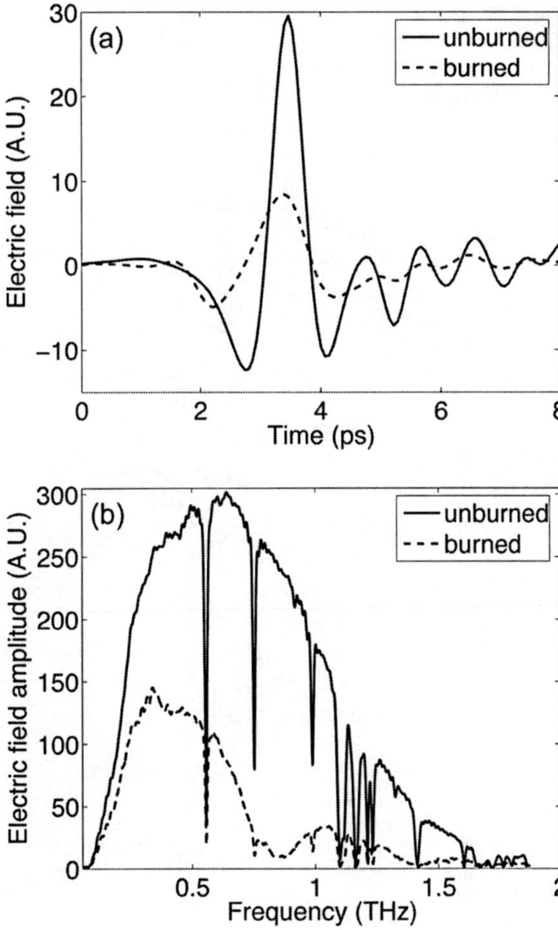

Fig. 8.20 THz pulses reflected from burned and unburned area on a carbon fiber composition material. (**a**) Waveforms of THz pulses and (**b**) spectra

Currently the industrial standard evaluation method for carbon fiber composite material is the bending test, which is a destructive method. In a 3-point bending test, two spots of the testing sample are fixed while a force is applied at middle of those two spots. The sample will be bended by the compression force. The slope of the compression point displacement as a function the compression force indicates rigidity of the sample. The curve growing until the broken threshold is achieved and where the curve reaches to a peak. If the sample contents multiple layers, the curve shows saw teeth structure with multiple peaks. The bending test can evaluate strength of the sample but cannot locate the damaging position. And since it is a destructive testing method, the sample cannot be used after testing. Figure 8.22 shows evaluation of two burned carbon fiber composite samples using bending test and THz wave inspection methods. Sample 1 is a single layer material, whose evaluation result is shown in Fig. 8.22a and b. Sample 2 contents multiple

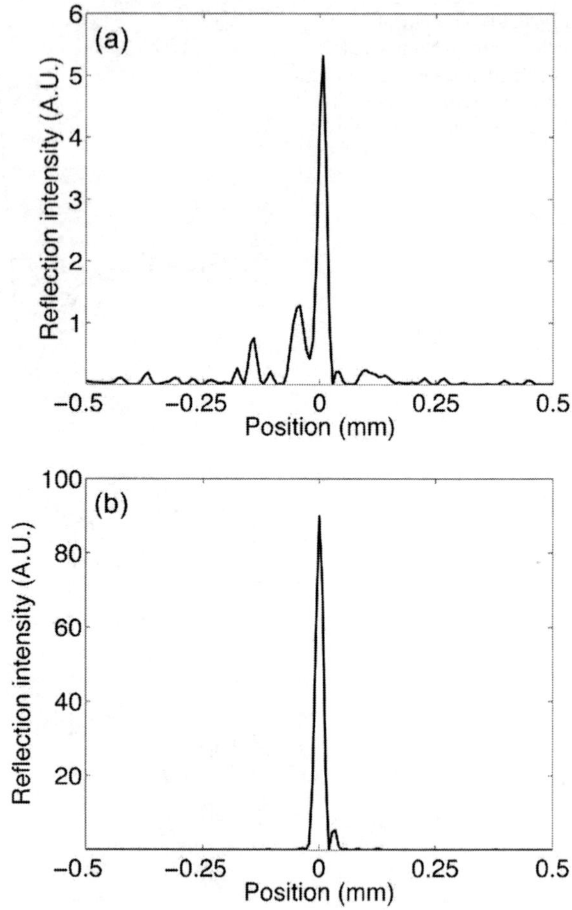

Fig. 8.21 The deconvolution results of THz pulse reflected from (**a**) burned areas and (**b**) unburned areas

layers, and its evaluation result is shown in Fig. 8.22c and d. The burning damage can be clearly identified by using THz wave time-domain measurement. It is worth to notice that, since carbon fiber composition material has high extinction coefficient to THz waves, using THz wave inspection usually limits the investigating depths less than 400 μm from the surface. THz wave imaging and spectroscopy are promising solutions to the problem of identifying evaluating damage to carbon fiber composite materials. While additional progress in THz technology will be required for some applications, currently existing tools are able to provide useful and quantifiable information regarding the extent and severity of damage. Such techniques could potentially increase safety and efficiency in the defense and aerospace industries.

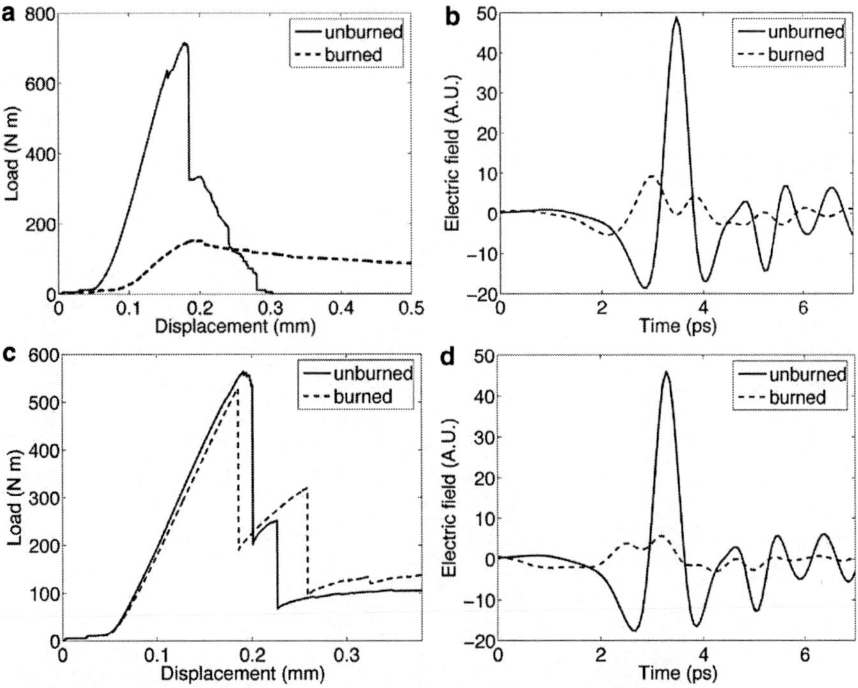

Fig. 8.22 Comparing THz wave inspection of two carbon fiber composition samples with bending test. Sample 1 is a single layer target, whose inspection result is in (**a**) and (**b**). Sample 2 contents multi-layers, and the inspection result is in (**c**) and (**d**). (**a**) and (**c**) are bending test results and (**b**) and (**d**) are THz wave inspection results

Chapter 9
THz Technology in Security Checks

The development of techniques for inspection of explosives and other hazardous materials has become more and more attractive as concerns about public security have increased considerably in the past years. Among all explosive devices, landmine is the most demanding target to be detected. Landmines were widely used in all kinds of battlefields and they are very difficult to eliminate once a conflict ends. The remaining landmines represent an enormous danger for the people, both military and civilian, that occupy the terrain affected by the presence of landmines. As of today, more than 100 million mines remain active and undetected in many fields around the world. Those mines claim more than 30,000 lives or injures each year. Although antilandmine technologies already exist and are being used in minefields, most of these technologies tend to give high false-positive results due to the presence of other objects present in the area.

There are other types of improvised explosive devices or IED that have been recently used in terrorism attacks. IEDs are more difficult to detect because there is no standard way to fabricate them. Up to now, no standard method to detect IED has been developed. Despite the variety of formats of explosive devices, they have one thing in common, which is that they contain explosives. Explosive compounds are usually organic compounds that have nitryl bonds. Most explosives have very low saturation vapor pressure, thus they are not easy for evaporating. On the one hand, it is difficult to detect them by they vapor in the ambient air. On the other hand, once an explosive contaminates a target, the residue will stay for a long time, which allows detecting the explosives by tracing the residue. Therefore, detection of explosive residue using their spectral features is very crucial in finding explosives and is of great help to enforce public security.

Spectra of Explosives in THz Band

Being organic compounds, most of explosive molecules have their unique spectral features in the THz band due to their rotation and collective vibration transitions. Those spectral features allow using THz wave spectroscopy to fingerprint explosives. The broadband THz radiation can be treated as sum of monochromatic waves,

X.-C. Zhang, J. Xu, *Introduction to THz Wave Photonics*,
DOI 10.1007/978-1-4419-0978-7_9,

each monochromatic wave described as:

$$\tilde{E}_0(\omega) = A(\omega)e^{-i(\omega t - kz + \phi_0(\omega))}, \tag{1}$$

where $k = \omega/c$ is the wave vector, ϕ_0 is the original phase, t and z are temporal and spatial position of the wave, respectively. The electric field of the transmitted wave is:

$$\begin{aligned}\tilde{E}_S(\omega) &= \tilde{t}_1\tilde{t}_2A(\omega)e^{-i(\omega t - kz + \phi_0(\omega))}e^{ik(\tilde{n}-1)d} \\ &= \tilde{t}_1\tilde{t}_2\tilde{E}_0(\omega)e^{-k\kappa d}e^{ik(n-1)d}\end{aligned}, \tag{2}$$

where $\tilde{t}_1$ and $\tilde{t}_2$ are the transmission coefficients of the THz field through both surfaces of the target, and $\tilde{n} = n + i\kappa$ is the complex refractive index of the composition. When the Fresnel loss is ignored, the effect of the target on the THz field is a decay of amplitude and a delay of the phase. The former is controlled by the imaginary part of the refractive index and the later is related to the real part. The decay of THz field can be measured through the absorption coefficient of the target $\alpha = k\kappa$. The absorption coefficient α may have different format depending on the characteristics of the samples. For instance, the absorption coefficient could be described based on thickness for some homogenous samples, for other samples, it may be described based on mass or concentration. It can also be categorized as intensity absorption or electric field absorption. In a laboratory environment, the absorption coefficient of an explosive sample is practically based on the mass of the sample, which is:

$$\alpha(\omega) = \ln\frac{I_0(\omega)}{I_S(\omega)}/m. \tag{3}$$

Here, I_0 is the original power of the THz wave, I_S is the THz power measured after the transmission through the target, and m is mass of the explosive sample. Absorption peak appears when there is transition in the explosive molecule. Figure 9.1 gives the absorption spectra of 4 explosives and their related compounds from 2 to 21 THz. The spectra were measured using a Fourier transform infrared (FTIR) spectrometer. Explosives and related compounds present rich spectral features in their THz spectra. Comparing to FTIR spectroscopy, THz wave time-domain spectroscopy (TDS) provides better data in the low frequency regime. Figure 9.2 shows spectra of 10 different explosives and their related compounds from 0.2 to 3 THz measured using THz TDS. Clear spectral features are detected for all those compounds in the lower THz band. Table 9.1 summarizes spectral features of the 15 most popular explosives and related compounds in the THz band. Using THz wave spectroscopy, it is possible to identify explosives through their spectral features and, therefore, it is possible to detect explosive devices such as landmines and IEDs.

Explosives can be identified through THz wave absorption spectroscopy. However, in reality, a THz wave might not be able to transmit through the target. For instance, in passenger screening, THz waves are not able to penetrate through

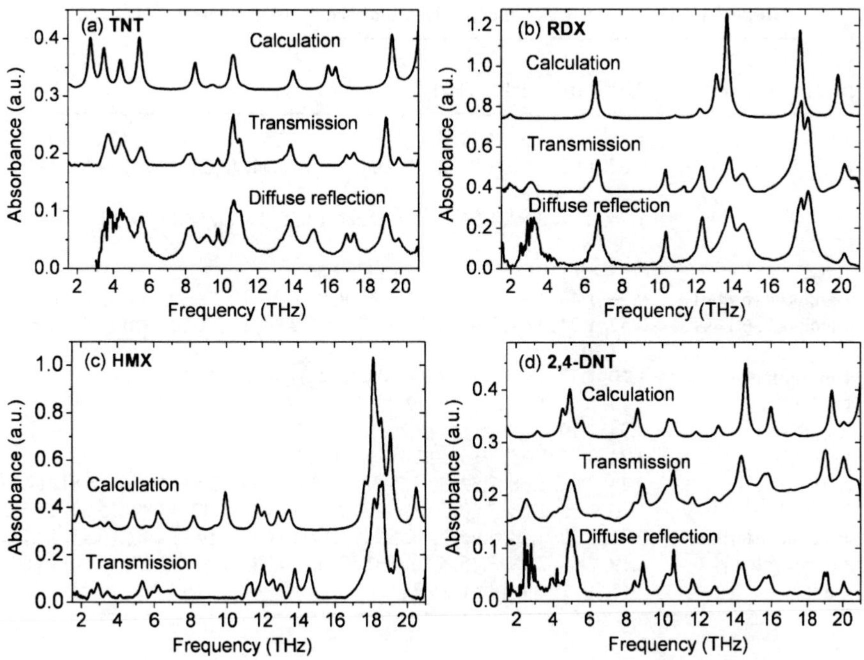

Fig. 9.1 Spectra of explosives and related compounds from 2 to 21 THz measured by FTIR. Transmission spectra and diffused reflection spectra are compared with calculation results

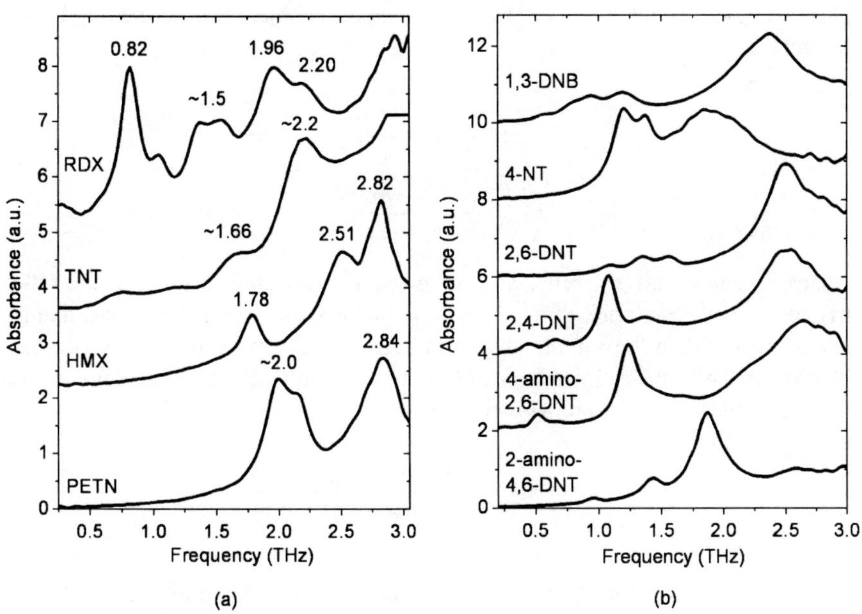

Fig. 9.2 Absorption spectra of explosives and related compounds measured using THz TDS (0.2–3 THz)

Table 9.1 Absorption peaks of explosives and related components in THz band

Explosives and related compounds	Absorption peaks (THz)
TNT	1.66, 2.20, 3.69, 4.71, 5.58, 8.16, 9.15, 9.75, 10.65, 13.89, 15.09, 19.17
RDX	0.82, 1.05, 1.50, 1.96, 2.20, 3.08, 6.73, 10.34, 12.33, 13.86, 14.54, 17.74, 18.12, 20.13
HMX	1.78, 2.51, 2.82, 6.06, 11.10, 11.97, 13.56, 14.52, 18.18, 18.51
PETN	2.0, 2.84
Tetryl	5.97, 10.11, 11.28, 14.67, 16.14, 18.36
2-amino-4, 6-DNT	0.96, 1.43, 1.87, 3.96, 5.07, 6.27, 8.49, 9.87, 10.77, 12.15, 13.44, 16.68
4-amino-2, 6-DNT	0.52, 1.24, 2.64, 3.96, 5.04, 5.82, 7.53, 9.30, 10.20, 11.13, 13.86, 14.97, 17.70
4-nitrotoluene	1.20, 1.37, 1.86, 6.75, 8.85, 10.83, 14.04, 15.66, 18.51
1,3,5-TNB	4.17, 4.62, 10.05, 11.19, 13.80, 15.75, 19.05
1,3-DNB	0.94, 1.19, 2.37, 10.56, 12.18, 15.33, 17.13
1,4-DNB	3.24, 3.96, 5.55, 10.38, 12.45, 13.29, 15.21, 15.54
2,4-DNT	0.45, 0.66, 1.08, 2.52, 5.01, 8.88, 10.56, 11.58, 12.81, 14.34, 15.81, 19.05
2,6-DNT	1.10, 1.35, 1.56, 2.50, 5.61, 6.75, 9.78, 11.43, 13.32, 13.89, 15.39, 17.25
3,5-dinitro aniline	0.96, 1.20, 3.18, 4.62, 5.04, 5.91, 7.44, 10.62, 10.98, 14.46, 16.41, 18.18
2-nitro diphenyl anine	2.19, 2.58, 2.88, 3.45, 5.13, 6.18, 7.56, 10.08, 12.33, 13.05, 15.00, 15.60, 16.29, 17.34, 18.51, 19.32

the body of the passenger. In such a case, reflection geometry is the only feasible choice. THz wave reflection spectroscopy analyzes the spectral feature of the specular reflection or the diffused reflection coming from a sample. Specular reflection is used when the target has a smooth surface and diffuse reflection is used for samples that have a rough surface. The reflectance of an object is described by Fresnel formula as:

$$\begin{aligned} \tilde{r}_{//} &= \frac{n\cos\theta_i - \cos\theta_t}{n\cos\theta_i + \cos\theta_t} \\ \tilde{r}_{\perp} &= \frac{\cos\theta_i - n\cos\theta_t}{\cos\theta_i + n\cos\theta_t} \end{aligned}, \tag{4}$$

where $r_{//}$ and $r_{\perp}$ are the reflectance of p and s polarization wave, n is the reflective index of the material, θ_i and θ_t are the incident angle and transmission angle, respectively. When the sample absorbs a THz wave, its reflective index is a complex value. Thus, the reflectance will have both amplitude and phase elements. For normal incidence, Equation (4) is simplified to:

$$\tilde{r}_{//} = -\tilde{r}_{\perp} = \frac{n^2 + \kappa^2 - 1}{(n+1)^2 + \kappa^2} + i\frac{2\kappa}{(n+1)^2 + \kappa^2} \tag{5}$$

Thus, spectral features of the target are reflected in both amplitude and phase changes in the reflection spectrum. Figure 9.3 shows the reflection spectrum of RDX

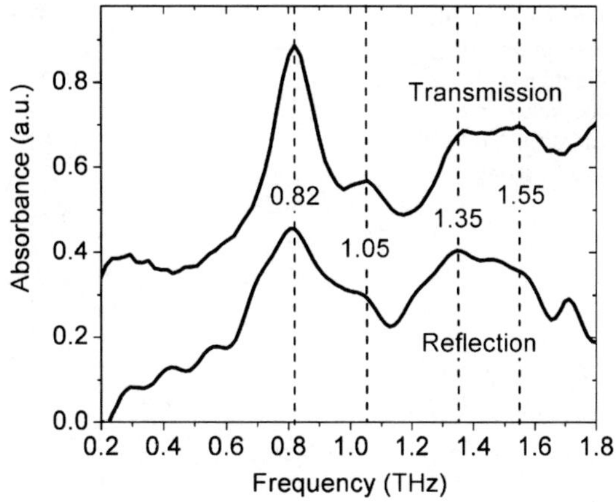

Fig. 9.3 Spectra of RDX measured by THz TDS in transmission and reflection geometries. The *dashed lines* indict locations of absorption peaks

with its absorption spectrum as comparison. Both spectra show the same spectral features.

If the target contains fine structures (for example powder) THz waves scatter once they penetrate into the target. Partial of the backscattered beam could be collected and analyzed. If certain approximations are satisfied, the problem can be solved as a pure diffuse spectroscopy analysis. These approximations are: (i) first of all, specular reflection from the surface of the target is ignored; (ii) secondly, the depth of the sample can be considered infinite compared to the penetration depth; and (iii) finally, the sample is homogenous and the illumination of the target is uniform. Under these conditions, the spectrum can be described using the Kubelka-Munk method [1] as:

$$F(R_\infty) = \frac{(1 - R_\infty)^2}{2R_\infty}, \tag{6}$$

where $R_\infty = R_{\text{Sig}}/R_{\text{Ref}}$ is defined as the diffuse factor of the target, R_{sig} denotes the reflectance of the target and R_{Ref} denotes the reflectance of a reference sample. The reference has similar physical characters as the target but it does not contain any spectral features. For instance, polyethylene powder can be used as a reference when THz wave spectroscopy is used in measuring explosive powder samples.

The diffused reflection spectra of 4 explosives and their related compounds are compared with their transmission spectra in Fig. 9.1. The diffused reflection spectra show similar spectral features as the absorption spectra. For certain weak absorption peaks, diffused reflection spectra show even a higher sensitivity. The modification of the frequency band caused by the target can be used as the effective absorption strength of the target, which can be modeled as the product of the absorption coefficient α and the effective interaction distance l_{Eff}. In the transmission spectrum, the effective absorption strength can be simply described as the product of the absorption coefficient of the target and its thickness. In diffuse reflection spectroscopy the

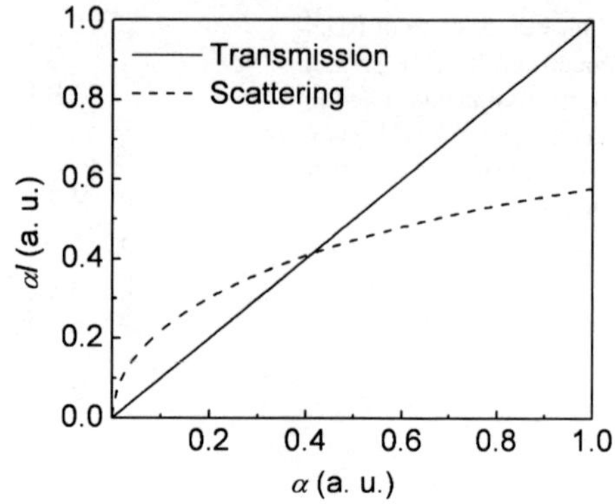

Fig. 9.4 The effective absorption strength as a function of the absorption coefficient in transmission and diffuse reflection spectroscopy

scattering of the material significantly increases the effective interaction distance of the THz waves inside the target. Additionally, the interaction length is different depending on the scattering paths. Therefore, the effective absorption needs to be calculated using a statistic method as [2]:

$$\alpha l_{\mathrm{Eff}} = (1 + 2s/\alpha)^{-1/2}, \quad (7)$$

where s is the scattering factor of the target, which is determined by the size of the fine structures and other properties of the target. Figure 9.4 compares the effective absorption strength in transmission and diffused reflection spectrum as a function of the absorption coefficient α. The figure shows that the diffused reflection spectrum shows a higher sensitivity of weaker absorption features.

The spectral features of explosives and related compounds are the result of the collective rotation or vibration modes of the molecules. The resonant structure of molecules can be calculated using density function methods once the structure of the molecule is determined. Figure 9.5 shows the molecular structure of 2,4-DNT, which is produced from the degradation of TNT, the most popular explosive.

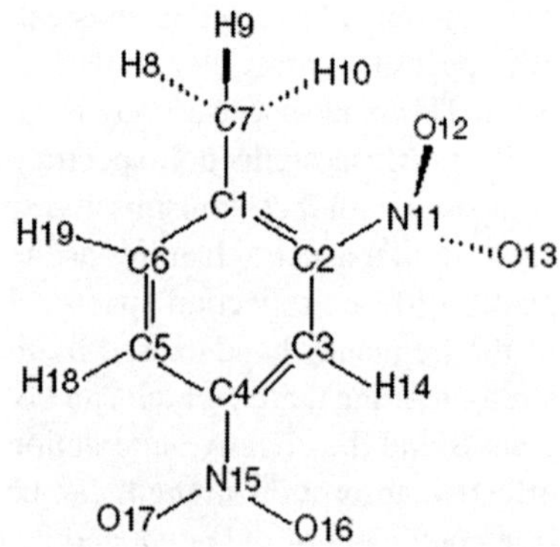

Fig. 9.5 Molecular structure of 2, 4-DNT

Therefore, it is considered as a hint for the presence of TNT. A molecule of 2,4-DNT has a benzene ring as the main frame, a methyl connecting with the first carbon atom C1, and two nitryls connecting to the second (C2) and the forth (C4) carbon atoms. Due to the asymmetry of the structure, the benzene ring is distorted in the 2,4-DNT molecule. The methyl acts as an electron donor and the nitryles act as electron acceptors. The interaction between methyl and the two nitryles prolongs C1-C2 and C1-C6 bonds, while C2-C3, C3-C4, C4-C5 and C5-C6 are compressed shorter than normal. This distortion causes the bond angle $\angle C2 - C1 - C6$ to be smaller than other bond angles. The forth nitryle, the benzene ring and the carbon atom in the methyl lie in the same plane. The second nitryle is pushed out of this plane 30° because of conflicts between its oxygen atom and the hydrogen atom in the methyl. Based on the previous analysis, its vibration transitions at ground state can be calculated using density function theory. Table 9.2 gives the calculated parameters of the 2,4-DNT molecule, including bond lengths, bond angles, the plane angle of the second nitryle, polarizations of the molecule as well as its energy structure. Figure 9.6a compares the theoretically predicted resonant features with the absorption spectrum measured using FTIR spectrometer. Most of the calculated features match the experimental result well, except for the absorption peak that appears at 2.52 THz in the experimental data, which is not present in the calculations. This peak is assumed to come from lattice vibration of crystalline 2,4-DNT. The calculation only considered single molecule, thus, lattice vibration modes are not considered. This assumption is confirmed by further experimental results. Figure 9.6b compares absorption spectra of a solid 2,4-DNT sample and the toluene solution sample. The absorption peak located at 2.52 THz disappears in the solution sample, where no crystalline structure exists. Table 9.3 compares calculation results and experimental data. The calculation does not only predict the location of those resonances but also provides the strength

Table 9.2 Parameters of 2,4-DNT molecular structure calculated using density function

Parameters	Calculation result
r (C1-C2)	1.406 Å
r (C2-C3)	1.390 Å
r (C3-C4)	1.384 Å
r (C4-C5)	1.390 Å
r (C5-C6)	1.388 Å
r (C1-C6)	1.403 Å
r (C1-C7)	1.506 Å
$\angle(C2 - C1 - C6)$	116.0°
$\angle(C1 - C2 - C3)$	123.0°
$\angle(C2 - C3 - C4)$	118.2°
$\angle(C3 - C4 - C5)$	121.5°
$\angle(C4 - C5 - C6)$	118.7°
$\angle(C5 - C6 - C1)$	122.5°
τ(O12-N11-C2-C1)	29.9°
τ(O16-N15-C4-C3)	0.8°
$\lvert\mu_t\rvert$	5.246 Debye
$-E$	680.8 Hartree

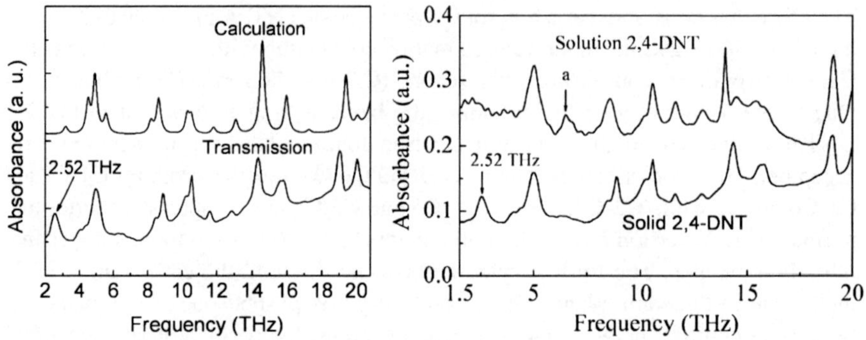

Fig. 9.6 (**a**) Comparing of absorption spectrum of 2,4-DNT in THz band with calculated result using density function. The absorption peak located at 2.52 THz does not appear in calculated result. (**b**) Absorption spectrum of 2,4-DNT sample in solid state and toluene solution. The absorption peak at 2.52 THz only appears in solid state sample

Table 9.3 Comparison of the absorption peaks and strength of 2,4-DNT between the experimental results and calculation results

Experimental result		Calculation result		
Frequency (THz)	Strength	Frequency (THz)	Strength	Cause of absorption
1.08	–	–	–	Phonon or intermolecular transition
2.52	–	–	–	Phonon or inter molecular transition
5.01	Strong	4.92	Strong	2,4 C-NO_2 in-plane bending vibration
8.88	Strong	8.61	Strong	Benzene ring out-of-plane bending vibration
10.56	Strong	10.53	Strong	Benzene ring in-plane bending vibration
11.58	Weak	11.82	Weak	Distortion of methyl and benzene ring out-of-plane bending vibration
12.81	Weak	13.08	Weak	C-CH_3 out-of-plane swing
14.34	Strong	14.58	Weak	4 C-N out-of-plane swing
15.81	Middle	15.96	Middle	4 C-N in plane bending and benzene ring distortion
19.05	Strong	19.38	Strong	Benzene ring distortion

information of those absorption structures. Calculation results of more explosive and related compounds are presented in Fig. 9.1.

Remote Sensing with THz Wave

One of the most attractive advantages of using THz waves to inspect hazardous materials is that the propagation of THz waves allows the operator to stay at a distance from the target. The propagation properties of THz wave in the ambient air are essential factors in the study of the remote sensing capabilities of the THz waves.

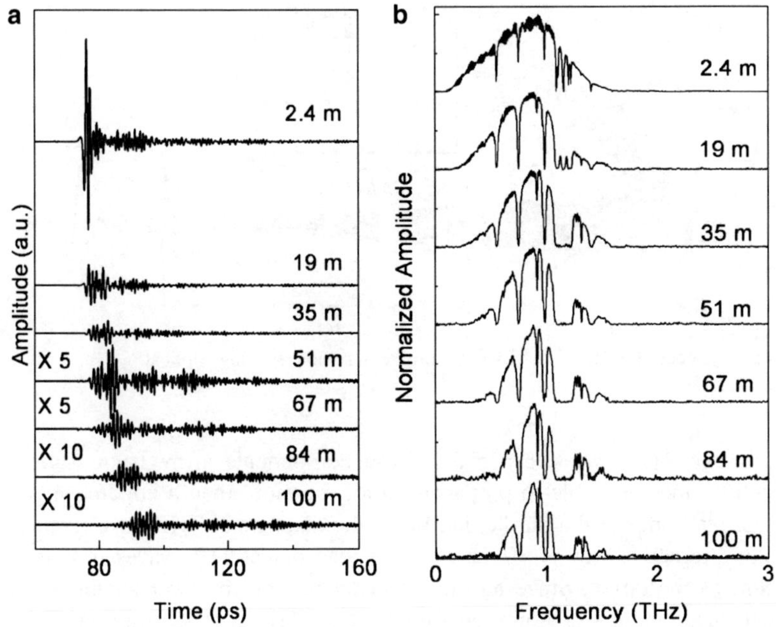

Fig. 9.7 Waveform and spectrum of THz pulse evolute with propagation distance in free space

Figure 9.7 shows the waveforms (a) and the spectra (b) of THz pulses after propagating through air for different distances. The amplitude of the THz field decreases for long propagation distances while the absorption due to water vapor becomes more severe. THz waveform is still detectable after 100 m of propagation, and it can still provide spectral measurement except the windows that are close to water vapor absorptions. Figure 9.8 shows the propagation windows for THz waves below 1.6 THz. There are 7 major windows in that region that can be used in remote sensing applications.

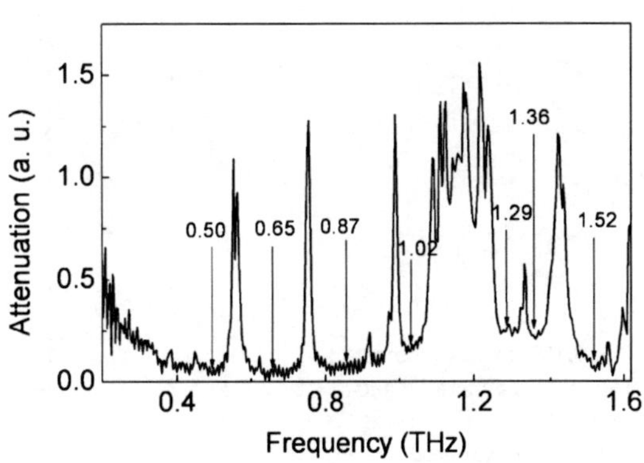

Fig. 9.8 THz propagation windows in air

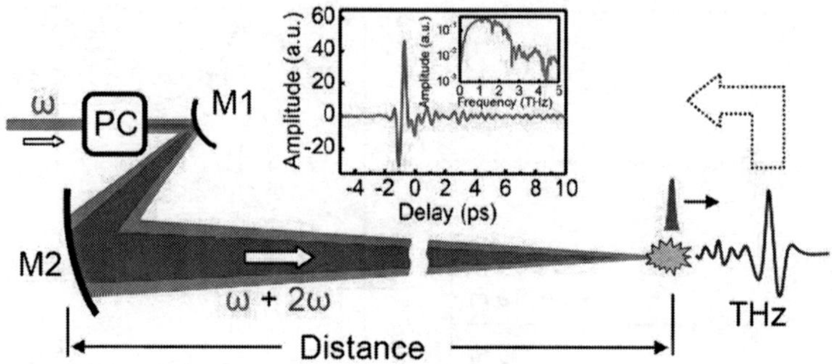

Fig. 9.9 Concept of using THz ABCD technique in remote sensing applications

Although THz waves, especially those components away from water vapor absorption lines, are able to propagate in air for more than a hundred meters, its high attenuation is still a hurdle that limits the distance of THz wave remote sensing. An alternative way is to generate and then detect THz waves in ambient air adjacent to the target. Since an optical beam has much lower attenuation in air (except dusty, smoky or foggy conditions) than THz waves, using air generation and detection promises longer sensing distance than directly send THz waves to the target. Figure 9.9 shows the concept of using THz wave air breakdown coherent detection (ABCD) technique in remote sensing. An ultrafast laser pulse and its double frequency pulses are focused next to the target, where they generate THz wave through the breakdown of air. THz waves coming from the target are detected locally using air as the sensor. Because a long propagation distance for THz waves has been avoided, the full spectral band and full dynamic range of the THz system can be used. Figure 9.10 shows a THz waveform generated and detected at 17 m away from the light source. Remote generation and detection of THz waves in air can also be performed using THz radiation enhanced-emission-of-fluorescence (REEF) technique, as presented in Fig. 9.11. Comparing to ABCD

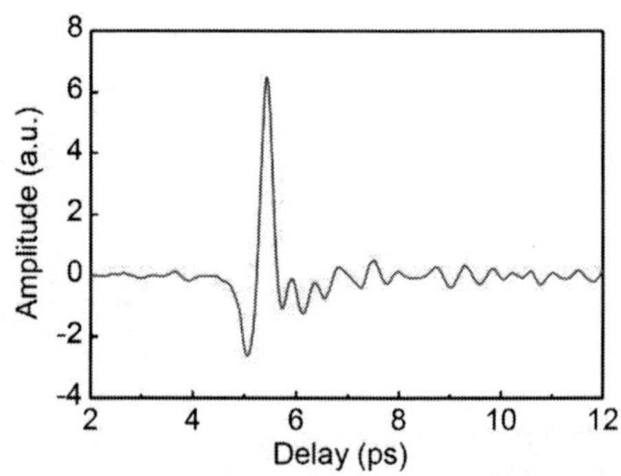

Fig. 9.10 THz waveform of THz pulses generated and detected in air 17 m away from the light source

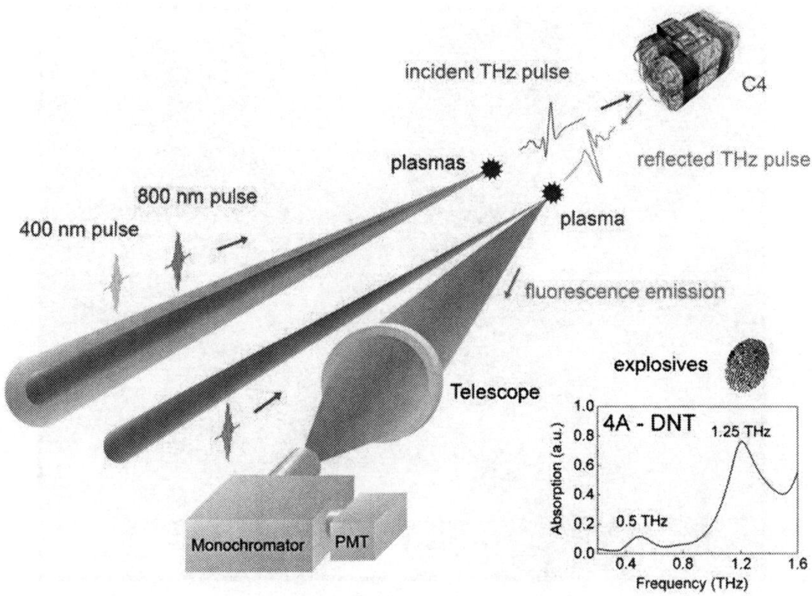

Fig. 9.11 Concept of using THz REEF technique in remote sensing applications

method, the REEF method detects fluorescence rather than second harmonic generation. Figure 9.12 compares THz waveforms detected using EO sampling method and REEF method. Using THz-TDS powered with REEF detection technique, it is possible to fingerprint explosives through their spectral features obtained locally besides the target.

The penetration capability of THz wave through several kinds of materials has been discussed in Chapter 8. THz waves have been used in NDE applications and its penetration capability has also lead using them in security inspection applications. Figure 9.13 shows a 0.094 THz wave image of a person [3]. A 1.2 m square shape

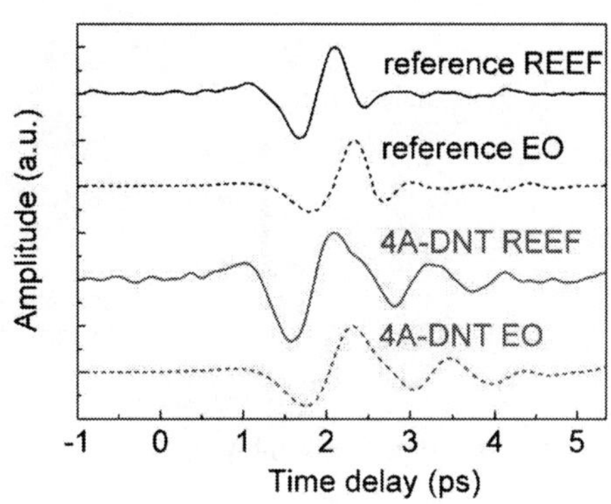

Fig. 9.12 THz waveforms detected using THz REEF methods comparing to those detected using EO sampling method

Fig. 9.13 THz wave (0.094 THz) image of a person. *Left* is a photo of the person as comparison and the *right* is his THz wave image (Courtesy of QinetiQ)

antenna placed at 7 m away from the person is used to take this image. THz wave image can unveil the knife hidden in the folding news paper and a pistol under cloth. THz wave imaging can be used in reflection geometry and, in some cases: transmission geometry is also possible, for example in package inspections. Figure 9.14 shows an experimental setup of a transmission THz wave imaging system operation in raster scanning configuration. Figure 9.15 shows THz wave image of a briefcase taken by 0.2 THz wave. The briefcase contenting items are clearly shown in its THz wave image, which included knife, CD, video and audiocassettes and pens,

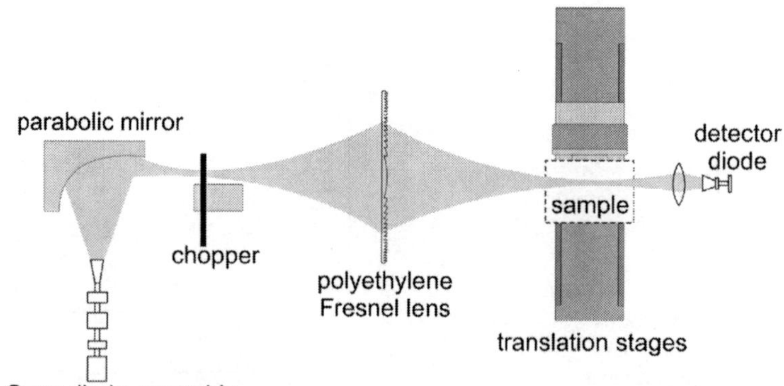

Fig. 9.14 Concept of THz wave transmission imaging

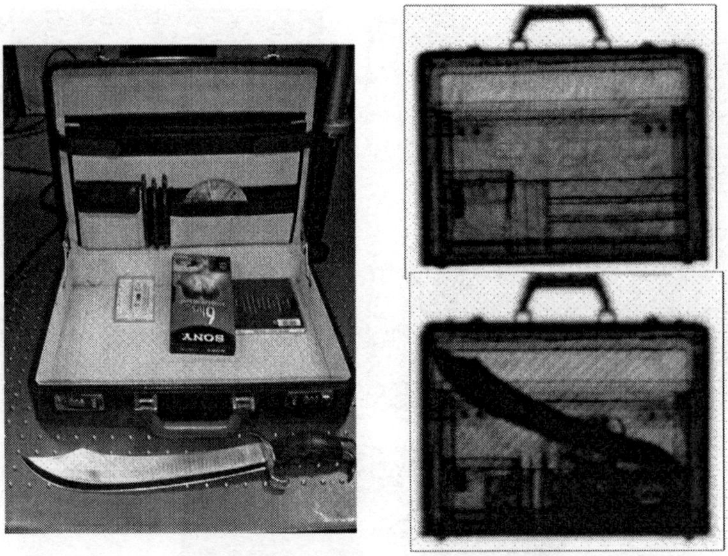

Fig. 9.15 THz wave image (0.2 THz) of a briefcase

etc. (Sub) THZ wave imaging system has already been tested in many airports for passenger screening, and the result is very permissing. It is very possible that this technology will be used in real security inspection in a near future.

When considering landmine detection, THz wave penetration into the soil is a critical factor. Since water highly absorbs THz waves, THz wave imaging can only be used in the detection of mines that are buried under dry soil, i.e. in the desert. According to Chapter 8, the penetration depth of THz wave in materials such as sand and rock is in the range of few centimeters. As a result, THz wave imaging is able to see a target buried under a few cm of soil. There are basically two kinds of landmines, which are the antipersonnel mines and the antitank mines. The former are usually buried within a very shallow depth, usually less than a few cm. The later can be buried as deep as ten centimeters or deeper. Thus THz wave image is appropriate to be used in detection of antipersonnel mines under dry soil. Figure 9.16 shows THz wave images of a metal target buried under dry sand at different depths. The result indicates that THz wave image can see metal target under sand up to 2 cm. Using THz wave inspection does not only see those hidden targets but also be able to identify those targets through their spectral features. Figure 9.17 shows THz wave transmission spectra of RDX sample hidden under paper, polyethylene film, polyester film and leather. The absorption peak at 0.82 THz can clearly be observed even with different covers.

THz Wave Stand-Off Detection

THz wave spectroscopy can fingerprint hazard materials, such as explosives, and it can be used in remote sensing condition as well as to detect targets under cover.

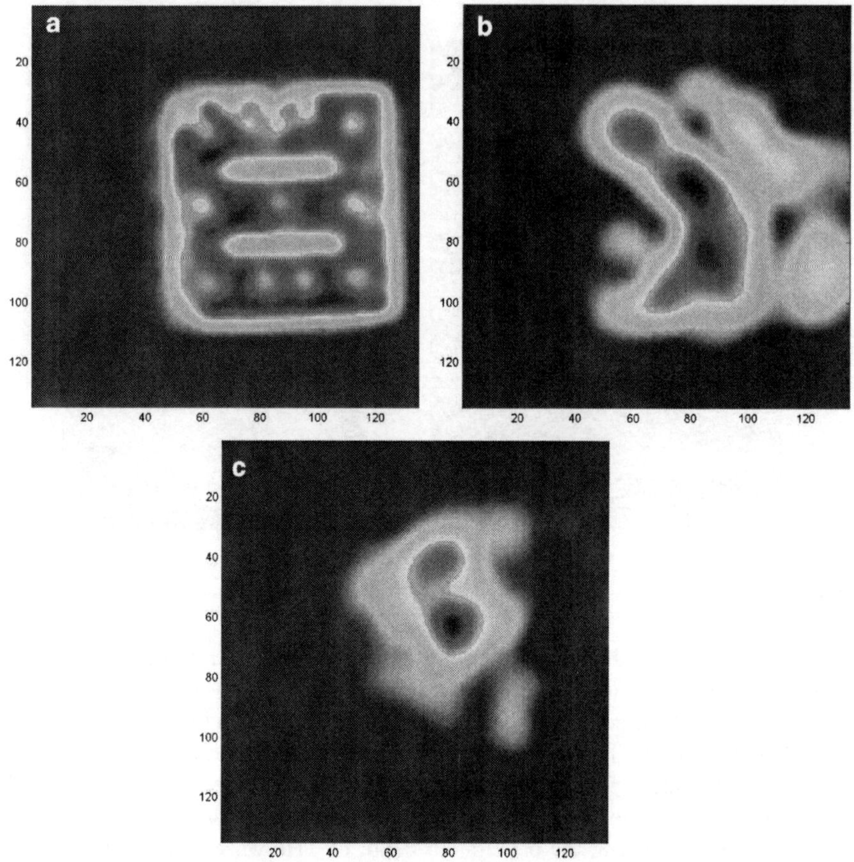

Fig. 9.16 Pulsed THz wave images of metal block buried under dry sand. The depths of metal: (**a**) 5 mm, (**b**) 10 mm, and (**c**) 20 mm

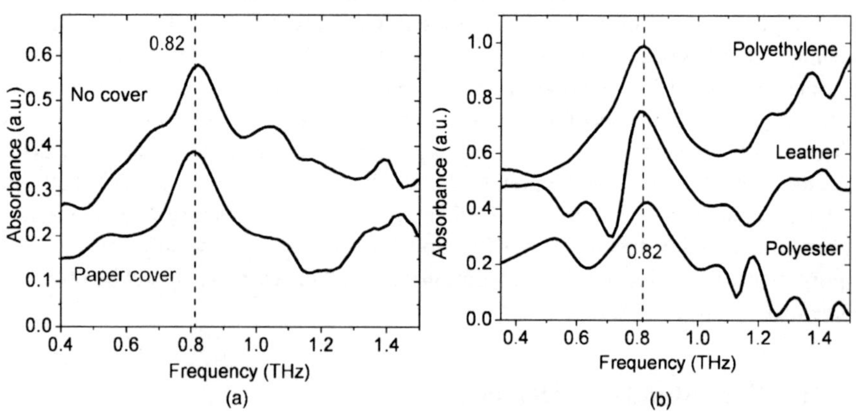

Fig. 9.17 THz TDS signal of a RDX sample under different covers

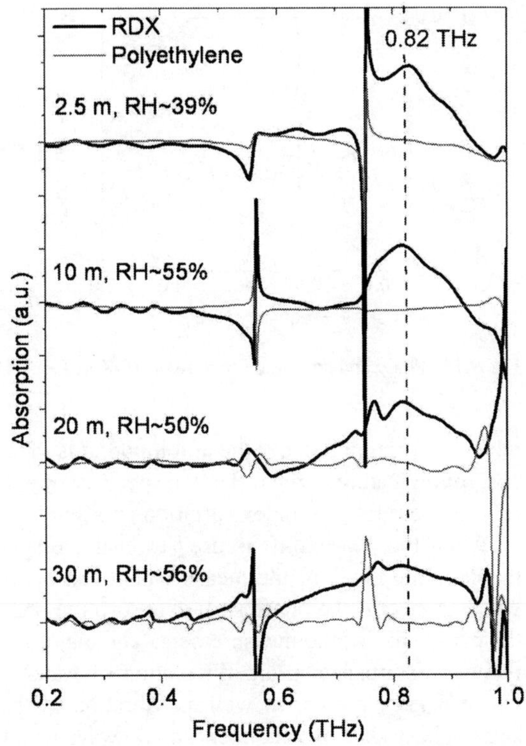

Fig. 9.18 THz TDS signal of a RDX sample when the spectrometer locating at different distances from the target. The reference spectra were taken using polyethylene as the reference sample

These features allow using THz wave in stand-off sensing, which allows the operator to stay out of the hazardous zone and interrogate the suspect target. Figure 9.18 shows THz wave spectroscopy of a RDX sample when the sample is located at different distances from the THz wave spectrometer. The experimental result shows that THz wave spectrometer can identify RDX even at 30 m away by catching the absorption peak at 0.82 THz. The reflection spectroscopy geometry is the primary choice in stand-off detection because the source and detector stay close to each other into a single unit. Therefore, the measured results will be analyzed using Equation (5) rather than Equation (3). In Equation (5), the reflectance has a real part and imaginary part. If $\kappa << 1$ is satisfied, the reflectance can be simplified to:

$$|\tilde{r}| \approx \frac{n-1}{n+1} \qquad \Delta\phi \approx \frac{2\kappa}{n^2-1} \tag{8}$$

Here $|\tilde{r}|$ and $\Delta\phi$ are the amplitude and phase change of the THz wave during reflection, respectively. Figure 9.19 shows the amplitude and phase spectra of a RDX sample measured in reflection spectroscopy. Unlike in transmission spectroscopy,

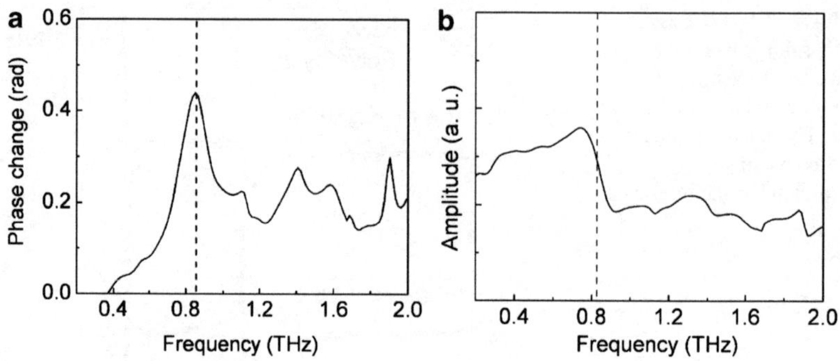

Fig. 9.19 Phase and amplitude spectra of RDX in THz-TDS

where absorption reduces the amplitude, it is the phase that changes, presenting the absorption feature in the reflection spectroscopy. The amplitude spectrum, however, reflects the refractive index variation adjacent to the absorption peak.

When the phase shift is used to evaluate spectrum in the stand-off detection, the dynamic range of the measurement can be described as $D = \Delta\phi_{\mathrm{Max}}/\Delta\phi_{\mathrm{Fluc}}$, where $\Delta\phi_{\mathrm{Max}}$ is the phase shift at absorption peak, and $\Delta\phi_{\mathrm{Fluc}}$ is the fluctuation of the phase shift of the measurement. The major sources of phase fluctuation (noise) include disturbance in air, fluctuation of water vapor concentration, timing jitter between laser pulses, as well as vibration of the optic components. Figure 9.20 shows phase shift of a monochromatic wave resulting from the measurement noise. The phase fluctuation of the THz wave is affected by the dynamic range of the system according to:

$$\delta\phi \propto \frac{1}{D_0}\sqrt{\delta t}. \tag{9}$$

Here D_0 denotes the dynamic range of the THz system and δt is the temporal resolution in the recording of the THz waveform. As a result, the detectable range of a THz wave stand-off detection system can be estimated according to the dynamic

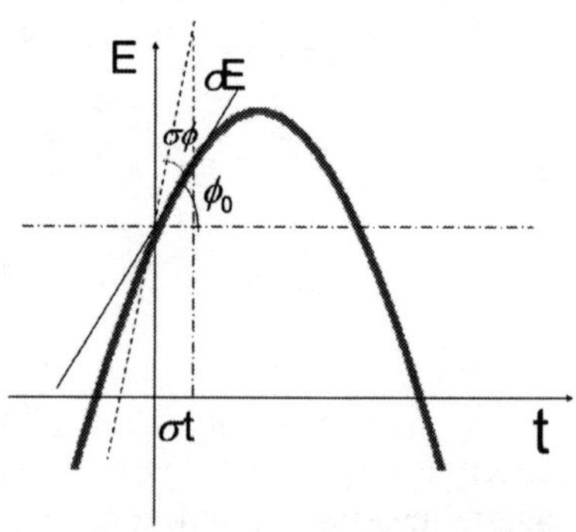

Fig. 9.20 Phase fluctuation in THz wave stand-off detection

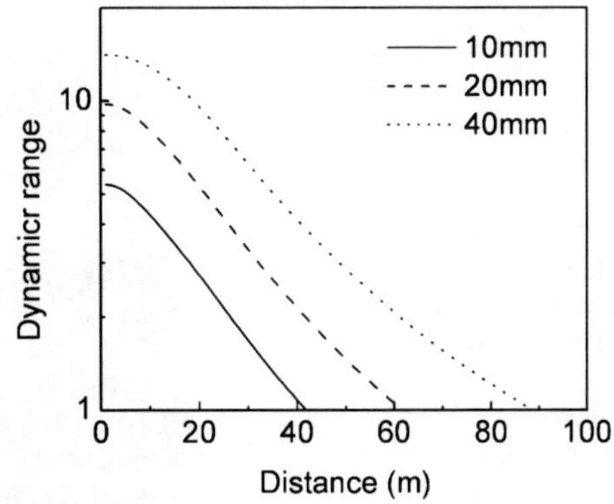

Fig. 9.21 Dynamic range of THz wave spectrum in measuring RDX sampling from different stand-off distances

range of the system, the beam emission and collection geometry, and the specific details of the target. Figure 9.21 provides the detectable range of a THz TDS system with a dynamic range of 15, which is used in the detection of RDX samples with different sizes. A collection aperture of 150 mm is used in the calculation.

THz Wave Spectroscopic Imaging

The combination of THz wave imaging technology with its spectroscopic capabilities results in THz wave spectroscopic imaging. THz wave spectroscopic image does not only see the profile of the target but also is able to identify its composition according to the spectral features. The example introduced in Chapter 8, where the spectral response of THz wave is used in detecting the defect in the carbon fiber composite sample. Following the unique spectral features of explosives, it is possible to identify a target containing explosives using THz wave spectroscopic imaging. Figure 9.22 shows spectroscopic images of three samples made by lactose, sugar, and RDX, respectively [4]. Those samples cannot be distinguished without using spectroscopic information. However, using their spectral features, each item can be distinguished from others.

THz wave spectroscopic imaging can be implemented with broadband pulsed THz wave, but it can also be implemented with several CW THz sources operating at selected wavelengths. Figure 9.23 shows a concept of using a serial of THz wave transceivers with multiwavelengths to provide THz wave spectroscopic imaging. Some of the wavelengths are located away from all absorption peaks in order to obtain a baseline. Other wavelengths are located on the absorption features of interest. Thus, by comparing images taken at different wavelengths, it is possible to distinguish the composition of the material. The simplest system uses just two wavelengths. One of the wavelengths is located away from the absorption peak and the other one located on the absorption peak or close to it. Figure 9.24 shows THz wave images of three targets made of polyethylene, TNT, and RDX acquired at

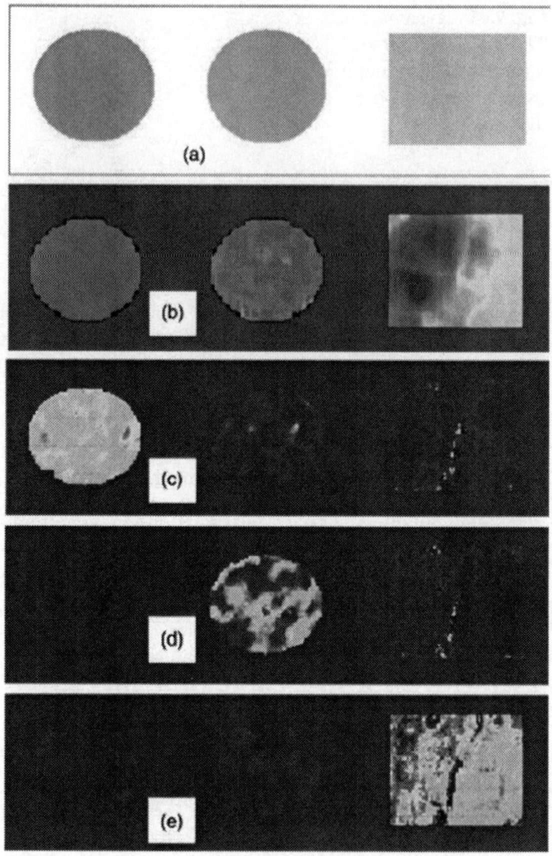

Fig. 9.22 THz wave spectroscopic images of targets made of lactose, sugar and RDX. (**a**) layout of samples. From the *left* to the *right*, lactose, sugar and RDX. (**b**) THz wave images of the three samples by measuring peak amplitude of THz pulses. (**c**, **d**, and **e**) are spectroscopic images of those three targets respectively (Courtesy of TeraView)

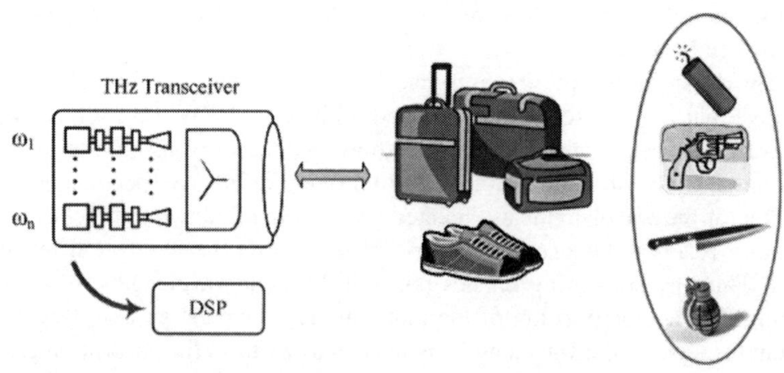

Fig. 9.23 THz wave spectroscopic imaging using cw THz wave transceivers

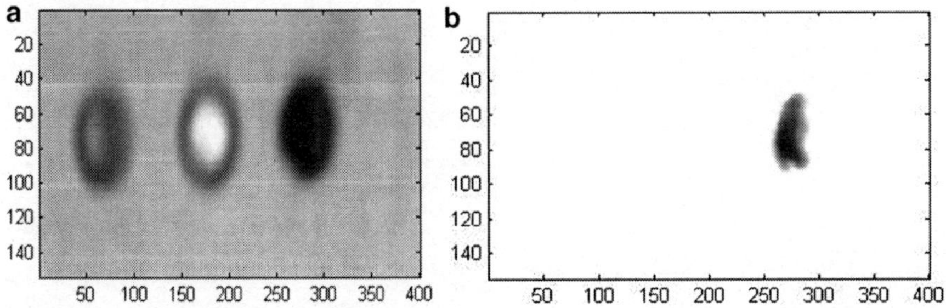

Fig. 9.24 (**a**) 0.6 THz wave images of three targets made of polyethylene, TNT, and RDX. (**b**) Identification of RDX from the other samples using spectroscopic imaging technique

0.2 and 0.6 THz frequencies. Figure 9.24a shows the images acquired at 0.6 THz. Although 0.6 THz does not overlap to the absorption peak of RDX (at 0.82 THz), the RDX sample still shows a higher absorption than the other samples at 0.6 THz. However, using 0.6 THz itself is not sufficient to highlight RDX. Suppressing the image acquired at 0.6 THz with the image acquired at 0.2 THz (Fig. 9.24b), the RDX sample is highlighted in the retrieved image while the other two samples are not presented because they do not have absorption feature close to 0.6 THz.

References

1. F. M. Mirabella, *Modern Techniques in Applied Molecular Spectroscopy*. Wiley, New York (1998).
2. M. Milosevic, and S. L. Berets, "A review of FT-IR diffuse reflection sampling considerations," *Appl. Spectr. Rev.* **37**, 347 (2002).
3. D. Clery, "Brainstorming their way to an imaging revolution," *Science*, **297**, 761 (2002).
4. Y. C. Shen, T. Lo, P. F. Taday, B. E. Cole, W. R. Tribe, and M. C. Kemp, "Detection and identification of explosives using terahertz pulsed spectroscopic imaging," *Appl. Phys. Lett.* **86**, 241116 (2005).

Chapter 10
THz Technology in Bio and Medical Applications

As THz waves interact with vibration and rotation transitions of organic molecules, they can be used to identify specific molecules based on their spectral features. In this way, THz technology can be used as a complement to other electromagnetic spectroscopy methods, such as visible and infrared. As THz photons have lower energy, they are unable to ionize biological samples under normal conditions. This makes THz spectroscopy an ideal tool for the examination of active biomedical samples. Due to the complexity of working with biological samples, biological applications are considered a mid- to long-term goal of THz research.

In this chapter, we will briefly discuss the use of THz waves in various biomedical applications. First, we will discuss the special challenges that biomedical applications present to THz and overview some techniques that could be used. Next, we will examine a few examples of the use of THz technology in pharmaceutical applications. Finally, we will discuss the use of THz in medical diagnostics. As with most of the work in the field, most examples discussed here are still at the proof-of-concept stage. As biomedical applications are relatively new to the THz field, more effort is needed to bring THz technology to the point at which it can be used in real biomedical applications.

THz Wave Spectra of Small Biomolecules

The THz wave spectra of small molecules usually exhibit clearer spectral features due to less broadening and overlap. As THz waves are strongly absorbed by liquid water and other polar liquids, THz wave spectroscopy typically requires samples to be solid-state or dissolved in nonpolar liquids. Figure 10.1 shows the THz wave absorption spectrum of glumatic acid taken by THz-TDS. The sample was made by mixing glumatic acid powder with polyethylene powder (ratio of 1:5 by mass) and then compressing it to form a chip with 0.5 mm thickness. Two absorption peaks located at 1.21 and 2.04 THz were observed in the THz spectrum. Those absorption peaks result from intermolecular collective vibrations. Other biomolecules, such as different amino acid and base molecules, also present spectral features in the THz band. Figure 10.2 shows the THz wave absorption spectra of a purine sample at

X.-C. Zhang, J. Xu, *Introduction to THz Wave Photonics*,
DOI 10.1007/978-1-4419-0978-7_10, © Springer Science+Business Media, LLC 2010

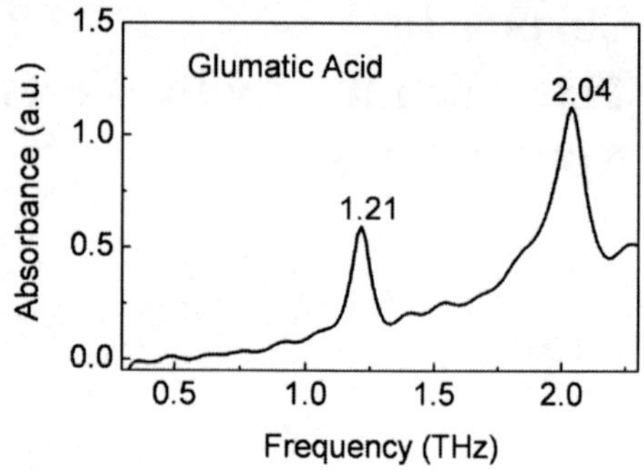

Fig. 10.1 THz wave absorption spectrum of glumatic acid in solid state

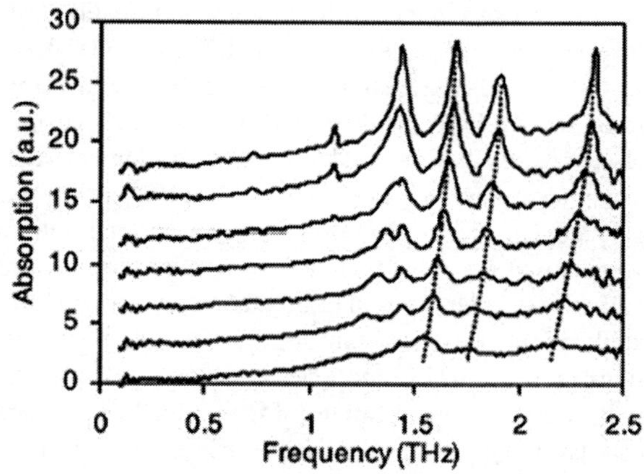

Fig. 10.2 THz wave absorption spectra of purine samples in solid state. Temperature of the sample changes from the top to the bottom at 4, 54, 105, 204, 253 and 295 K. The *dashed curve* indicates a shift of absorption peaks due to temperature change (courtesy of TeraView Corp.)

different temperatures [1], where the sample was made by mixing multicrystalline purine powder with polyethylene powder (1:10) and the mixture was compressed to form a chip with 1.3 mm thickness. Similar to the explosives discussed in Chapter 9, the absorption features of biomolecules also result from the rotation, vibration and collective vibration of those molecules. Since biomolecules usually contain more atoms than explosives, they have denser collective vibration modes. Due to the interaction among atoms, vibration modes in biomolecules typically differ from the simple harmonic format. The spectra of these biomolecules also suffer from inhomogeneous broadening, causing the absorption features to be indistinguishable from each other. Measuring the spectrum at a low temperature helps to minimize spectral broadening and thus narrows absorption features; they can then be distinguished with THz spectroscopy (as presented in Fig. 10.2).

Monomers, such as amino acid molecules, show spectral features in the THz band. In addition, small polymers consisting of a few monomers, like short polypeptides, also have responses in the THz band. Figure 10.3 compares the terahertz absorption spectra of polypeptides consisting of 1–4 glycine molecules (Gly)n (n = 1–4) measured using an FTIR spectrometer [2]. The THz absorption spectra

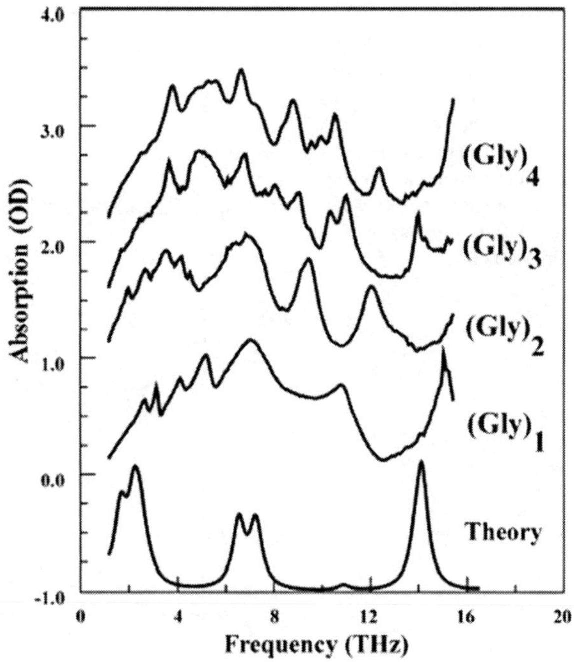

Fig. 10.3 THz wave absorption spectra of a polypeptide consisting of 1–4 glycine molecules. A calculation result is used as the comparison (courtesy of Dr. Heilweil)

exhibit increasing mode density in lower frequencies as the chain length increases. All four samples show clear spectral features in the THz band. Polypeptides with more monomers show a more complicated structure within the THz wave absorption spectra and that structure is thus predicted and calculated differently. Biomolecules that contain even more monomers, such as proteins or DNA molecules, usually do not give clear spectral features in the THz band due to broadening and overlapping of spectral features. Even in this case, THz wave spectroscopy can still be used to inspect the general properties of these molecules.

THz Wave Spectra of Biomacromolecules

Although biomacromolecules, such as protein or DNA, do not show clear spectral features in the THz band, they still draw attention in research due to their vital importance to human lives. When a terahertz wave interacts with a molecule, part of the wave can be transmitted, and a phase change can be induced. Different molecules, or the same molecules at different conformations or configurations, may present different absorption coefficients and refractive indexes in the THz band. Therefore, THz wave spectroscopy can be used to evaluate the aberration of these molecules. Figure 10.4 shows the THz absorption spectra of DNA, bovine serum albumin, and collagen in the 0.1–2 THz range [3]. The spectra were taken from

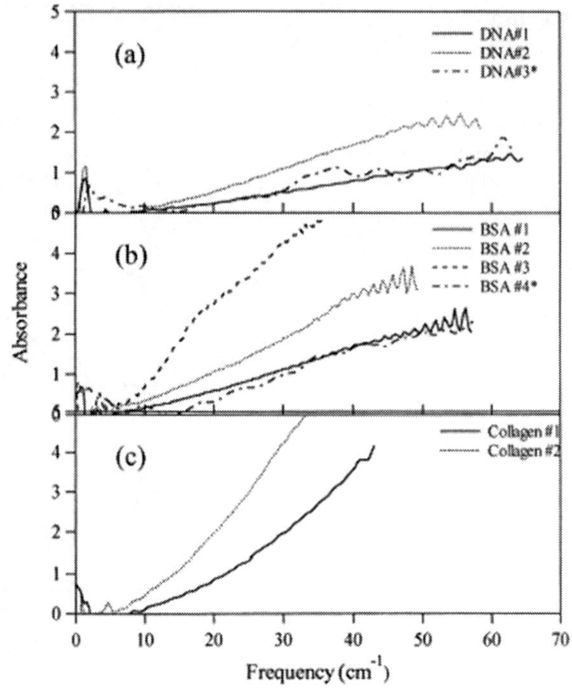

Fig. 10.4 THz wave absorption spectra of DNA, bovine serum albumin, and collagen samples. Except DNA #3 and BSA #4 sample, which are pure samples, all other samples are mixed with polyethylene powder as a buffer (courtesy of Dr. Markelz)

solid samples. Although there was no clear spectral feature that appeared under 2 THz for any the samples, each material shows consistent molar absorbtivity – the format of each sample may be different or even mixed with buffer materials. This unique molar absorbtivity can be used to a limited extent in distinguishing different materials.

Figure 10.5 shows a THz wave spectroscopy system that can be used to detect trace amounts of material [4]. This micro-THz-spectrometer is made by connecting a THz waveguide with a band pass filter. The filter acts as a THz-wave oscillator. It is used to enhance the reaction between the THz wave and the sample. This allows the THz spectrometer to detect trace amounts of the material. It was used for measuring DNA samples, where the DNA sample (in water solution) was dropped on the filter chip and left a layer of DNA coating about 40–80 nm thick after the water evaporated. Figure 10.6 shows the THz wave transmission spectra of denatured DNA molecules and hybridized DNA molecules [4]. Due to existence of DNA molecules, which change the local dielectric constant, the transmission spectrum of the band pass filter is changed. Two different kinds of DNA molecules present different spectral response in the THz band, and thus one can use the THz spectrum to distinguish between those DNA molecules. There are several other techniques that can be used to enhance the interaction between the THz wave and the target, which include total internal reflection and surface plasmon oscillation.

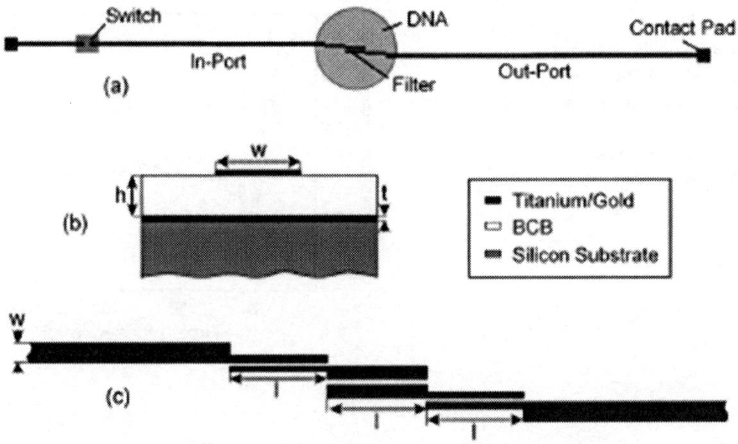

Fig. 10.5 Concept of THz wave micro spectrometer. (**a**) *top view* of the spectrometer, (**b**) cross section of the transmission line, where BCB is benzocyclobutene, which is a low k material in the transmission line, and (**c**) A zoomed-in *top view* of the filter, where $w = 16\ \mu m$, $l = 85\ \mu m$ (courtesy of Dr. Nagel.)

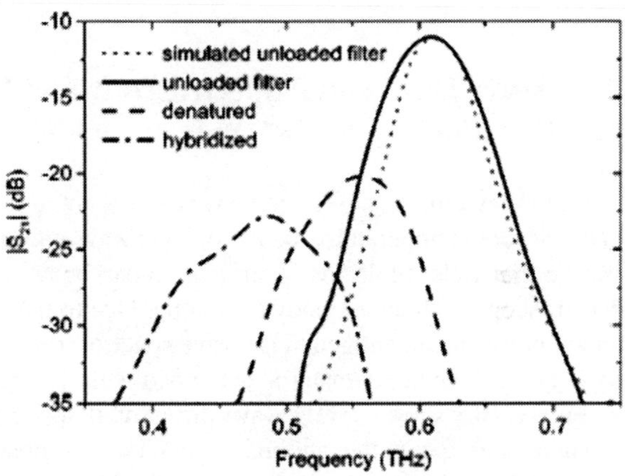

Fig. 10.6 Transmission spectra of the band pass filter when it was coated with denatured and hybridized DNA molecules, respectively. The calculation and measured transmission spectra of the filter itself are used as comparisons (courtesy of Dr. Nagel)

Protein is the most important functional material in human life. The function and activity of a protein is determined not only by its molecular structure but also by conformation and configuration of the molecule. As proteins are extremely complex, physical structure and properties arise from interactions between different amino acids in the chain, even if they are not directly neighbors; when a protein is formed, it folds into a unique shape that determines its function. If a protein molecule is excited by far infrared light, the excitation changes its vibration structure and thus causes a change in its THz wave absorption. The change of absorption recovers after a short interval referred to as the relaxation time. Utilizing THz wave pump and probe spectroscopy, one can investigate the dynamics of a protein molecule's

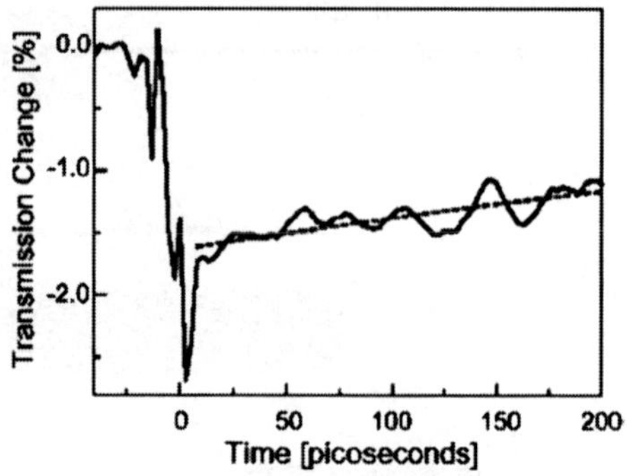

Fig. 10.7 THz wave transmission evolution of bacteriorhodopsin after being excited by pulse lasers with a wavelength of 87 μm and a pulse duration of 10 ps (courtesy of Dr. Xie)

collective vibration evolution after excitation. Figure 10.7 presents the relaxation process of a collective vibration mode located at 3.45 THz after a membrane protein has been excited [5]. A free-electron laserwith a wavelength of 87 μm and pulse duration of 10 ps was used to excite and detect the protein sample coated onto a piece of polyethylene film.

THz Wave Differential Spectroscopy and Biomolecule Identification

Since biomacromolecules do not have clear spectral features in the THz band, especially at room temperature, the use of THz wave spectroscopy alone cannot identify certain materials. Molecule identification can be realized by combining THz wave spectroscopy with an antibody technique. Due to the specific combination between an antibody and an antigen, THz wave spectroscopy can identify certain molecules by judging if the target material is bonded with the antibody.

Figure 10.8 shows a THz wave differential spectroscopy system used to detect biomolecules using the antibody technique. Similar to what was discussed in Chapter 3, a chip coated with an antibody, whose size is greater than twice the size of the THz focal spot, is placed at the focal point of the THz beam. Half of the chip is dipped into a solution with the interesting target and the solvent is evaporated to dry.Thus, half the chip will have the sample, and the other half will have a reference to compare the sample against. To avoid contamination by a solute that does not bond to the antibody, the chip is then washed using a solvent. If there is antigen material in the target-of-interest, it will bond to the antibody and cause a difference between the parts of the chip that have and have not been dipped into the solution. On the other hand, if there is no antigen material, these two parts will remain the same. A shaker is then used to alternately set those two parts of the chip. A lock-in technique is used to measure the differences in THz wave amplitude and phase between the two parts of the chip. If there is antigen material in the solution, it will bond to the antibody and the THz wave differential spectrum records the difference.

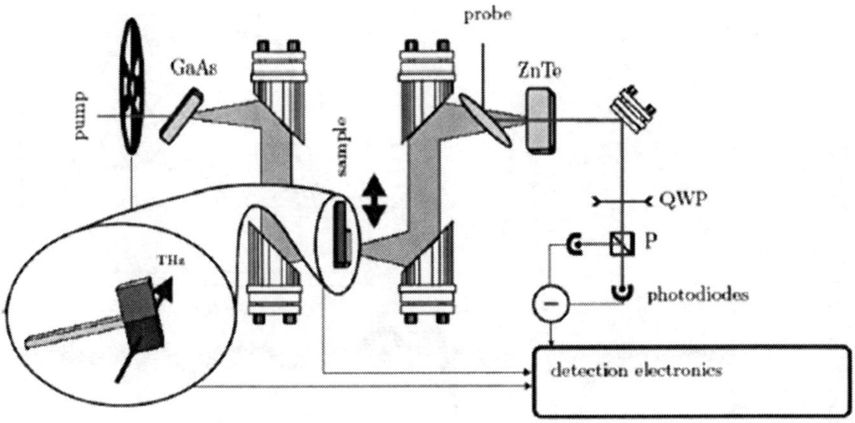

Fig. 10.8 Concept of a THz wave differential spectrometer combined with antibody technique

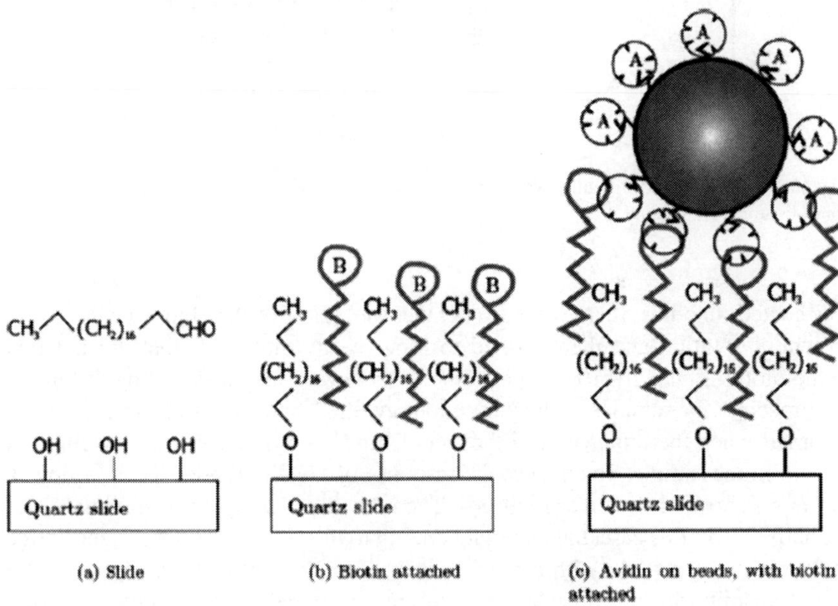

Fig. 10.9 Preparation of antibody chip and bonding between the antibody and the antigen. (**a**) coat octadecanol on the fused silica substrate, (**b**) bond biotin single molecule layer on the chip, and (**c**) avidin (combining with agarose bead) were bonded onto biotin

On the other hand, if there is no antigen material, then the differential spectroscopy records no signal. Figure 10.9 shows the concept of an antibody bonding with an antigen.

In the experiment, avidin was used as the antigen and biotin (Vitamin H) was used as the antibody. The chip was made using a fused silica wafer. The biotin

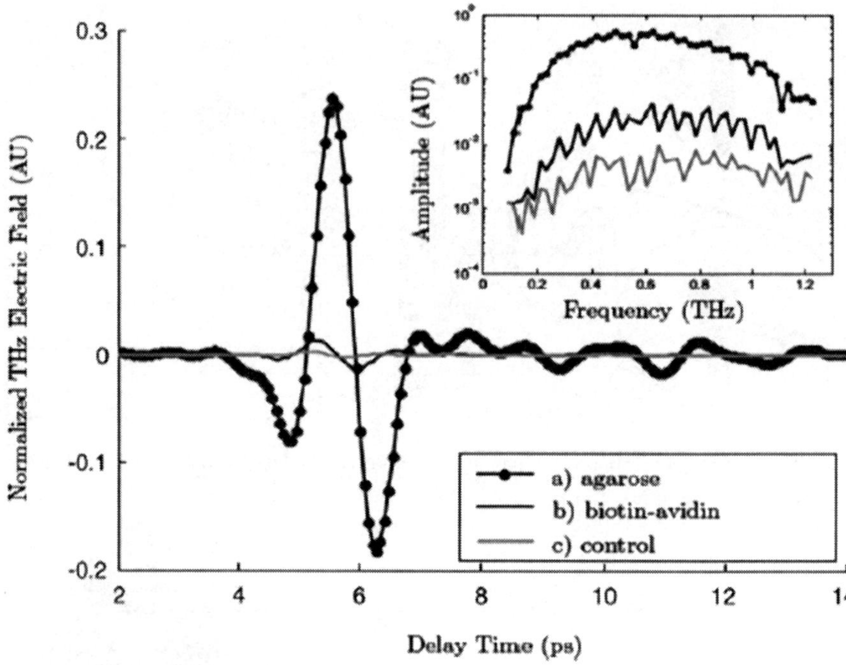

Fig. 10.10 THz wave differential waveforms and spectra (*inset*) of sample containing avidin. The control signals are used as a comparison

was coated onto the silica wafer via octadecanol. Half of the chip was dipped into an avidin solution and then washed to remove any molecules that did not bond to the biotin. Figure 10.10 compares the THz waveforms measured by differential spectroscopy for samples with and without avidin. There is a detectable difference recorded when there is avidin in the sample and no difference when avidin is not present in the sample. To enhance the response of the differential spectroscopy, the avidin was bonded with agarose beads. Then the mixed sample was used to bond to the antibody. In this case, the material bonding to the antibody is not just the antigen molecule but a particle with a much larger size. As a result, it changes the THz wave transmittance in a much larger scale and dramatically enhances the differential signal.

Hydration of Molecules and Its Response in THz Wave Spectroscopy

THz waves strongly interact with water, which is the essential material of all life. Most living things need to associate with water in order to remain active. The existence of water reduces the dynamic range when using THz wave spectroscopy to investigate those materials. On the other hand, the high sensitivity of THz spectroscopy to water molecules can be employed to verify the existence of water in biological samples as well as to identify the formation of water molecules in the bio samples. Hydration is a very common method by which water molecules exist in other materials. Hydration and dehydration of a material can dramatically change its properties, and therefore it is important to know the hydration condition of materials in applications such as quality control of pharmaceuticals.

Mixing caffeine with water, some of the water molecules will bond to caffeine molecules via hydrogen bonds, and the bonding water molecules are not easily removed even after the sample are dried. Figure 10.11 shows the THz wave absorption spectra of a caffeine molecule and hydrated caffeine molecules. Hydration induces an interaction between caffeine molecules and water molecules, and changes the original interaction among caffeine molecules as well as its crystalline structure. Those changes result in different spectral structures, which can be observed in THz spectroscopy. The change in the caffeine crystalline structure by hydration has been confirmed via X-ray diffraction. Heating the hydrated material may cause dehydration of the material, where a portion of the bonding water molecules is removed from the material.

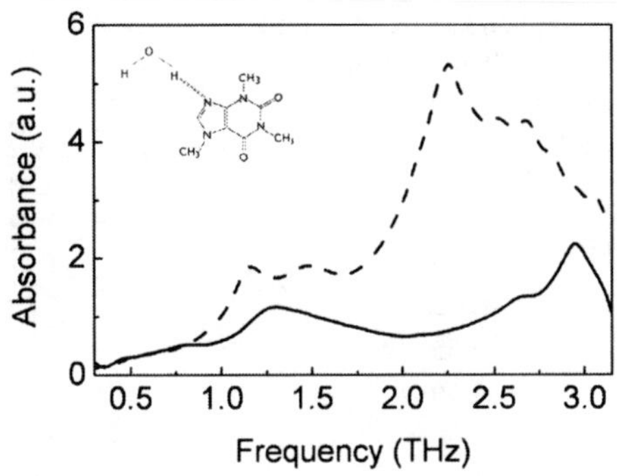

Fig. 10.11 THz wave absorption spectra of caffeine (*solid curve*) and hydrated caffeine (*dashed curve*). *Inset* shows structure of hydrated caffeine

Figure 10.12 shows the THz wave absorption spectra of mono-hydrated dextrose being heated to 45°C over time. Due to the loss of bonding water molecules at high temperature, the spectrum of dextrose changed. The obvious changes include an increase in the strength of the absorption peak at 1.44 THz, the disappearance of absorption peaks at 1.80 THz and 1.95 THz, and the appearance of an absorption peak located at 2.07 THz. These changes in the THz wave absorption spectrum allow for the evaluation of the hydration ratio of dextrose.

Hydration is one example of bonding between different molecules. There are also other formats of molecular bonding. For instance, sulfamethoxazole (SMA), which is an important composition of popular antibacterial drugs, can easily bond with other molecules such as caffeine or phylline via hydrogen bonds. Figure 10.13 compares the THz wave absorption spectra of two samples containing SMZ and caffeine, where in one sample the SMZ and caffeine are just mixed and in the other sample

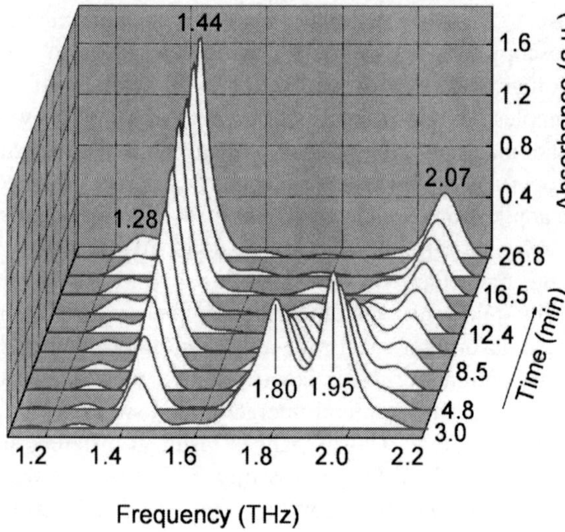

Fig. 10.12 THz wave absorption spectra of mono-hydrated dextrose being heated at 45°C for different periods of time

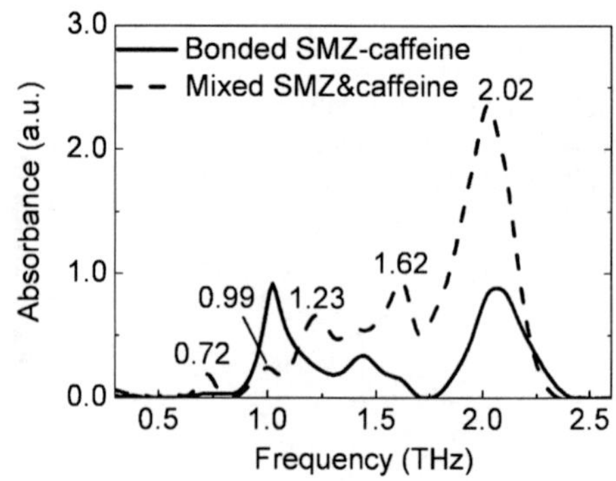

Fig. 10.13 THz wave absorption spectra of bonded SMZ-caffeine sample and mixed SMZ and caffeine sample

those two components are bonded to each other. THz wave spectroscopy shows a clear change in the absorption spectra due to the bonding between molecules.

Using THz Technologies in Quality Control of Pharmaceutical Products

THz wave spectroscopy can be used to identify molecular compositions if the compositions have spectral features in the THz band or to distinguish a change in molecular compositions if there are no features. THz wave technologies can be used in the quality control of pharmaceutical products to inspect if the drug meets the product specifications, such as concentration of effective composition, degradation

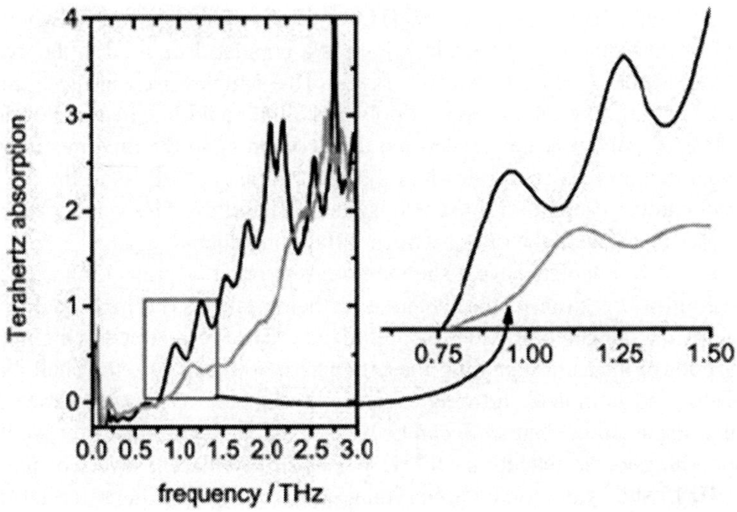

Fig. 10.14 THz wave absorption spectra of two isomers of ranitidine hydrochloride (courtesy of TeraView Corp.)

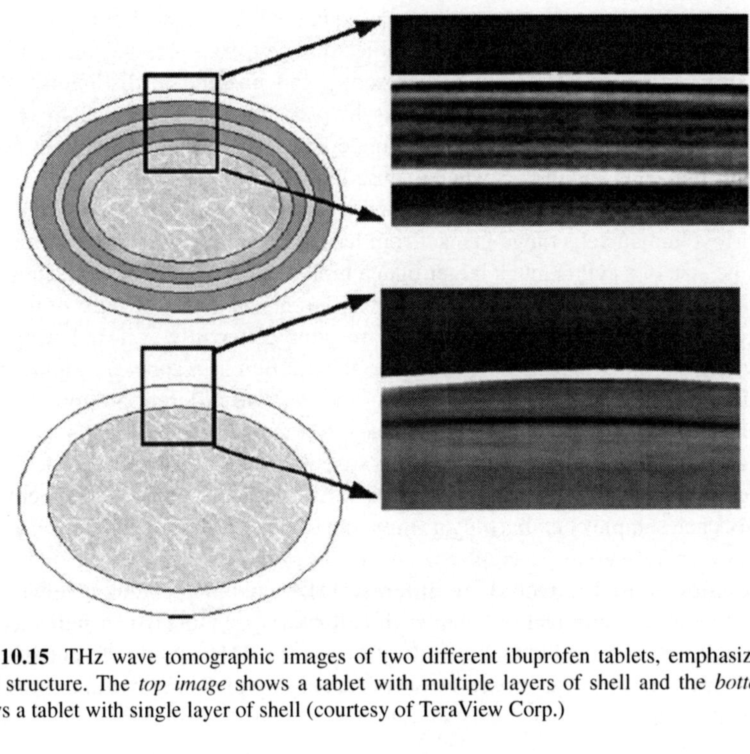

Fig. 10.15 THz wave tomographic images of two different ibuprofen tablets, emphasizing the shell structure. The *top image* shows a tablet with multiple layers of shell and the *bottom one* shows a tablet with single layer of shell (courtesy of TeraView Corp.)

level, etc. Figure 10.14 compares the THz wave absorption spectra of two different isomers of ranitidine hydrochloride, which is a popular drug used in the treatment of stomach diseases, e.g., gastric ulcers [6]. The samples used in the experiment were made by mixing each isomer of polycrystalline ranitidine hydrochloride powder with 25% polyethylene powder and then compressing the mixtures into chips. The experimental result indicates that the absorption coefficients in the THz band are significantly different for those two isomers. Therefore, THz wave spectroscopy can be used to measure the ratio between different isomers.

Most medicine tablets have a shell structure in order to protect the active agent and to control digestion of the tablets after being taken. The high quality of the shell helps the tablets to provide the optimized effect for treatment. One important quality control measure regarding these tablets is to make sure the shell structure is uniform and complete, forming layers in the tablet. THz wave time-of-flight imaging, discussed in Chapter 3, can be used to map a 3D profile of the shell structure according to the reflection of THz pulses from different layers of the tablet. Figure 10.15 shows the tomographic image of two different ibuprofen tablets [7]. The shell structure can be different for products made by different companies.

THz Wave Spectroscopy of Cells and Tissues

THz technologies have the potential to be used in medical diagnostics due to the sensitivity of THz waves to water and biomolecules. For example, one very attractive application is to use THz wave spectroscopy and imaging to distinguish abnormal tissues from healthy ones, and thus help to diagnose diseases. In order to develop such applications, one needs to understand the difference between normal and abnormal cells and tissues when exposed to THz waves.

A cell is the fundamental unit of life, and it can individually play some functional roles. Most human cells range in size from less than one micron to tens of microns. Since the size of a cell is much larger than a biomolecule, it does not present a clear spectral feature in the THz band. However, different categories of cells, or the same kinds of cells at different conditions may respond differently to THz waves. As a result, one can distinguish those cells via their different responses. Figure 10.16 compares the THz wave differential waveforms of two different groups of bovine lung microvascular endothelial cells. One group contains natural cells and in the other group, the cells were treated using vascular endothelial growth factor. The samples were made by growing a single layer of cells on a piece of polyethylene chip. As each sample contains only a single layer of cells, it only gives a weak modulation to the THz wave. To emphasize the effect of the cells, THz wave differential spectroscopy is used to record the different THz wave transmittance between the bare polyethylene chip and the chip with cell coating. Figure 10.16 indicates that the aberration of the cells dramatically changes the THz wave differential waveform. Thus, THz wave spectroscopy has the capability to distinguish abnormal cells from healthy ones.

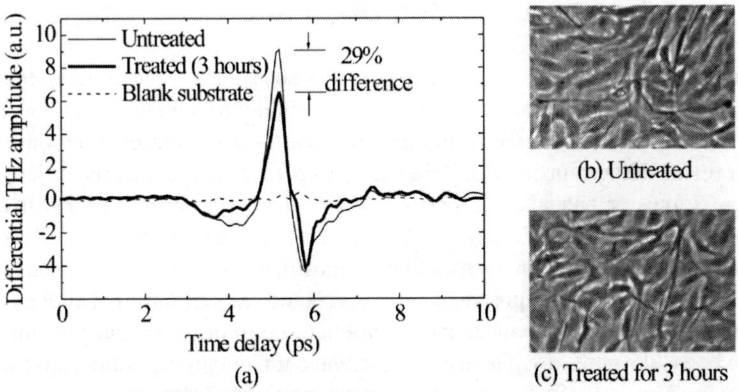

Fig. 10.16 (**a**) THz wave differential waveforms of treated and untreated bovine lung microvascular endothelial cells. (**b** and **c**) are microscopic images of untreated and treated cell samples

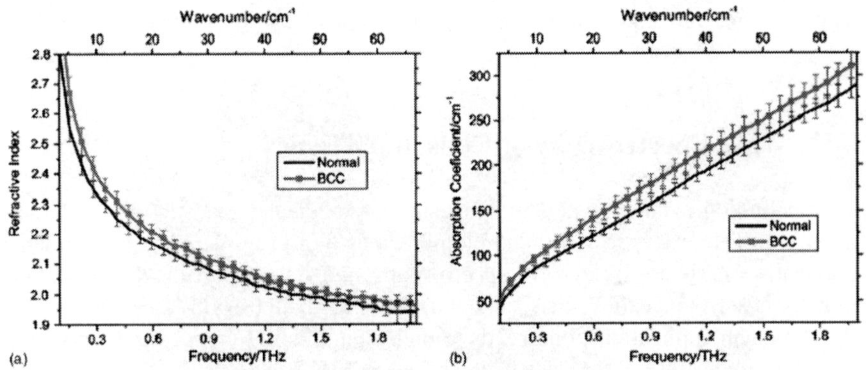

Fig. 10.17 Refractive index and absorption coefficient of healthy and cancer skin tissues in the THz band (courtesy of TeraView Corp.)

Tissues are an ensemble of similar cells and form an intermediate stage between cells and organisms. The syndromes of most illnesses are present at the tissue level. As a result, the identification of sick tissues is very important in diagnosing diseases. Figure 10.17 compares the refractive index and absorption coefficient of healthy skin tissue and cancer tissue (basal cell carcinoma) in the THz band [8]. The presented results were measured from tissues of 10 patients. The statistical analysis indicates that the diagnostic accuracy rate is larger than 95%. Consistent differences between the healthy tissues and the cancer tissues were observed in the THz wave spectra for both refractive index and absorption coefficient. This makes it possible to use THz wave spectroscopy to identify cancerous tissues. The reason different tissues have different responses to THz waves is not yet known, however, a common understanding is that this may be due to different water concentrations in different tissues.

THz Wave Imaging in Medical Diagnostics

Due to the different THz spectra of different tissues, THz wave imaging technologies can be used in medical diagnostics. However, the following two factors should be noted: the penetration capability and limited transmission of THz waves. The former refers to the capability of THz waves to penetrate through lots of daily items such as clothes or bandages. Therefore, THz waves can be used to investigate an illness or a wound concealed by those materials. Limited transmission refers to the high absorption of THz waves by water in most tissues. Since most human tissues (such as muscle) are composed of water, THz waves can only penetrate into the human body a shallow distance, and cannot be used to inspect organisms inside the human body like an X-ray. The use of THz wave imaging (or spectroscopy) in medical diagnostics is limited to the following conditions: THz wave imaging can be used to diagnose skin diseases, THz wave imaging can be used to investigate slices of tissues, and THz wave imaging can be used to inspect inside the human body via an endoscope. There are some human tissues that contain less water than others,

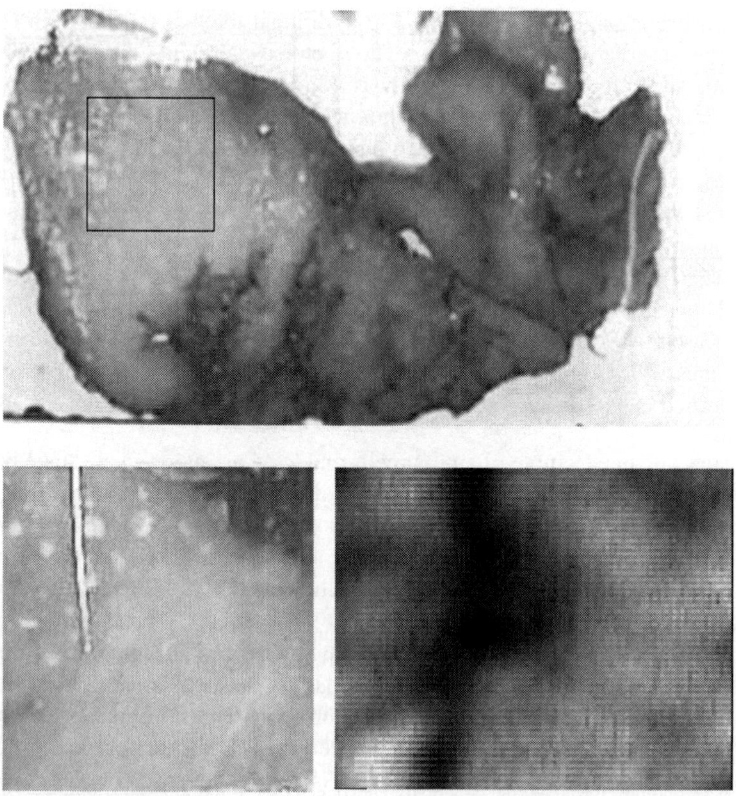

Fig. 10.18 THz wave images of breast tissue and fiber buried in the breast tissue

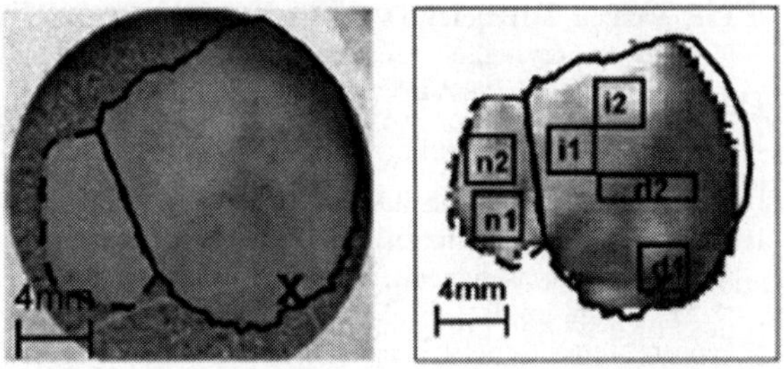

Fig. 10.19 Optical image and THz wave image of skin cancer. The images of the cancerous area (boundary with *solid curve*) are compared with the healthy area (boundary with *dashed curve*) (courtesy of TeraView Corp.)

and THz wave can penetrate a greater distance into those tissues. For example, breast tissue contains a lot of fat and thus has better transmittance for THz waves than muscles. Figure 10.18 shows THz wave images of breast tissue. The THz wave image is able to see the hidden artificial fiber under the breast tissue. Besides breast tissue, bone and teeth are other elements of the human body that have higher transmittance for THz waves.

An additional example of using THz wave imaging in medical diagnostics is THz wave imaging of skin cancer. The traditional method to treat skin cancer is based on observation by the doctor. The doctor usually cuts the suspect area and takes the sliced tissue for further analysis. Since a portion of the cancerous tissues can be buried under the top layer of skin, which is difficult to observe in an optical image, in a clinic the doctor needs to make a series of operations in which all the cancerous tissues are removed. A series of operations not only takes more time and is more expensive, but can also be more painful to the patient. THz wave time-of-flight imaging is able to observe THz waves reflected from different layers of the tissues, and thus is able to inspect the cancerous tissues under the top layer of skin. This technique can help the doctor to evaluate size, distribution, and depth of the diseased tissues before the operation. Figure 10.19 shows THz wave images of skin cancer [9]. Compared to the optic image, the THz wave image not only sees the exposed cancer tissue but also sees cancerous tissue underneath the top layer of skin.

References

1. Y. C. Shen, P. C. Upadhya, H. Linfield, and A. Ga. Davies, "Temperature-dependent low-frequency vibrational spectra of purine and adenine," *Appl. Phys. Lett.* **82**, 2350 (2003).
2. M. R. Kutteruf, C. M. Brown, L. K. Iwaki, M. B. Campbell, T. M. Korter, and E. J. Heilweil, "Terahertz spectroscopy of short-chain polypeptides," *Chem. Phys. Lett.* **375**, 337 (2003).
3. A. G. Markelz, A. Roitberg, and E. J. Heilweil, "Pulsed terahertz spectroscopy of DNA, bovine serum albumin and collagen between 0.1 and 2.0 THz," *Chem. Phys. Lett.* **320**, 42 (2000).
4. M. Nagel, P. H. Bolivar, M. Brucherseifer, H. Kurz, A. Bosserhoff,, and R. Buttner, "Integrated THz technology for label-free genetic diagnostics," *Appl. Phys. Lett.* **80**, 154 (2002).
5. A. Xie, A. F. G. van der Meer, and R. H. Austin, "Excited-state lifetimes of far-infrared collective modes in proteins," *Phys. Rev. Lett.* **88**, 018102 (2002).

6. P. F. Taday, I. V. Bradley, D. D. Arnone, and M. Pepper, "Using terahertz pulse spectroscopy to study the crystalline structure of a drug: a case study of the polymorphs of ranitidine hydrochloride," *J. Pharm. Sci.* **92**, 831 (2003).
7. A. J. Fitzgerald, B. E. Cole, and P. F. Taday, "Nondestructive analysis of tablet coating thicknesses using terahertz pulsed imaging," *J. Pharm. Sci.* **94**, 177 (2006).
8. E. Pickwell, A. J. Fitzgerald, B. E. Cole, P. F. Taday, R. J. Pye, T. Ha, M. Pepper, and V. P. Wallace, "Simulating the response of terahertz radiation to basal cell carcinoma using ex vivo spectroscopy measurements," *J. Biomed. Opt.* **10**, 064021 (2005).
9. R. M. Woodward, B. E. Cole, V. P. Wallace, R. J. Pye, D. D. Arnone, E. H. Linfield, and M. Pepper, "Terahertz pulse imaging in reflection geometry of human skin cancer and skin tissue," *Phys. Med. Biol.* **47**, 3853 (2002).

Additional Reference

Sensing with Terahertz Radiation, by Daniel Mittleman (Editor). Springer (2002) ISBN: 978-3-540-43110-7

Principles of Terahertz Science and Technology, by Yun-Shik Lee. Springer (2009) ISBN-13: 978-0387095394

THz Sensing Technology Volume 2, by Dwight L. Woolard, William R. Loerop, and Michael S. Shur (Editors). World Scientific (2004) ISBN-13: 978-9812386113

THz Frequency Detection and Identification of Materials and Objects, by R. E. Miles, X.-C. Zhang, H. Eisele, and A. Krotkus (Editors). Springer (2007) ISBN-13: 978-1402065026

THz Science and Technology for Military and Security Applications, by Dwight L. Woolard, James O. Jensen, R. Jennifer Hwu, and Michael S. Shur (Editors). World Scientific (2007) ISBN-13: 978-9812771797

Terahertz Spectroscopy: Principles and Applications (Optical Science and Engineering Series), by Susan L. Dexheimer (Editor). Taylor and Francis (2007) ISBN-13: 978-0849375255

Terahertz Optoelectronics, by Kiyomi Sakai (Editor). Springer (2005) ISBN-13: 978-3540200130

Intense Terahertz Excitation of Semiconductors (Series on Semiconductor Science and Technology), by S.G. Ganiche and W. Prettl. Oxford University Press (2006) ISBN-13: 978-0198528302

Terahertz Plasmonics: Metamaterials and Guided-wave Devices, by Amit Agarwal and Ajay Nahata. VDM Verlag (2009) ISBN-13: 978-3639129533

Assessment of Millimeter-Wave and Terahertz Technology for Detection and Identification of Concealed Explosives and Weapons, by Committee on Assessment of Security Technologies for Transportation National Research Council National. Academies Press (2007) ISBN-13: 978-0309104692

Terahertz Sources and Systems (NATO Science Series II: Mathematics, Physics and Chemistry) by R. E. Miles, P. Harrison, D. Lippens (Editors). Springer (2001) ISBN-13: 978-0792370970

X.-C. Zhang, J. Xu, *Introduction to THz Wave Photonics*,
DOI 10.1007/978-1-4419-0978-7,

Index

Index

Note: Locators followed by "f" and "t" refer figures and tables respectively.

Lightning Source UK Ltd.
Milton Keynes UK
11 July 2010

156819UK00010B/81/P

9 781441 909770